Advances in Principal Component Analysis

Ganesh R. Naik
Editor

Advances in Principal Component Analysis

Research and Development

Springer

Editor
Ganesh R. Naik
BENS Research Group, MARCS Institute
Western Sydney University
Kingswood
Australia

ISBN 978-981-10-6703-7 ISBN 978-981-10-6704-4 (eBook)
https://doi.org/10.1007/978-981-10-6704-4

Library of Congress Control Number: 2017957664

Printed on acid-free paper

This Springer imprint is published by Springer Nature
The registered company is Springer Nature Singapore Pte Ltd.
The registered company address is: 152 Beach Road, #21-01/04 Gateway East, Singapore 189721, Singapore

Preface

Principal component analysis (PCA) is one of the widely used matrix factorization techniques for dimensionality reduction and revealing hidden factors that underlie sets of random variables, signals, or measurements. PCA is essentially a method for extracting individual signals from mixtures of signals. Its power resides in the physical assumptions that the different physical processes generate unrelated signals. The main aim of PCA is to reduce the dimensionality of a data set in which there are a large number of interrelated variables, while retaining as much as possible of the variation present in the data set. This reduction is achieved by transforming to a new set of variables, the principal components (PCs), which are uncorrelated, and are ordered so that the first few retain most of the variation present in all of the original variables.

PCA research can be motivated by the open problems and continuing research on these problems, and hence a need to edit this book to report latest results on the topic. These challenges motivate the further effort on the study of PCA, and the book intends to report the new results of these efforts. This book aims to disseminate timely to the scientific community the new developments in PCA spanning from theoretical frameworks, algorithmic developments, to a variety of applications. The book covers the emerging techniques in PCA, especially those developed recently, offering academic researchers and practitioners with a complete update about the new development in this field. The book provides a forum for researchers to exchange their ideas and to foster a better understanding of the state of the art of the subject.

This book is intended for both computer science and electronics engineers (researchers and graduate students) who wish to get novel research ideas and some training in PCA, dimensional reduction, artificial intelligence, and image processing and signal processing applications. Furthermore, the research results previously scattered in several high-quality scientific journals papers will be methodically collected and presented in the book in a unified form. As a result of its twofold characteristics, the book is likely to be of interest to university researchers, R&D engineers, and graduates wishing to learn the core principles, methods, algorithms, and applications of PCA. Furthermore, the book may also be of interest to researchers working in different areas of science, as a number of results and concepts will be included which may be useful for their further research.

For this book, we have collected nine chapters with several novel contributions, namely the idea of sparse PCA by *Zhenfang Hu, Gang Pan, Yueming Wang,* and *Zhaohui Wu*, Kernel PCA and dimensionality reduction in hyperspectral images by *Aloke Datta, Susmita Ghosh*, and *Ashish Ghosh*, PCA in the presence of missing data by *Marco Geraci* and *Alessio Farcomeni*, robust PCA using generalized mean by *Jiyong Oh* and *Nojun Kwak*, PCA techniques for visualization of volumetric data by *Salaheddin Alakkari* and *John Dingliana*, outlier-resistant data processing with L1-norm PCA by *P. P. Markopoulos, S. Kundu, S. Chamadia, N. Tsagkarakis,* and *D. A. Pados*, damage and fault detection of structures using PCA by *Francesc Pozo* and *Yolanda Vidal*, PCA for exponential family data by *Meng Lu*, *Kai He*, *Jianhua Z. Huang*, and *Xiaoning Qian*, and application and extension of PCA concepts to blind unmixing of hyperspectral data with intra-class variability by *Yannick Deville, Charlotte Revel, V′eronique Achard*, and *Xavier Briottet*.

I would like to thank the authors for their excellent submissions (chapters) to this book and their significant contributions to the review process which have helped to ensure the high quality of this publication. Without their contributions, it would have not been possible for the book to come successfully into existence.

Kingswood, Australia
September 2017

Ganesh R. Naik

Contents

Sparse Principal Component Analysis via Rotation and Truncation

Zhenfang Hu, Gang Pan, Yueming Wang and Zhaohui Wu

Abstract This chapter begins with the motivation of sparse PCA–to improve the physical interpretation of the loadings. Second, we introduce the issues involved in sparse PCA problem that are distinct from PCA problem. Third, we briefly review some sparse PCA algorithms in the literature, and comment their limitations as well as problems unresolved. Forth, we introduce one of the state-of-the-art algorithms, SPCArt Hu et al. (IEEE Trans. Neural Networks Learn. Syst. 27(4):875–890, 2016), including its motivating idea, formulation, optimization solution, and performance analysis. Along with the introduction, we describe how SPCArt addresses the unresolved problems. Fifth, based on the Eckart-Young Theorem, we provide a unified view to a series of sparse PCA algorithms including SPCArt. Finally, we make a concluding remark.

1 Motivation of Sparse PCA

Commonly, the dimensions of the data have physical explanations. For example, in financial or biological applications, each dimension may correspond to a specific asset or gene [3]. However, the loadings obtained by PCA are usually dense, so the principal component, obtained by inner product, is a mixture of all dimensions,

Z. Hu · G. Pan (✉) · Z. Wu
Computer Science and Technology, Zhejiang University,
Hangzhou 310027, China
e-mail: gpan@zju.edu.cn

Z. Hu
e-mail: fancij@139.com

Z. Wu
e-mail: wzh@zju.edu.cn

Y. Wang
Qiushi Academy for Advanced Studies, Zhejiang University,
Hangzhou 310027, China
e-mail: ymingwang@zju.edu.cn

G. R. Naik (ed.), *Advances in Principal Component Analysis*,
https://doi.org/10.1007/978-981-10-6704-4_1

which makes it difficult to interpret. If most of the entries in the loadings are zeros (sparse), each principal component becomes a linear combination of a few non-zero entries. This facilitates the understanding of the physical meaning of the loadings as well as the principal components [10]. Further, the physical interpretation would be clearer if different loadings have different non-zero entries, corresponding to different dimensions. This is the motivation of sparse PCA.

2 Involved Issues

Sparse PCA attempts to find a sparse basis to make the result more interpretable [11]. At the same time, the basis is required to represent the data distribution faithfully. Thus, there is a tradeoff between statistical fidelity and interpretability.

During the past decade, a variety of methods for sparse PCA have been proposed. Most have considered the tradeoff between sparsity and explained variance. However, three points have not received sufficient attention: orthogonality between loadings, balance of sparsity among loadings, and the pitfall of deflation algorithms.

- Orthogonality. PCA loadings are orthogonal. But in pursuing sparse loadings, this property is easily lost. Orthogonality is desirable in that it indicates the independence of the physical meaning of the loadings. When the loadings are sufficiently sparse, orthogonality usually implies non-overlapping of their supports. So under the background of improving the interpretation of PCA, each sparse loading is associated with distinctive physical variables, so are the principal components. This makes interpretation simpler. If the loadings are not an orthogonal basis, the inner products between the data and the loadings, used to compute the components, do not constitute an exact projection. As an extreme example, if two loadings were very close, the two components would be similarly close, which would be meaningless.
- Balance of sparsity. There should not be any member of the loadings that is highly dense, particularly those leading ones that account for most variance. We emphasize this point, because quite a few existing algorithms yield highly dense leading loadings (close to those of PCA) while the minor ones are sparse. Thus, sparsity is achieved by the minor loadings, whereas variance is explained by the dense loadings. This is unreasonable, since sparse PCA aims to find sparse loadings that explain as much variance as possible.
- Pitfall of deflation. Existing work can be categorized into deflation and block groups. To obtain r sparse loadings, the deflation group computes one loading at a time, with others calculated via removing components that have been computed [17]. This follows traditional PCA. The block group finds all loadings together. However, in general, the optimal loadings found when we restrict the subspace

to be of dimension r may not overlap with the $r + 1$ optimal loadings when the dimensionality increases to $r + 1$ [12]. This does not occur for PCA, whose loadings successively maximize the variance, and the loadings found via deflation are always globally optimal for any r. But it is not the case for sparse PCA, the deflation method is greedy and may not find optimal sparse loadings, whereas the block group has the potential to obtain optimal solutions and should be preferred.

Finally, it should be mentioned that the loadings obtained by deflation are nearly orthogonal, while the block group usually does not ensure orthogonality.

3 Related Work

First we give a brief review of various sparse PCA methods proposed previously, and then introduce a new method, called SPCArt [9], that has achieved the state-of-the-art performance recently.

- Post-processing PCA. Originally, interpretability was gained by post-processing the PCA loadings. Loading rotation (LR) [12] applied various criteria to rotate the PCA loadings so that 'simple structure' appeared, e.g., varimax criterion drives the entries to be either small or large, which is close to a sparse structure. Simple thresholding (ST) [2] obtains sparse loadings via directly setting entries of PCA loadings below a small threshold to zero.
- Covariance matrix maximization. More recently, systematic approaches based on solving explicit objectives have been proposed, starting from SCoTLASS [11], which optimizes the classical objective of PCA, i.e., maximizing the quadratic form of the covariance matrix, while imposing a sparsity constraint on each loading.
- Matrix approximation. SPCA [29] formulates the problem as a regression-type optimization, to facilitate the use of LASSO [22] or elastic-net [28] techniques to solve the problem. Recently, it has been extended to the tensor context [14]. rSVD [21] and SPC [24] obtain sparse loadings by solving a sequence of rank-1 matrix approximations, with an imposed sparsity penalty or constraint.
- Semidefinite convex relaxation. Most proposed methods are local, which suffer from being trapped in local minima. DSPCA [4] transforms the problem into a semidefinite convex relaxation problem, so that global optimality of the solution is guaranteed. This distinguishes it from most local methods. Unfortunately, its computational complexity is as high as $O(p^4\sqrt{\log p})$ (p is the number of variables), which is expensive for most applications. A variable elimination method [27] of complexity $O(p^3)$ was later developed to make the application feasible on large scale problems.

- Greedy methods. In [19], greedy search and branch-and-bound methods were used to solve small instances of the problem exactly. Each step of the algorithm has complexity $O(p^3)$, leading to a total complexity of $O(p^4)$ for a full set of solutions (solutions of cardinality ranging from 1 to p). This bound was improved for classification [18]. In contrast, another greedy algorithm PathSPCA [3] was proposed to further approximate the solution process of [19], resulting in complexity of $O(p^3)$ for a full set of solutions. For a review of DSPCA, PathSPCA, and their applications, please refer to [26].
- Power methods. The GPower method [13] formulates the problem as maximization of a convex objective function, and the solution is obtained by generalizing the power method [8] used to compute the PCA loadings. Recently, a new power method, TPower [25], and a somewhat different but related power method, ITSPCA [16], which aims at recovering sparse principal subspace, have been proposed.
- Augmented Lagrangian optimization. ALSPCA [15] solves the problem based on an augmented Lagrangian optimization. The most special feature of ALSPCA is that it simultaneously considers the explained variance, orthogonality, and correlation among principal components.

Among these methods, only LR [12], SCoTLASS [11], and ALSPCA [15] have considered the orthogonality of loadings. SCoTLASS, rSVD [21], SPC [24], the greedy methods [3, 19], one version of GPower [13], and TPower [25] belong to the deflation group. Only the solution of [4] is guaranteed to be globally optimal.

3.1 SPCArt

Recently a new approach: Sparse PCA via rotation and truncation (SPCArt) has been proposed [9]. Distinct from most traditional work, which are based on adding a sparsity penalty on the PCA objective, SPCArt looks for a rotation matrix and a sparse basis such that the sparse basis approximates the loadings of PCA after the rotation. The resulting algorithm consists of three alternative steps: rotating PCA loadings, truncating small entries, and updating the rotation matrix.

SPCArt resolves or alleviates the three issues discussed above. It has the following merits. (1) SPCArt is able to explain as much variance as the PCA loadings, since the sparse basis spans almost the same subspace as the PCA loadings. (2) The new basis is close to orthogonal, since it approximates the rotated PCA loadings. (3) The truncation tends to produce more balanced sparsity, since vectors of the rotated PCA loadings are of equal length. (4) SPCArt belongs to the block group, it is not greedy compared with the deflation group.

We list the computational complexities of some of the above algorithms in Table 1.

Table 1 Time complexities for computing r loadings from n samples of dimension p. m is the number of iterations, and k is the cardinality of a loading. Preprocessing and initialization overheads are omitted (ST and SPCArt have the additional cost of PCA). The complexities of SPCArt listed below are of the truncation types T-ℓ_0 and T-ℓ_1. Those of T-sp and T-en are $O(rp\log p + r^2p + r^3)$. "GPowerB" refers to the block version of GPower, while "GPower" the deflation version

	PCA [10]	ST [2]	SPCA [29]	PathSPCA [3]	ALSPCA [15]	GPower [13], TPower [25]	GPowerB [13]	SPCArt [9]
$n > p$	$O(np^2)$	$O(rp)$	$mO(r^2p + rp^3)$	$O(rkp^2 + rk^3)$	$mO(rp^2)$	$mO(rp^2)$	$mO(rpn + r^2n)$	$mO(r^2p + r^3)$
$n < p$	$O(pn^2)$	$O(rp)$	$mO(r^2p + rnp)$	$O(rknp + rk^3)$	$mO(rnp)$	$mO(rnp)$	$mO(rpn + r^2n)$	$mO(r^2p + r^3)$

4 SPCArt: Sparse PCA via Rotation and Truncation

We first explain the basic idea of SPCArt, then introduce the motivation, the objective and optimization, and the truncation types. Finally, we introduce performance analysis. The major notations are listed in Table 2.

The basic idea of SPCArt is as follows. Any rotation of the r PCA loadings, $[V_1, \ldots, V_r] \in \mathbb{R}^{p\times r}$, constitutes an orthogonal basis spanning the same subspace, denoted with $X = VR$ ($R \in \mathbb{R}^{r\times r}$, $R^T R = I$). SPCArt wants to find a rotation matrix, R, through which V is transformed to a sparse basis, X. It is difficult to solve this problem directly, so SPCArt seeks a rotation matrix and a sparse basis such that the sparse basis approximates the PCA loadings after the rotation $V \approx XR$.

4.1 Motivation

SPCArt was motivated by the Eckart-Young theorem [6]. This theorem considers the problem of approximating a matrix by the product of two low-rank matrices.

Theorem 1 (Eckart-Young Theorem) *Assume the SVD of a matrix* $A \in \mathbb{R}^{n\times p}$ *is* $A = U\Sigma V^T$, *where* $U \in \mathbb{R}^{n\times m}$, $m \le min\{n, p\}$, $\Sigma \in \mathbb{R}^{m\times m}$ *is diagonal with*

Table 2 Major notations

Notation	Note
$A \in \mathbb{R}^{n\times p}$	A data matrix with n samples of p variables
$V = [V_1, V_2, \ldots]$	PCA loadings arranged column-wise. V_i denotes the ith column. $V_{1:r}$ denotes the first r columns
R	The rotation matrix
Z	The rotated PCA loadings, i.e., VR^T
X	Sparse loadings arranged column-wise, similar to V
$Polar(\cdot)$	For a matrix $B \in \mathbb{R}^{n\times p}$, $n \ge p$, let the thin SVD be WDQ^T, $D \in \mathbb{R}^{p\times p}$, then $Polar(B) = WQ^T$
$S_\lambda(\cdot)$	$0 \le \lambda < 1$. For a vector x, $S_\lambda(x)$ is entry-wise soft thresholding: $S_\lambda(x_i) = sign(x_i)(\|x_i\| - \lambda)_+$, where $[y]_+ = y$ if $y \ge 0$ and $[y]_+ = 0$ otherwise
$H_\lambda(\cdot)$	$0 \le \lambda < 1$. For a vector x, $H_\lambda(x)$ is entry-wise hard thresholding: $H_\lambda(x_i) = x_i[sign(\|x_i\| - \lambda)]_+$, i.e., $H_\lambda(x_i) = 0$ if $\|x_i\| \le \lambda$, $H_\lambda(x_i) = x_i$ otherwise
$P_\lambda(\cdot)$	$\lambda \in \{0, 1, 2, \ldots\}$. For a vector x, $P_\lambda(x)$ sets the smallest λ entries (absolute value) to zero
$E_\lambda(\cdot)$	$0 \le \lambda < 1$. For a vector x, $E_\lambda(x)$ sets the smallest k entries, whose energy accounts for at most λ proportion, to zero. k is found as follows. Sort $\|x_1\|, \|x_2\|, \ldots$ in ascending order: $\bar{x}_1, \bar{x}_2, \ldots$, then $k = \max_i i,\ s.t. \sum_{j=1}^{i} \bar{x}_j^2/\\|x\\|_2^2 \le \lambda$

$\Sigma_{11} \geq \Sigma_{22} \geq \cdots \geq \Sigma_{mm}$, *and* $V \in \mathbb{R}^{p \times m}$. *A rank-r* $(r \leq m)$ *approximation of* A *is to solve the following problem:*

$$\min_{Y,X} \|A - YX^T\|_F^2, \ s.t. X^T X = I, \tag{1}$$

where $Y \in \mathbb{R}^{n \times r}$, *and* $X \in \mathbb{R}^{p \times r}$. *A solution is*

$$X^* = V_{1:r}, \ Y^* = AX^*, \tag{2}$$

where $V_{1:r}$ *is the first r columns of* V.

Alternatively, the solution can be expressed as

$$Y^* = U_{1:r}\Sigma_{1:r}, \ X^* = Polar(A^T Y^*), \tag{3}$$

where $Polar(\cdot)$ is the orthonormal component of the polar decomposition [13]. From the SVD perspective, its equivalent definition is provided in Table 2.

Note that if A is a mean-removed data matrix with n samples, then $V_{1:r}$ are the loadings obtained by PCA. Clearly, $X^* = V_{1:r}R$ and $Y^* = AX^* = U_{1:r}\Sigma_{1:r}R$ are also a solution of (1), $\forall$ rotation matrix R. This implies that any rotation of the r orthonormal leading eigenvectors, $V_{1:r}$, is a solution of the best rank-r approximation of A. That is, any orthonormal basis in the corresponding eigensubspace is capable of representing the original data distribution as well as the original basis. Thus, a natural idea for sparse PCA is to find a rotation matrix, R, so that $X = V_{1:r}R$ becomes sparse, i.e.,

$$\min_{R \in \mathbb{R}^{r \times r}} \|V_{1:r}R\|_0, \ s.t. \ R^T R = I, \tag{4}$$

where $\|\cdot\|_0$ denotes the sum of ℓ_0 (pseudo) norm of the columns of a matrix, i.e., the count of non-zeros of a matrix.

4.2 Objective and Optimization

Unfortunately, the above problem is difficult to solve, so SPCArt approximates it. Since $X = V_{1:r}R \Leftrightarrow V_{1:r} = XR^T$, SPCArt wants to find a rotation matrix, R, through which a sparse basis, X, approximates the PCA loadings. Without confusion, we use V to denote $V_{1:r}$ hereafter. Let us first consider the ℓ_1 version:

$$\begin{aligned} &\min_{X,R} \frac{1}{2}\|V - XR\|_F^2 + \lambda \sum_i \|X_i\|_1, \\ &s.t. \forall i, \ \|X_i\|_2 = 1, \ R^T R = I, \end{aligned} \tag{5}$$

where $\|\cdot\|_1$ is the ℓ_1 norm of a vector, i.e., the sum of absolute values. The ℓ_0 version will be introduced in the next section. It is well-known that ℓ_1 norm is sparsity inducing, which is a convex surrogate of the ℓ_0 norm [5]. Under this objective, the solution may not be orthogonal, and may deviate from the eigensubspace spanned by V. However, if the approximation is accurate enough, i.e., $V \approx XR$, then $X \approx VR^T$ would be nearly orthogonal and explain similar variance as V. Note that the above objective turns out to be a matrix approximation problem of the Eckart-Young theorem. The key difference is that a sparsity penalty is added, but the solutions still share some common features.

There are no closed-form solutions for R and X simultaneously. To find a local minimum, SPCArt solves the problem by fixing one and optimizing the other alternately, i.e., the block coordinate descent [23]. Fortunately, both subproblems have closed-form solutions.

4.2.1 Fixing *X* and Solving *R*

When X is fixed, it becomes a Procrustes problem [29]:

$$\min_R \|V - XR\|_F^2, \ s.t. \ R^T R = I. \tag{6}$$

The solution is $R^* = Polar(X^T V)$. It has the same form as the right part of (3).

4.2.2 Fixing *R* and Solving *X*

When R is fixed, it becomes

$$\min_X \frac{1}{2}\|VR^T - X\|_F^2 + \lambda \sum_i \|X_i\|_1, \ s.t. \forall i, \ \|X_i\|_2 = 1. \tag{7}$$

There are r independent subproblems, one for each column: $\min_{X_i} 1/2\|Z_i - X_i\|_2^2 + \lambda\|X_i\|_1, \ s.t. \ \|X_i\|_2 = 1$, where $Z = VR^T$. It is equivalent to $\max_{X_i} Z_i^T X_i - \lambda \|X_i\|_1, \ s.t. \ \|X_i\|_2 = 1$. The solution is $X_i^* = S_\lambda(Z_i)/\|S_\lambda(Z_i)\|_2$ [13]. $S_\lambda(\cdot)$ is the entry-wise soft thresholding, defined in Table 2. This is the truncation type T-ℓ_1, i.e., soft thresholding.

The solution has the following physical explanations. Z contains the rotated PCA loadings, so it is orthonormal. X is obtained via truncating small entries of Z. On one hand, because of the unit length of each column in Z, a single threshold $0 \le \lambda < 1$ is feasible to make the sparsity distribute evenly among the columns in X; otherwise we have to apply different thresholds for different columns, which would be difficult to determine. On the other hand, because of the orthogonality of Z and small truncations, X is still possible to be nearly orthogonal. These are the most distinctive features of SPCArt, which enable simple analysis and parameter setting.

Algorithm 1 SPCArt

1: **Input:** Data matrix $A \in \mathbb{R}^{n \times p}$, number of loadings r, truncation type T, and truncation parameter λ.
2: **Output:** Sparse loadings $X = [X_1, \ldots, X_r] \in \mathbb{R}^{p \times r}$.
3: Initialize R: $R = I$.
4: PCA: compute rank-r SVD of A: $U \Sigma V^T$, $V \in \mathbb{R}^{p \times r}$.
5: **repeat**
6: Rotation: $Z = V R^T$.
7: Truncation: $\forall i$, $X_i = T_\lambda(Z_i)/\|T_\lambda(Z_i)\|_2$.
8: Update R: thin SVD of $X^T V$: $W D Q^T$, $R = W Q^T$.
9: **until** convergence

The algorithm of SPCArt is presented in Algorithm 1, where the truncation in line 7 can be any type, including T-ℓ_1 and the others that will be introduced in next section.

The computational complexity of SPCArt is shown in Table 1. Except for the computational cost of PCA loadings, SPCArt scales linearly with the data dimension. When the number of loadings is not too large, it is efficient.

4.3 Truncation Types

Given rotated PCA loadings, Z, we introduce the truncation operation of SPCArt, $T_\lambda(Z_i)$, where T_λ is one of the following four types: T-ℓ_1, soft thresholding S_λ; T-ℓ_0, hard thresholding H_λ; T-sp, truncation by sparsity P_λ; and T-en, truncation by energy E_λ. T-ℓ_1 was introduced in the previous section, we now introduce the remaining three types.

T-ℓ_0: hard thresholding. Set the entries below threshold λ to zero: $X_i^* = H_\lambda(Z_i)/\|H_\lambda(Z_i)\|_2$. $H_\lambda(\cdot)$ is defined in Table 2. It is resulted from ℓ_0 penalty:

$$\min_{X,R} \|V - XR\|_F^2 + \lambda^2 \sum_i \|X_i\|_0, \; s.t. \; R^T R = I. \tag{8}$$

The optimization is similar to the ℓ_1 case. Fixing X, $R^* = Polar(X^T V)$. Fixing R, the problem becomes $\min_X \|V R^T - X\|_F^2 + \lambda^2 \|X\|_0$. Let $Z = V R^T$, it can be decomposed to $p \times r$ entry-wise subproblems, and the solution is apparent: if $|Z_{ji}| \le \lambda$, then $X_{ji}^* = 0$, otherwise $X_{ji}^* = Z_{ji}$. Hence the solution can be expressed as $X_i^* = H_\lambda(Z_i)$.

There is no normalization for X^* compared with the ℓ_1 case. This is because, if the unit length constraint, $\|X_i\|_2 = 1$, is added, there will be no closed-form solution. However, in practice, SPCArt still uses $X_i^* = H_\lambda(Z_i)/\|H_\lambda(Z_i)\|_2$ for consistency, since empirically no significant difference is observed.

Note that both ℓ_0 and ℓ_1 penalties only result in thresholding operations on Z. Hence, we may devise other heuristic truncation types irrespective of explicit objectives.

T-sp: truncation by sparsity. Truncate the smallest λ entries: $X_i = P_\lambda(Z_i)/\|P_\lambda(Z_i)\|_2, \lambda \in \{0, 1, \ldots, p-1\}$. Table 2 gives the precise definition of $P_\lambda(\cdot)$. It can be shown that this heuristic type is resulted from the ℓ_0 constraint:

$$\min_{X,R} \|V - XR\|_F^2, \tag{9}$$
$$s.t. \forall i, \ \|X_i\|_0 \le p - \lambda, \ \|X_i\|_2 = 1, \ R^T R = I.$$

When X is fixed, it is the same as the ℓ_0 and ℓ_1 cases. When R is fixed, the solution is $X_i^* = P_\lambda(Z_i)/\|P_\lambda(Z_i)\|_2$, where $Z = VR^T$ (refer to [9] for the proof).

T-en: truncation by energy. Truncate the smallest entries whose energy (sum of square) accounts for λ proportion: $X_i = E_\lambda(Z_i)/\|E_\lambda(Z_i)\|_2$. E_λ is described in Table 2. Unlike previous cases, the objective associated with this type is not clear.

Algorithm 1 describes the complete SPCArt algorithm. SPCArt promotes the seminal ideas of simple thresholding (ST) [2] and loading rotation (LR) [12]. When using T-ℓ_0, the first iteration of SPCArt, i.e., $X_i = H_\lambda(V_i)$, corresponds to the ad-hoc ST, which is frequently used in practice and sometimes achieved good results [19, 29]. The motivation of SPCArt, i.e., (4), is similar to LR, whereas SPCArt explicitly seeks sparse loadings via ℓ_0 pseudo-norm, LR seeks a 'simple structure' via various criteria. For example, the varimax criterion maximizes the variances of squared loadings, $\sum_i[\sum_j Z_{ji}^4 - 1/p(\sum_k Z_{ki}^2)]$, where $Z = VR$. It drives the entries to distribute unevenly, either small or large (see Sect. 7.2 in [10]).

4.4 *Performance Analysis*

This section discusses the performance bounds for each truncation type. For $X_i = T_\lambda(Z_i)/\|T_\lambda(Z_i)\|_2$, the following problems are studied:

1. How much sparsity of X_i is guaranteed?
2. How much does X_i deviate from Z_i?
3. What is the orthogonality degree of X?
4. How much variance is explained by X?

The derived performance bounds are functions of λ. The sparsity, orthogonality, and explained variance can be directly or indirectly controlled via λ.[1] We first introduce some definitions.

[1]Theorem 13 is specific to SPCArt, which concerns the important explained variance. The other results are applicable to more general situations: Propositions 6–11 are applicable to any orthonormal Z, Theorem 12 is applicable to any matrix X. To obtain results specific to SPCArt, some assumptions of the data distribution are needed.

Definition 2 $\forall x \in \mathbb{R}^p$, the **sparsity** of x is the proportion of zero entries: $s(x) = 1 - \|x\|_0/p$.

Definition 3 $\forall z \in \mathbb{R}^p$, $z \neq 0$, $x = T_\lambda(z)/\|T_\lambda(z)\|_2$, the **deviation** of x from z is $\sin(\theta(x, z))$, where $\theta(x, z)$ is the included angle between x and z, $0 \leq \theta(x, z) \leq \pi/2$. If $x = 0$, $\theta(x, y)$ is defined to be $\pi/2$.

Definition 4 $\forall x,\ y \in \mathbb{R}^p$, $x \neq 0$, $y \neq 0$, the **nonorthogonality** between x and y is $|\cos(\theta(x, y))| = |x^T y|/(\|x\|_2 \cdot \|y\|_2)$, where $\theta(x, y)$ is the included angle between x and y.

Definition 5 Given data matrix $A \in \mathbb{R}^{n\times p}$ containing n samples of dimension p, $\forall$ basis $X \in \mathbb{R}^{p\times r}$, $r \leq p$, the **explained variance** by X is $EV(X) = tr(X^T A^T AX)$. Let U be any orthonormal basis in the subspace spanned by X, then the **cumulative percentage of explained variance** is $CPEV(X) = tr(U^T A^T AU)/tr(A^T A)$ [21].

Intuitively, larger λ leads to higher sparsity and larger deviation. When two truncated vectors deviate from their originally orthogonal vectors, in the worst case, the nonorthogonality degenerates as the 'sum' of their deviations. On the other side, if the deviations of a sparse basis from the rotated loadings are small, we may expect the sparse basis still represents the data well, and the explained variance maintains a similar level to that of PCA. In a word, both the nonorthogonality and explained variance depend on the deviation, and the deviation and sparsity in turn are controlled by λ. We now go into details. For the proofs of the results, please refer to [9].

4.4.1 Orthogonality

Proposition 6 *The relative upper bound of nonorthogonality between X_i and X_j, $i \neq j$, is*

$$|\cos(\theta(X_i, X_j))| \leq \begin{cases} \sin(\theta(X_i, Z_i) + \theta(X_j, Z_j)) & , \theta(X_i, Z_i) + \theta(X_j, Z_j) \leq \frac{\pi}{2}, \\ 1 & , \textit{otherwise}. \end{cases} \tag{10}$$

The bounds can be obtained by considering the two conical surfaces generated by the axes Z_i and Z_j, with rotational angles $\theta(X_i, Z_i)$ and $\theta(X_j, Z_j)$. The proposition implies the nonorthogonality is determined by the sum of deviation angles. When the deviations are small, the orthogonality is good. The deviation depends on λ, which is introduced below.

4.4.2 Sparsity and Deviation

The following results only concern a single vector of the basis. We will denote Z_i by z, and X_i by x for simplicity, and derive bounds of sparsity, $s(x)$, and deviation,

$\sin(\theta(x, z))$, for each T. They depend on a key value, $1/\sqrt{p}$, which is the entry value of a uniform vector.

Proposition 7 *For T-ℓ_0, the sparsity bounds are*

$$\begin{cases} 0 \leq s(x) \leq 1 - \frac{1}{p} & , \lambda < \frac{1}{\sqrt{p}}, \\ 1 - \frac{1}{p\lambda^2} < s(x) \leq 1 & , \lambda \geq \frac{1}{\sqrt{p}}. \end{cases} \tag{11}$$

The deviation is $\sin(\theta(x, z)) = \|\bar{z}\|_2$, *where* $\bar{z}$ *is the truncated part:* $\bar{z}_i = z_i$ *if* $x_i = 0$, *and* $\bar{z}_i = 0$ *otherwise. The absolute bounds of deviation are:*

$$0 \leq \sin(\theta(x, z)) \leq \begin{cases} \sqrt{p-1}\lambda & , \lambda < \frac{1}{\sqrt{p}}, \\ 1 & , \lambda \geq \frac{1}{\sqrt{p}}. \end{cases} \tag{12}$$

All the above bounds are achievable.

Since when $\lambda < 1/\sqrt{p}$, there is no sparsity guarantee, λ is usually set to be $1/\sqrt{p}$ in practice. It generally works well.

Proposition 8 *For T-ℓ_1, the bounds of* $s(x)$ *and lower bound of* $\sin(\theta(x, z))$ *are the same as T-ℓ_0's. In addition, there are relative deviation bounds*

$$\|\bar{z}\|_2 \leq \sin(\theta(x, z)) < \sqrt{\|\bar{z}\|_2^2 + \lambda^2 \|x\|_0}. \tag{13}$$

It is still an open question whether T-ℓ_1 has the same upper bound of deviation as T-ℓ_0. By the relative lower bounds, we have

Corollary 9 *The deviation of soft thresholding is always larger than that of hard thresholding, if the same* λ *is applied.*

This implies that results obtained by T-ℓ_1 have potentially greater sparsity and less explained variance than those of T-ℓ_0.

Proposition 10 *For T-sp,* $\lambda/p \leq s(z) < 1$, *and*

$$0 \leq \sin(\theta(x, z)) \leq \sqrt{\lambda/p}\ . \tag{14}$$

Generally $s(z) = \lambda/p$, except for the unusual case that x originally has many zeros. The main advantage of T-sp lies in its direct control on the sparsity.

Proposition 11 *For T-en,* $0 \leq \sin(\theta(x, z)) \leq \sqrt{\lambda}$. *In addition,*

$$\lfloor \lambda p \rfloor / p \leq s(x) \leq 1 - 1/p. \tag{15}$$

If $\lambda < 1/p$, *there is no sparsity guarantee. When* p *is moderately large,* $\lfloor \lambda p \rfloor / p \approx \lambda$.

Due to the discrete nature of the operand, the actually truncated energy may be less than λ. However, in practice, and especially when p is moderately large, the effect is negligible. Thus, usually $\sin(\theta(x, z)) \approx \sqrt{\lambda}$. The main advantage of T-en is that it allows direct control of deviation. Recall that the deviation has direct influence on the explained variance. Thus, if it is desirable to gain specific explained variance, T-en is preferable. Besides, if p is moderately large, T-en also provides control on sparsity.

4.4.3 Explained Variance

Finally, we introduce bounds for the explained variance $EV(X)$. Two results are obtained. The first is general and applicable to any basis X, not limited to sparse ones. The second is tailored to SPCArt.

Theorem 12 *Let rank-r SVD of* $A \in \mathbb{R}^{n\times p}$ *be* $U\Sigma V^T$, $\Sigma \in \mathbb{R}^{r\times r}$. *Given* $X \in \mathbb{R}^{p\times r}$, *assume the SVD of* $X^T V$ *to be* WDQ^T, $D \in \mathbb{R}^{r\times r}$, $d_{min} = \min_i D_{ii}$, *then*

$$d_{min}^2 \cdot EV(V) \leq EV(X), \tag{16}$$

and $EV(V) = \sum_i \Sigma_{ii}^2$.

The theorem can be interpreted as follows. If X is a basis that approximates the rotated PCA loadings well, then d_{min} will be close to one, and so the variance explained by X is close to that explained by PCA. Note that the variance explained by PCA loadings is the largest value that is possible to be achieved by an orthonormal basis. Conversely, if X deviates greatly from the rotated PCA loadings, then d_{min} tends to zero, so the variance explained by X is not guaranteed to be large. Thus, the less the sparse loadings deviate from the rotated PCA loadings, the more variance they explain.

When SPCArt converges, i.e., $X_i = T_\lambda(Z_i)/\|T_\lambda(Z_i)\|_2$, where $Z = VR^T$, and $R = Polar(X^T V)$ hold simultaneously, there is another estimation (mainly valid for T-en).

Theorem 13 *Let* $C = Z^T X$, *i.e.,* $C_{ij} = \cos(\theta(Z_i, X_j))$, *and let* $\bar{C}$ *be the diagonal-removed version. Assume* $\forall i$, $\theta(Z_i, X_i) = \theta$ *and* $\sum_j^r \bar{C}_{ij}^2 \leq 1$, *then*

$$(\cos^2(\theta) - \sqrt{r-1}\sin(2\theta)) \cdot EV(V) \leq EV(X). \tag{17}$$

When θ *is sufficiently small,*

$$(\cos^2(\theta) - O(\theta)) \cdot EV(V) \leq EV(X). \tag{18}$$

Since the sparse loadings are obtained by truncating small entries of the rotated PCA loadings, and θ is the deviation angle, the theorem implies that if the deviation is small then the explained variance is close to that of PCA, as $\cos^2(\theta) \approx 1$. For example, if the truncated energy $\|\bar{z}\|_2^2 = \sin^2(\theta)$ is approximately 0.05, then 95% EV of PCA loadings is guaranteed.

The assumptions $\theta(Z_i, X_i) = \theta$ and $\sum_j^r C_{ij}^2 \leq 1$, $\forall i$, are broadly satisfied by T-en using small λ. Uniform deviation $\theta(Z_i, X_i) = \theta \; \forall i$ can be achieved by T-en, as indicated by Proposition 11. $\sum_j^r C_{ij}^2 \leq 1$ means the sum of projected length is less than 1 when Z_i is projected onto each X_j. It is satisfied if X is exactly orthogonal, whereas it is likely satisfied if X is nearly orthogonal (note Z_i may not lie in the subspace spanned by X), which can be achieved by setting small λ according to Proposition 6. In this case, about $(1 - \lambda)EV(V)$ is guaranteed.

In practice, we prefer CPEV [21] to EV. CPEV measures the variance explained by subspace rather than basis. Since it is also the projected length of A onto the subspace spanned by X, the higher CPEV, the better X represents the data. If X is not an orthogonal basis, EV may overestimate or underestimate the variance. However, if X is nearly orthogonal, the difference is small, and it is nearly proportional to CPEV.

5 A Unified View to Some Prior Work

A series of methods: PCA [10], SCoTLASS [11], SPCA [29], GPower [13], rSVD [21], TPower [25], SPC [24], and SPCArt, although proposed independently and formulated in various forms, can be derived from the common source of Theorem 1, the Eckart-Young Theorem. Most of them can be seen as the problems of matrix approximation (1), with different sparsity penalties. Most of them have two matrix variables, and the solutions of them can usually be obtained by an alternating scheme: fixing one and solving the other. Similar to SPCArt, the two subproblems are a sparsity penalized/constrained regression problem and a Procrustes problem.

PCA [10]. Since $Y^* = AX^*$, substituting $Y = AX$ into (1) and optimizing X, the problem is equivalent to

$$\max_X tr(X^T A^T AX), \;\; s.t. \, X^T X = I. \tag{19}$$

By the Ky Fan theorem [7], $X^* = V_{1:r}R, \forall R^T R = I$. If A is a mean-removed data matrix, the special solution $X^* = V_{1:r}$ contains exactly the r loadings obtained by PCA.

SCoTLASS [11]. Constraining X to be sparse in (19), we get SCotLASS

$$\max_X tr(X^T A^T AX), \; s.t. \, X^T X = I, \; \forall i, \; \|X_i\|_1 \leq \lambda. \tag{20}$$

Unfortunately, this problem is not easy to solve.

SPCA [29]. If we substitute $Y = AX$ into (1) and separate the two X's into two independent variables X and Z (so as to solve the problem via alternating), and then impose some penalties on Z, we obtain SPCA

$$\min_{Z,X} \|A - AZX^T\|_F^2 + \lambda\|Z\|_F^2 + \sum_i \lambda_{1i}\|Z_i\|_1, \tag{21}$$
$$s.t.\ X^T X = I,$$

where Z is the target sparse loadings, and λ's are weights. When X is fixed, the problem is equivalent to r elastic-net problems: $\min_{Z_i} \|AX_i - AZ_i\|_F^2 + \lambda\|Z_i\|_2^2 + \lambda_{1i}\|Z_i\|_1$. When Z is fixed, it is a Procrustes problem: $\min_X \|A - AZX^T\|_F^2$, $s.t.$ $X^T X = I$, and $X^* = Polar(A^T AZ)$.

GPower [13]. Except for some artificial factors, the original GPower solves the following ℓ_0 and ℓ_1 versions of objectives:

$$\max_{Y,W} \sum_i (Y_i^T AW_i)^2 - \lambda_i\|W_i\|_0,\ s.t. Y^T Y = I, \forall i,\ \|W_i\|_2 = 1, \tag{22}$$

$$\max_{Y,W} \sum_i Y_i^T AW_i - \lambda_i\|W_i\|_1,\ s.t.\ Y^T Y = I,\ \forall i,\ \|W_i\|_2 = 1. \tag{23}$$

They can be seen as derived from the following more fundamental cases (see [9] for details).

$$\min_{Y,X} \|A - YX^T\|_F^2 + \sum_i \lambda_i\|X_i\|_0,\ \ s.t.\ Y^T Y = I, \tag{24}$$

$$\min_{Y,X} \frac{1}{2}\|A - YX^T\|_F^2 + \sum_i \lambda_i\|X_i\|_1,\ \ s.t.\ Y^T Y = I. \tag{25}$$

These two objectives can be seen as derived from a mirror version of (1): $\min_{Y,X}$ $\|A - YX^T\|_F^2$, $s.t.$ $Y^T Y = I$, where $A \in \mathbb{R}^{n\times p}$ is still a data matrix containing n samples of dimension p. The solution is $X^* = V_{1:r}\Sigma_{1:r}R$ and $Y^* = Polar(AX^*) = U_{1:r}R$. Adding sparsity penalties to X, we get (24) and (25).

Following the alternating optimization scheme, when X is fixed, in both cases $Y^* = Polar(AX)$. When Y is fixed, the ℓ_0 case becomes $\min_X \|A^T Y - X\|_F^2 + \sum_i \lambda_i\|X_i\|_0$. Let $Z = A^T Y$, then $X_i^* = H_{\sqrt{\lambda_i}}(Z_i)$. The ℓ_1 case becomes $\min_X 1/2\|A^T Y - X\|_F^2 + \sum_i \lambda_i\|X_i\|_1$, $X_i^* = S_\lambda(Z_i)$. The ith loading is obtained by normalizing X_i to unit length.

The iterative steps combined together produce essentially the same solution processes as the original ones in [13]. However, the matrix approximation formulation makes the relations of GPower to SPCArt and others apparent. The methods rSVD, TPower, and SPC below can be seen as special cases of GPower.

rSVD [21]. rSVD can be seen as a special case of GPower, i.e., the single component case ($r = 1$). $Polar(\cdot)$ reduces to unit-length normalization. More loadings can be obtained via deflation [17, 21], e.g., updating $A \leftarrow A(I - x^*x^{*T})$ and running the procedure again. Since $Ax^* = 0$, the subsequent loadings obtained are nearly orthogonal to x^*.

If the penalties in rSVD are replaced with constraints, we obtain TPower and SPC.

TPower [25]. The ℓ_0 case leads to TPower

$$\min_{y\in\mathbb{R}^n, x\in\mathbb{R}^p} \|A - yx^T\|_F^2, \; s.t. \; \|x\|_0 \leq \lambda, \; \|y\|_2 = 1. \tag{26}$$

By alternating optimization, $y^* = Ax/\|Ax\|_2$, $x^* = P_{p-\lambda}(A^T y)$. $P_{p-\lambda}(\cdot)$ sets the smallest $p - \lambda$ entries to zero.[2] By iteration, $x^{(t+1)} \propto P_{p-\lambda}(A^T Ax^{(t)})$, which indicates equivalence to the original TPower algorithm.

SPC [24]. The ℓ_1 case is $\min_{y,d,x} \|A - ydx^T\|_F^2$, $s.t.$ $\|x\|_1 \leq \lambda$, $\|y\|_2 = 1$, $\|x\|_2 = 1$, $d \in \mathbb{R}$. d serves as the length of x in (26). If the other variables are fixed, $d^* = y^T Ax$. If d is fixed, the problem is: $\max_{y,x} tr(y^T Ax)$, $s.t.$ $\|x\|_1 \leq \lambda$, $\|y\|_2 = 1$, $\|x\|_2 = 1$. A small modification leads to SPC:

$$\max_{y,x} tr(y^T Ax), \; s.t. \; \|x\|_1 \leq \lambda, \; \|y\|_2 \leq 1, \; \|x\|_2 \leq 1,$$

which is biconvex. $y^* = Ax/\|Ax\|_2$. However, there is no analytic solution for x, it is solved by linear searching.

SPCArt shares a close relation with GPower, and based on the ideas of SPCArt, an improved version of GPower called rSVD-GP has been developed, please refer to [9] for detailed discussions.

6 Conclusion

According to the experimental results in [9], SPCArt, rSVD-GP, and PathSPCA generally perform well. PathSPCA consistently explains most variance, but it is the most computational expensive among the three. rSVD-GP and SPCArt perform similarly on sparsity, explained variance, orthogonality, and balance of sparsity. However rSVD-GP is more sensitive to parameter setting (except rSVD-GP(T-sp), i.e., TPower), and it is a greedy deflation algorithm. SPCArt belongs to the block group, its solution improves with the target dimension, and it has the potential to obtain a globally optimal solution.

When the sample size is larger than the dimension, the time cost of PathSPCA and rSVD-GP go nonlinearly with the dimension, while that of SPCArt increases much slower. They can deal with high dimensional data under different situations,

---

[2] [21] did implement this version for rSVD, but using a heuristic approach.

SPCArt: when the number of loadings is small; rSVD-GP: when the sample size is small; PathSPCA: when the target cardinality is small.

The four truncation types of SPCArt work well in different aspects: T-ℓ_0 performs well overall; T-ℓ_1 provides the best sparsity and orthogonality; T-sp directly controls the sparsity; T-en guarantees explained variance.

There are still two open questions unresolved. (1) Under what conditions can SPCArt recover the underlying sparse basis? Efforts have been made recently in [1, 16, 20, 25]. (2) Is there any explicit objective formulation for T-en?

References

1. Amini, A., Wainwright, M.: High-dimensional analysis of semidefinite relaxations for sparse principal components. Ann. Stat. **37**(5B), 2877–2921 (2009)
2. Cadima, J., Jolliffe, I.: Loading and correlations in the interpretation of principle components. J. Appl. Stat. **22**(2), 203–214 (1995)
3. d'Aspremont, A., Bach, F., Ghaoui, L.: Optimal solutions for sparse principal component analysis. J. Mach. Learn. Res. **9**, 1269–1294 (2008)
4. d'Aspremont, A., El Ghaoui, L., Jordan, M., Lanckriet, G.: A direct formulation for sparse pca using semidefinite programming. SIAM Rev. **49**(3), 434–448 (2007)
5. Donoho, D.L.: For most large underdetermined systems of linear equations the minimal ℓ1-norm solution is also the sparsest solution. Commun. Pure Appl. Math. **59**(6), 797–829 (2006)
6. Eckart, C., Young, G.: The approximation of one matrix by another of lower rank. Psychometrika **1**(3), 211–218 (1936)
7. Fan, K.: A generalization of Tychonoff's fixed point theorem. Math. Ann. **142**(3), 305–310 (1961)
8. Golub, G., Van Loan, C.: Matrix Computations, vol. 3. Johns Hopkins University Press, Baltimore (1996)
9. Hu, Z., Pan, G., Wang, Y., Wu, Z.: Sparse principal component analysis via rotation and truncation. IEEE Trans. Neural Networks Learn. Syst. **27**(4), 875–890 (2016)
10. Jolliffe, I.: Principal Component Analysis. Springer, Berlin (2002)
11. Jolliffe, I., Trendafilov, N., Uddin, M.: A modified principal component technique based on the lasso. J. Comput. Graphical Stat. **12**(3), 531–547 (2003)
12. Jolliffe, I.T.: Rotation of ill-defined principal components. Appl. Stat. pp. 139–147 (1989)
13. Journée, M., Nesterov, Y., Richtárik, P., Sepulchre, R.: Generalized power method for sparse principal component analysis. J. Mach. Learn. Res. **11**, 517–553 (2010)
14. Lai, Z., Xu, Y., Chen, Q., Yang, J., Zhang, D.: Multilinear sparse principal component analysis. IEEE Trans. Neural Networks Learn. Syst. **25**(10), 1942–1950 (2014)
15. Lu, Z., Zhang, Y.: An augmented Lagrangian approach for sparse principal component analysis. Math. Program. **135**(1–2), 149–193 (2012)
16. Ma, Z.: Sparse principal component analysis and iterative thresholding. Ann. Stat. **41**(2), 772–801 (2013)
17. Mackey, L.: Deflation methods for sparse PCA. Adv. Neural Inf. Process. Syst. **21**, 1017–1024 (2009)
18. Moghaddam, B., Weiss, Y., Avidan, S.: Generalized spectral bounds for sparse LDA. In: Proceedings of the 23rd International Conference on Machine Learning, pp. 641-648. ACM, New York (2006)
19. Moghaddam, B., Weiss, Y., Avidan, S.: Spectral bounds for sparse PCA: exact and greedy algorithms. Adv. Neural Inf. Process. Syst. **18**, 915 (2006)
20. Paul, D., Johnstone, I.M.: Augmented sparse principal component analysis for high dimensional data. arXiv preprint arXiv:1202.1242, (2012)

21. Shen, H., Huang, J.: Sparse principal component analysis via regularized low rank matrix approximation. J. Multivar. Anal. **99**(6), 1015–1034 (2008)
22. Tibshirani, R.: Regression shrinkage and selection via the lasso. J. Royal Stat. Soc. Series B (Methodol.), pp. 267–288 (1996)
23. Tseng, P.: Convergence of a block coordinate descent method for nondifferentiable minimization. J. Optim. Theory Appl. **109**(3), 475–494 (2001)
24. Witten, D., Tibshirani, R., Hastie, T.: A penalized matrix decomposition, with applications to sparse principal components and canonical correlation analysis. Biostatistics **10**(3), 515 (2009)
25. Yuan, X., Zhang, T.: Truncated power method for sparse eigenvalue problems. J. Mach. Learn. Res. **14**, 899–925 (2013)
26. Zhang, Y., d'Aspremont, A., Ghaoui, L.: Sparse pca: convex relaxations, algorithms and applications. In: Handbook on Semidefinite, Conic and Polynomial Optimization, pp. 915–940 (2012)
27. Zhang, Y., Ghaoui, L.E.: Large-scale sparse principal component analysis with application to text data. In: Advances in Neural Information Processing Systems (2011)
28. Zou, H., Hastie, T.: Regularization and variable selection via the elastic net. J. Royal Stat. Soc.: Series B (Stat. Methodol.) **67**(2), 301–320 (2005)
29. Zou, H., Hastie, T., Tibshirani, R.: Sparse principal component analysis. J. Comput. Graphical Stat. **15**(2), 265–286 (2006)

PCA, Kernel PCA and Dimensionality Reduction in Hyperspectral Images

Aloke Datta, Susmita Ghosh and Ashish Ghosh

Abstract In this chapter an application of PCA, kernel PCA with their modified versions are discussed in the field of dimensionality reduction of hyperspectral images. Hyperspectral image cube is a set of images from hundreds of narrow and contiguous bands of electromagnetic spectrum from visible to near-infrared regions, which usually contains large amount of information to identify and distinguish spectrally unique materials. In hyperspectral image analysis, reducing the dimensionality is an important step where the aim is to discard the redundant bands and make it less time consuming for classification. Principal component analysis (PCA), and the modified version of PCA, i.e., segmented PCA are useful for reducing the dimensionality. A brief detail of these PCA based methods in the field of hyperspectral images with their advantages and disadvantages are discussed here. Also, dimensionality reduction using kernel PCA (one of the non linear PCA) and its modification i.e., clustering oriented kernel PCA in this field are elaborated in this chapter. Advantages and disadvantages of all these methods are experimentally evaluated over few hyperspectral data sets with different performance measures.

1 Introduction

Development of hyperspectral sensors [1] is a significant breakthrough in remote sensing. Hyperspectral sensors acquire a set of images from hundreds of narrow and contiguous bands of the electromagnetic spectrum from visible to infrared regions. Images captured by hyperspectral sensors have ample spectral information to identify and distinguish spectrally unique materials. There are various applications of hyperspectral images [2–5] like target detection, material identification, mineral mapping,

A. Datta
Department of CSE, NIT Meghalaya, Shillong, India

S. Ghosh
Department of CSE, Jadavpur University, Kolkata, India

A. Ghosh (✉)
Center for Soft Computing Research, Indian Statistical Institute, Kolkata, India
e-mail: ash@isical.ac.in

G. R. Naik (ed.), *Advances in Principal Component Analysis*,
https://doi.org/10.1007/978-981-10-6704-4_2

vegetation species identification, mapping details of surface properties etc. To perform these tasks, homogeneous pixels with defined similarity have to be grouped together (recognition/classification) in hyperspectral images.

Recognition/Classification of patterns is either of the two tasks: supervised classification (or simply known as classification) and unsupervised classification (also known as clustering) [6]. Classification task of hyperspectral images is a very challenging task in recent days due to the presence of a large number of features for each pixel. The performance of a classifier depends on the interrelationship between sample sizes, number of features and classifier complexity. The minimum number of training patterns required for proper training may be an exponential function of the number of features present in a data set [7]. It has been often observed that more features may not increase the performance of a classifier, if the number of training samples is small relative to the number of features. This phenomenon is termed as "curse of dimensionality" [6, 8]. Another fact of hyperspectral images is that the neighboring bands are generally strongly correlated. As a result, it is possible that very less relevant information is actually being added by increasing the spectral resolution. Thus, it can be concluded that large number of features is not always needed. In case of analysis of hyperspectral images, dimensionality reduction is an important issue [9–11].

The main two approaches of dimensionality reductions in hyperspectral images are feature selection and feature extraction [8, 12]. In brief, feature selection [6, 13–19] is nothing but selecting a subset of features from the original set of features to preserve crucial information and reduce redundancy among information. Feature selection methods preserve the original physical meaning of the features; whereas, transforming the original features into a reduced set of features, which preserves the class separability as much as possible in the transformed space, is called feature extraction [20–24]. The extracted features lose the meaning of the original features, but each of the original features may contribute to make a transformed feature. The main advantages of performing feature selection and feature extraction are to improve the classification accuracy by avoiding the "curse of dimensionality" and to reduce the computational cost for classification or clustering of data. Depending on the availability of labeled patterns, feature selection/extraction is categorized into supervised and unsupervised ones. Supervised methods use class label information of patterns and, when no labeled patterns are available, unsupervised method is used for dimensionality reduction.

In this chapter, our main aim is to represent principal component analysis and its various modifications in respect to feature extraction in hyperspectral images. Principal component analysis (PCA) [10], and the modified version of PCA, i.e., segmented PCA [20] are useful for reducing the dimensionality. A brief detail of these PCA based methods in the field of hyperspectral images with their advantage and disadvantages are discussed here. Also, dimensionality reduction using kernel PCA (one of the non linear PCA) [22, 25] and its modification i.e., clustering oriented kernel PCA [26] in this field are elaborated in this chapter. Advantages and disadvantages of all these methods are experimentally evaluated over few hyperspectral data sets in terms of different performance measures.

2 Principal Component Analysis (PCA) Based Feature Extraction Method

Principal component analysis (PCA) [10, 12, 27] is an orthogonal basis transformation with the advantage that the first few principal components preserve most of the variance of the data set. This method [27], initially, calculates the covariance matrix of the given data set, and then finds the eigenvalues and eigenvectors of this matrix. Next it selects a few eigenvectors whose eigenvalues are more to form the transformation matrix to reduce the dimensions of the data set.

Suppose, there are D number of band images. So, a pixel has D number of different responses over different wavelengths. As a consequences, a pixel may be treated as a pattern of D attributes. The main target is to reduce the dimensionality from D to d $(d \ll D)$of hyperspectral image pixel.

Let, there be a set of pattern x_i, where $x_i \in \Re^D$, $i = 1, 2, ..., N$. Assume that the data are centered, i.e., $x_i \Longleftarrow x_i - E\{x_i\}$. Conventional PCA formulates the eigenvalue problem by

$$\lambda V = \Sigma_x V \tag{1}$$

where λ is eigenvalue, V is eigenvector, Σ_x is the corresponding covariance matrix over data set x which is calculated by the following equation

$$\Sigma_x = \frac{1}{N}\sum_{i=1}^{N} x_i x_i^T. \tag{2}$$

The projection on the eigenvector V^k is calculated as

$$x_{pc}^k = V^k.x. \tag{3}$$

The principal component based transformation is defined as

$$y_i = W^T x_i; \tag{4}$$

where W is the matrix of first d normalized eigenvectors of highest eigenvalues of the image covariance matrix Σ_x. T denotes the transpose operation.

Here, a pattern x_i from original D-dimensional space is transformed into y_i, a pattern in reduced d-dimensional space by choosing only the first d components (eigenvectors of highest d eigenvalues).

The transformed data set has two main properties which are significant to the application here. The variance in the original data set has been rearranged and reordered so that first few components contain almost all of the variance in the original data, and the components in the new feature space are uncorrelated in nature [20].

3 Segmented Principal Component Analysis (SPCA) Based Feature Extraction Method

In hyperspectral images, the correlations between neighboring spectral bands are generally higher than for bands further apart. If conventional PCA based method is modified so that the transformation is carried out by avoiding the low correlations between the highly correlated blocks, the efficiency of PCA will be improved. Also, the computational load is a major consideration in the case of hyperspectral data transformation, i.e., it is inefficient to transform the complete data set. So, a segmented principal component analysis comes into picture.

In this scheme [20], the complete data set is first partitioned into several subgroups, depending on the correlations of neighboring features of hyperspectral images. Highly correlated features are selected as subgroups. Then, PCA based transformation is conducted separately on each subgroup of data.

At the onset, the D number of bands of a hyperspectral images is partitioned into a few number of contiguous intervals with constant intensities (i.e., K subgroups). Highly correlated bands should be in a subgroup. Let I_1, I_2, ..., I_k, be the number of bands in the 1st, 2nd, and Kth group, correspondingly. The purpose is to obtain a set of K breakpoints $P = \{p_1, p_2, \ldots, p_K\}$, which defines the contiguous intervals $I_k = [p_k, p_{k+1})$. The partition should follow the principle that each band should be inside one block.

Let Γ be a correlation matrix of size $D \times D$, where D is the number of bands present in a hyperspectral image. Each element of Γ is γ_{ij}, where γ_{ij} represents the correlation between band images B_i and B_j. Let the size of each band image be $M \times N$. The correlation coefficient between B_i and B_j is defined as

$$\gamma_{i,j} = \frac{\Sigma_{x=1}^{M}\Sigma_{y=1}^{N}|B_i(x, y) - \mu_i||B_j(x, y) - \mu_j|}{\sqrt{(\Sigma_{x=1}^{M}\Sigma_{y=1}^{N}[B_i(x, y) - \mu_i]^2)(\Sigma_{x=1}^{M}\Sigma_{y=1}^{N}[B_j(x, y) - \mu_j]^2)}} \tag{5}$$

where μ_i and μ_j are the mean of band images B_i and B_j, respectively. $|B_i(x, y) - \mu_i|$ measures the difference between the reflectance value of pixel (x, y) from the mean value of the total image.

It is observed that the correlation between neighboring spectral bands are generally higher than for bands further apart. Partitioning is performed based on the results obtained by first considering only correlations whose absolute value exceeds a given threshold, and simultaneously searching for edges in the "image" of the correlation matrix [20]. Each value of the correlation matrix is compared with a threshold (correlation). If the magnitude is greater than the threshold value (i.e., denoted by Θ), then replace it by 1; otherwise by 0. The value of Θ has been determined depending on the value of average correlation (μ_{corr}) and standard deviation (σ_{corr}) of correlation matrix Γ as

$$\Theta = \mu_{corr} + \sigma_{corr}; \tag{6}$$

Fig. 1 Gray scale image representation of the correlation matrix of Indian data set

where,

$$\mu_{corr} = \frac{1}{D^2} \Sigma_{i=1}^{D} \Sigma_{j=1}^{D} \gamma_{i,j}; \tag{7}$$

and

$$\sigma_{corr} = sqrt(\frac{1}{D^2} \Sigma_{i=1}^{D} \Sigma_{j=1}^{D} (\gamma_{i,j} - \mu_{corr})). \tag{8}$$

The image of the thresholded correlation matrix will be a binary image with the square blocks of white color in diagonal direction. These square blocks of white color are treated as a subgroup or partition of bands. An example of the correlation matrix of AVIRIS Indian data in image form is shown in Fig. 1.

Now, PCA based transformation is conducted on each subgroup of data. Selection over obtained principal components from each subgroup is performed based on pairwise separability measure, such as the Bhattacharyya distance [20].

4 Kernel Principal Component Analysis (KPCA) Based Feature Extraction Method

PCA, basically, rotates the original axes, so that the new coordinate system aligns with the orientation of maximum variability of data. Rotation is a linear transformation and the new coordinate axes are then a linear combination of the original axes. So, PCA as a linear algorithm is inadequate to extract the non linear structures of the data. Also, PCA only considers variance between patterns which is a second order statistics, that may limit the effectiveness of the method. So, a non-linear version of PCA is considered, which is called kernel PCA (KPCA). It is capable of capturing a part of higher order statistics. So it is useful for representing the information from the original data set which is more useful to discriminate among themselves.

Kernel principal component analysis [22], a nonlinear version of the PCA is capable of capturing a part of higher order statistics, which may represent the information in a better way from the original data set to reduced data set [25]. This technique is used for reducing the dimensionality of hyperspectral images. Here, the data of the input space $\Re^D$ is mapped into another space, called feature space F, to capture higher-order statistics. A non-linear mapping function Φ is used to transfer the data from input feature space to a new feature space by

$$\Phi : \Re^D \rightarrow F;$$

$$x \rightarrow \Phi(x). \tag{9}$$

The non-linear function Φ transforms a pattern x from D-dimensional input space to another feature space F. The covariance matrix in this feature space is calculated as

$$\Sigma_{\Phi(x)} = \frac{1}{N}\sum_{i=1}^{N} \Phi(x_i)\Phi(x_i)^T. \tag{10}$$

The principal components are then computed by solving the eigenvalue problem

$$\lambda V = \Sigma_{\Phi(x)} V = \frac{1}{N}\sum_{i=1}^{N}(\Phi(x_i).V)\Phi(x_i). \tag{11}$$

Furthermore, all eigenvectors with nonzero eigenvalue must be in the span of mapped data, i.e., $V \in span\{\Phi(x_1), ..., \Phi(x_N)\}$, and there exists coefficients α_i $(i = 1, 2, ..., N)$ such that

$$V = \sum_{i=1}^{N} \alpha_i \Phi(x_i). \tag{12}$$

Here, V denotes the eigenvector and x_i denotes the ith pattern. Multiplying Eq. 11 by $\Phi(x_k)$ from left and substituting Eq. 12 into it, we get

$$\lambda \sum_{i=1}^{N} \alpha_i (\Phi(x_k)\Phi(x_i)) = \frac{1}{N}\sum_{i=1}^{N} \alpha_i \left(\Phi(x_k). \sum_{j=1}^{N}(\Phi(x_j).\Phi(x_i))\Phi(x_j) \right); \tag{13}$$

for $k = 1, ..., N$.

Calculation of principle components in feature space F is computationally prohibitive. It is possible to work implicitly in F while all computations is done in the input space using kernel trick. Using kernel function, the product in feature space is reduced to a possibly nonlinear function (denoted by ψ) in the input space

$$\Phi(x_i).\Phi(x_j) = \psi(x_i, x_j). \tag{14}$$

Now, the NXN matrix, termed as kernel matrix Ψ, is defined as

$$\Psi = \begin{pmatrix} \psi(x_1, x_1) & \psi(x_1, x_2) & \cdots & \psi(x_1, x_N) \\ \psi(x_2, x_1) & \psi(x_2, x_2) & \cdots & \psi(x_2, x_N) \\ \vdots & \vdots & \ddots & \vdots \\ \psi(x_N, x_1) & \psi(x_N, x_2) & \cdots & \psi(x_N, x_N) \end{pmatrix}.$$

Using the kernel matrix Ψ, Eq. 13 becomes

$$\lambda\alpha = \Psi\alpha; \tag{15}$$

where, $\alpha = (\alpha_1, ..., \alpha_N)^T$, T denotes the transpose operation, and one computes an eigenvalue for the expansion coefficient α_i, which is solely dependent on kernel function.

Like PCA algorithm, the data needs to be centered in F and it is done by substituting the kernel matrix Ψ by

$$\Psi_c = \Psi - 1_N\Psi - \Psi 1_N + 1_N\Psi 1_N; \tag{16}$$

where 1_N is a square matrix such as $(1_N)_{ij} = 1/N$.

For extracting features of a new pattern x with KPCA, one simply projects the mapped pattern $\Phi(x)$ into kth eigen vector V^k by

$$(V^k.\Phi(x)) = \sum_{i=1}^{M} \alpha_i^k(\Phi(x_i).\Phi(x)) = \sum_{i=1}^{M} \alpha_i^k \psi(x_i, x). \tag{17}$$

The KPCA incorporates nonlinearity in the calculation of the matrix elements of Ψ and the evaluation of the expansion.

The function ψ is a positive semi-definite function on $\Re^D$ which incorporates nonlinearity into processing. This is usually called a kernel. Selecting an appropriate kernel is a new scope of research. It is better to use Gaussian kernel if there are assumptions of the nature of clusters of data as Gaussian. Hyperspectral remote sensing data are known to be well approximated by a Gaussian distribution [28]. So, in this article, Gaussian kernel is used, which is described by following equation

$$\psi(x_i, x_j) = exp\left(-\frac{||x_i - x_j||}{2\sigma^2}\right). \tag{18}$$

In the Gaussian kernel, the parameter σ, controls the width of the exponential function. For a very small value of σ, each sample is considered as an individual cluster, and vice-versa. The value of σ depends on data set [25]. This KPCA based feature extraction method selects some percent of data from the total data set randomly to calculate the kernel matrix, i.e., value of σ [22]. The minimum distance of all representative patterns x_i with other patterns is calculated. Thus, if there are N patterns, then there will be N minimum distances. The average of this N minimum distances is calculated. σ is taken as five times of this minimum value. Thus, σ value is dependent on the nature of data set.

5 Clustering Oriented Kernel Principal Component Analysis (KPCA) Based Feature Extraction Method

The clustering oriented KPCA based feature extraction method [26] performs kernel principal component analysis to transform the original data set of dimension D into d dimensional space. The KPCA is non linear in nature and uses higher order statistics of data set to discriminate the classes. The most important thing is to select the proper training set for calculating kernel matrix for KPCA. A randomly selected training pattern may not represent the overall data set properly. Also, it should not be too large so that the method becomes computationally prohibitive. So, a proper subset of original hyperspectral data set which can represent the total data set properly should be selected and this training set should not contain any noisy data. DBSCAN clustering technique is used for choosing the proper representative training set. In this section, selection of N representative patterns using DBSCAN clustering technique is described and then discuss about the KPCA based transformation using these data.

KPCA shares the same properties as the PCA, but in a different space. Both PCA and KPCA need to solve eigenvalue problem, but the dimensions of the problem are different, $D \times D$ for PCA and $N \times N$ for KPCA, where D is the dimensions of data set and N is number of representative patterns required to calculate kernel matrix Ψ. The size of the matrix becomes problematic for large N. Number of pixel points (N) in hyperspectral images is huge, so it is difficult to perform KPCA by taking all the pixels. If some percentage of total pixels are selected randomly, then the selected pixels may not represent the characteristics of total data. So, it is better to make small group of pixels according to their similarity, and then take some representative pixels from each group to make the representative pattern set for KPCA.

Selecting of N Representative Pixels using DBSCAN Clustering

Pixels on homogeneous region have similar properties and make group or region in hyperspectral images by clustering. Each pixel of a hyperspectral image can be treated as a pattern with D attributes, where D represents the total number of features present in the images. In the proposed investigation a density based spatial clustering technique (DBSCAN) [29] is applied to obtain the region types. It does not require prior information regarding the number of clusters. DBSCAN treats a noisy pattern as an isolated point, rather than including it into any cluster. The main concept of DBSCAN clustering technique is that within each cluster, density of points is considerably higher than outside the cluster, whereas, the density around the noisy area is lower than the density in any of the clusters. So, if the neighborhood of a given radius of a pattern, contains at least a minimum number of patterns, i.e. the density in the neighborhood exceeds some threshold, then that pattern is in a cluster.

It requires two user-defined parameters, neighborhood distance (*Eps*) and the minimum number of points (*MinPts*). For a given point, the points within an Eps distance are called neighbors of that point. DBSCAN labels the data points as core points, border points, and outlier points. Core points are those which have at least *MinPts* number of points within the *Eps* distance in all directions. Border points can be defined as points that are not core points, but are the neighbors of core points. Outlier points are those which are neither core points nor border points.

The algorithm starts with an arbitrary starting point and then finds all the neighboring points within Eps distance of the starting point. If the number of points of its neighborhood is greater than or equal to *MinPts*, a cluster is formed. The starting point and its neighbors are added to this cluster and the starting point is marked as visited. The algorithm then repeats this process for all the neighbors iteratively. If the number of neighbors is less than *MinPts*, the point is marked as noise (i.e., isolated point). If a cluster is fully expanded, then the algorithm proceeds to iterate through the remaining unvisited points in the data set. The steps of DBSCAN are given in Algorithm 1.

Algorithm 1 Pseudo Code of DBSCAN Algorithm

```
Let S = {x_1, x_2, ..., x_n}
Let class(x) = -1, ∀x ∈ S
Choose Eps and MinPts
class_no = 1
for i = 1 to n do
  A_i = {x ∈ S : d(x, x_i) ≤ Eps}
  if (| A_i |≥ MinPts) then
    if (class(x_i) == -1) then
      if (max(class(x : x ∈ A_i)) > -1) then
        new_class_no = min(class(x : x ∈ A_i and class(x : x ∈ A_i) > -1))
        class(x_i) = new_class_no
        class(x : ∀x ∈ A_i) = new_class_no
      else
        class(x_i) = class_no
        class(x : ∀x ∈ A_i) = class_no
        class_no = class_no + 1
      end if
    else
      new_class_no = class(x_i)
      class(x : ∀x ∈ A_i) = new_class_no
    end if
  end if
end for
return class
```

Let DBSCAN clustering technique produce C clusters. The isolated pixels identified by DBSCAN algorithm are discarded considering them as noise. Number of clusters (C) does not lie on any predefined range, it is dependent on the data set. Basically it is better that C be close to the number of regions/ land cover types present on the hyperspectral image. The value of C gives an approximation on the number of land cover types/ groups present in the images. DBSCAN clustering technique gives only the clusters present in the data set, but not the cluster centers.

From each cluster, a certain percentage of pixels are selected as representative patterns for calculating the kernel matrix of the KPCA based method. For example, if a cluster C_1 has N_1 pixels, then $N_1/10$ number of pixels are selected from that cluster. The first selected pixels of each cluster is the mean of all the pixels present in a cluster. If a cluster mean does not represent a physical pixel in that cluster, then the nearest pixel of cluster mean is selected from that cluster. Then the next pixels from another cluster is selected which has the maximum distance from other selected pixels of that cluster. The isolated pixels or noisy pixels, which is far away from any cluster (DBSCAN clustering technique detects them and considers them separately) would not be included in the representative pattern set, because KPCA is susceptible to noise.

Now, the KPCA based transformation is performed to reduce the dimensionality from D to d, as described in Sect. 4, where the set of representative patterns are selected by DBSCAN clustering technique to properly represent the characteristics of whole data set. This technique is called as clustering oriented KPCA based feature extraction method of hyperspectral images. An outline of the clustering oriented KPCA based feature extraction method is given in Algorithm 2.

Algorithm 2 Clustering oriented KPCA based feature extraction algorithm

1. Selecting N representative pixels
   - Perform DBSCAN clustering technique over pixels of hyperspectral images which is in D-dimensional space using Algorithm 1.
   - Choose some percentage of exemplar pixels from each cluster to make N representative pixels.
   - These N pixels are used as representative pixels for calculating the kernel matrix in KPCA.
2. Using kernel PCA, transform data into reduced d-dimensional space
   - Compute kernel matrix, Ψ, using Eq. 18
   - Center Ψ, using Eq. 16
   - Solve eigen value problem of Eq. 15
   - Extract the d first principal components using Eq. 17

6 Experimental Evaluation

6.1 Description of Data Sets

Experiments are carried out to evaluate the effectiveness of these feature extraction methods on three hyperspectral remotely sensed images namely, Indian Pine [30], KSC [31], and Botswana [31] images corresponding to the geographical areas of Indian Pine test site of Northwest Indiana, Kennedy Space Center of Florida and Okavango Delta of Botswana. The data sets are described here.

Indian Pine data:
Indian Pine image [30] data was captured by AVIRIS (Airborne Visible/Infrared Imaging Spectrometer) over an agricultural portion of northwest Indiana's Indian Pine test site in the early growing season of 1992. The data has been taken within the spectral range from 400 to 2500 nm with spectral resolution of about 10 nm and has 220 spectral bands.

The size of the image is 145 × 145 pixels and spatial resolution is 20 m. Twenty water absorption bands (numbered 104–108, 150–163 and 220) and 15 noisy bands (1–3, 103, 109–112, 148–149, 164–165 and 217–219) were removed, resulting in a total of 185 bands. There are 16 classes in this image. Class name and the number of labeled samples for each class are given in Table 1. Among the 16 classes, seven classes contain fewer samples. For more details and ground truth information, see [30] and visit http://dynamo.ecn.purdue.edu/biehl/ (Figs. 2, 3, and 4).

Table 1 Indian Pine data: class names and the number of samples

Class no	Class name	No. of samples
C1	Corn	191
C2	Corn-min	688
C3	Corn-notill	1083
C4	Soybean-clean	541
C5	Soybean-min	2234
C6	Soybean-notill	860
C7	Wheat	211
C8	Alfalfa	51
C9	Oats	20
C10	Grass/ Trees	605
C11	Grass/ Pasture	351
C12	Grass/ Pasture-mowed	17
C13	Woods	1293
C14	Hay-windrowed	477
C15	Bldg-Grass-Tree-Drives	380
C16	Stone-steel-towers	86

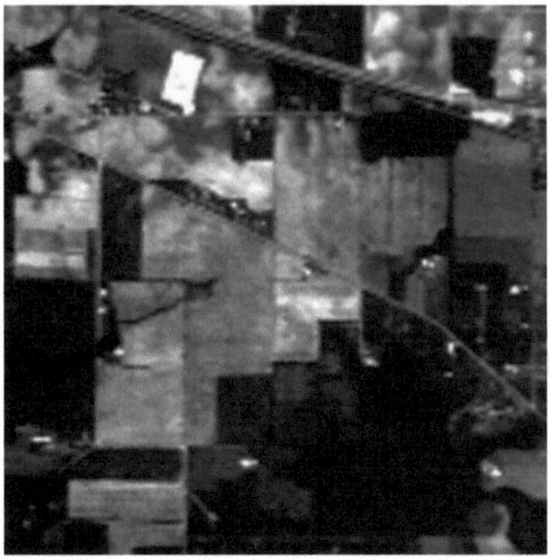

Fig. 2 Band 11 image of Indian Pine data

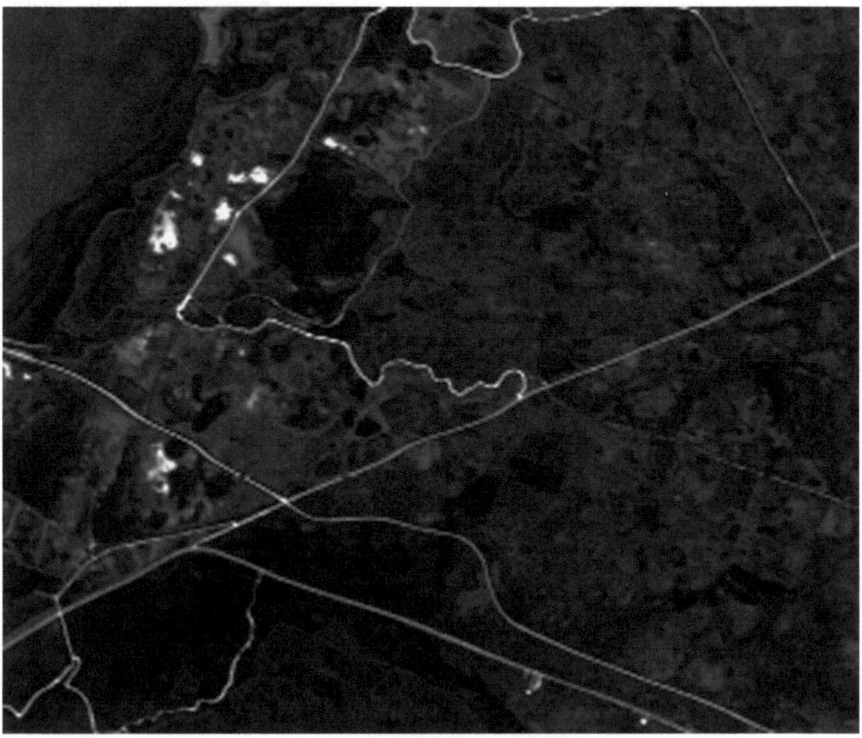

Fig. 3 Band 11 image of KSC data

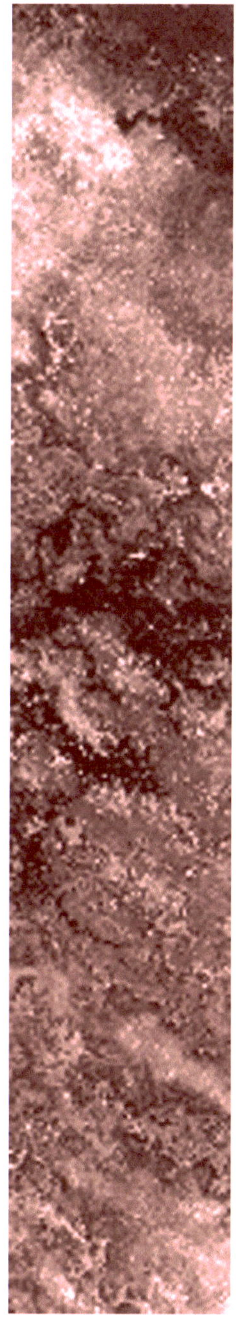

Fig. 4 Band 11 image of Botswana data

KSC data:
The KSC [31] images, acquired over Kennedy Space Center (KSC), Florida on March 23, 1996 by NASA AVIRIS, is of size 512 × 614. AVIRIS acquires data in 224 bands of 10 nm width with wavelengths ranging from 400 to 2500 nm. The data is acquired from an altitude of approximately 20 km with a spatial resolution of 18 m. After removing the bands disturbed due to water absorption or with low signal-to-noise-ratio (SNR) value (numbered 1–4, 102–116, 151–172 and 218–224), 176 bands are used for analysis. Training data were selected using land cover maps derived from color infrared photography provided by the Kennedy Space Center and Landsat Thematic Mapper (TM) imagery. The vegetation classification scheme was developed by KSC personnel in an effort to define functional types that are discernable at the spatial resolution of Landsat and the AVIRIS data [31]. Discrimination of land cover for this environment is difficult due to similarity of spectral signatures for certain vegetation type. Details of the 13 land cover classes considered in the KSC data area are listed in Table 2. For more details and ground truth information, see [31] and visit http://www.csr.utexas.edu/.

Botswana data:
The NASA Earth Observing 1 (EO-1) satellite acquired a sequence of 1476 × 256 pixels over the Okavango Delta, Botswana in 2001–2004 [31]. The Hyperion sensor on EO-1 acquired data at 30 m pixel resolution over a 7.7 km × 44 km surface are in 242 bands from the 400–2500 nm portion of the spectrum in 10 nm windows. Preprocessing of the data was performed by the UT Center for Space Research to mitigate the effects of bad detectors, inter-detector miscalibration, and intermittent anomalies. Uncalibrated and noisy bands which cover water absorption features were removed, and the remaining 145 bands were included as candidate features: [10–55, 82–97, 102–119, 134–164, 187–220]. This data was acquired on May 31, 2001 and

Table 2 KSC data: class names and the number of samples

Class no	Class name	No. of samples
C1	Scrub	761
C2	Willow swamp	243
C3	Cabbage palm hammock	256
C4	Cabbage palm/oak hammock	252
C5	Slash pine	161
C6	Oak/broadleaf hammock	229
C7	Hardwood swamp	105
C8	Graminoid marsh	431
C9	Spartina marsh	520
C10	Cattail marsh	404
C11	Salt marsh	419
C12	Mud flats	503
C13	Water	927

Table 3 Botswana data: class names and the number of samples

Class no	Class name	No. of samples
C1	Water	270
C2	Hippo Grass	101
C3	FloodPlain Grasses 1	251
C4	FloodPlain Grasses 2	215
C5	Reeds	269
C6	Riparian	269
C7	Firescar	259
C8	Island Interior	203
C9	Acacia Woodlands	314
C10	Acacia Shrublands	248
C11	Acacia Grasslands	305
C12	Short Mopane	181
C13	Mixed Mopane	268
C14	Exposed Soils	95

consists of observations from 14 identified classes representing the land cover types in seasonal swamps, occasional swamps, and drier woodlands located in the distal portion of the Delta [31]. These classes were chosen to reflect the impact of flooding on vegetation in the study area. Class names and corresponding number of ground truth observations used in our experiment are listed in Table 3. For more details and ground truth information, see [31] and visit http://www.csr.utexas.edu/.

6.2 Performance Measures

In this section, four feature evaluation indices namely, class separability (S) [8], overall classification accuracy (OA) [32], kappa coefficient (κ) [32] and entropy (E) [33], have been described which are considered for evaluating the effectiveness of the extracted features. The first three measuring indices need class label information of the samples while the last one does not require the same. The details of the evaluation indices used in this thesis, are given below.

Overall Accuracy (OA):
Overall accuracy [32] represents the ratio between the number of samples correctly recognized by the classification algorithm and the total number of test samples. To measure the overall accuracy, initially, confusion matrix is determined. The confusion matrix is a square matrix of size $C \times C$, where C represents the number of classes of the given data set. The element n_{ij} of the matrix denotes the number of samples of the jth ($j = 1, 2, ..., C$) category which are classified into ith ($i = 1, 2, ..., C$) category. Let N be the total number of samples; where $N = \sum_{i=1}^{C} \sum_{j=1}^{C} n_{ij}$. The

overall accuracy (*OA*) is defined as

$$OA = \frac{\sum_{i=1}^{C} n_{ii}}{N}. \tag{19}$$

***Kappa* Coefficient (κ):**
The kappa coefficient (κ) [32] is a measure defined on the difference between the actual agreement in the confusion matrix and the chance agreement, which is indicated by row and column totals of the confusion matrix. The kappa coefficient is widely adopted, as it also takes into consideration the off-diagonal elements of the confusion matrix and compensates for chance agreement. The value of κ lies in the range $[-1\ , +1]$. Closer the value of κ to $+1$, better is the classification.

Let, in the confusion matrix, the sum of the elements of ith row be denoted as n_{i+} (where, $n_{i+} = \sum_{j=1}^{C} n_{ij}$) and the sum of the elements of column j be n_{+j} (where $n_{+j} = \sum_{i=1}^{C} n_{ij}$). The kappa coefficient is then defined as

$$\kappa = \frac{N \sum_{i=1}^{C} n_{ii} - \sum_{i=1}^{C} n_{i+} n_{+i}}{N^2 - \sum_{i=1}^{C} n_{i+} n_{+i}}; \tag{20}$$

where N denotes the total number of samples and C denotes the number of classes of the given data set.

Class Separability:
Our aim is to look for a feature space where the inter-class distance is large and at the same time the intra-class distance is as small as possible. Let there be C classes $\omega_1, \omega_2, \ldots, \omega_C$. Assume S_w and S_b to be the intra-class and inter-class scatter matrices, respectively and can be defined as

$$S_w = \sum_{i=1}^{C} p_i \Xi\{(\mathbf{x} - \mu_i)(\mathbf{x} - \mu_i)^T \mid \omega_i\} = \sum_{i=1}^{C} p_i \Sigma_{\omega_i}; \tag{21}$$

$$S_b = \sum_{i=1}^{C} p_i (\mu - \mu_i)(\mu - \mu_i)^T; \tag{22}$$

where p_i is the a priori probability that a pattern belongs to class ω_i, $\mathbf{x}$ is a pattern vector, μ_i represents the sample mean vector of class ω_i, Σ_{ω_i} is the sample covariance matrix of class ω_i, and $\Xi\{\cdot\}$ calculates the expectation value. The overall mean vector (μ) for the entire data set is defined as

$$\mu = \sum_{i=1}^{C} p_i \mu_i. \tag{23}$$

Class separability [8], S, of a data set is defined as

$$S = trace(S_b^{-1} S_w). \tag{24}$$

A lower value of the separability measure S ensures that the classes are well separated.
Entropy:
The distance L_{pq} between the patterns $\mathbf{x}_p$ and $\mathbf{x}_q$ can be defined as:

$$L_{pq} = \left(\sum_{j=1}^{D} \left(\frac{x_{p,j} - x_{q,j}}{max_j - min_j} \right)^2 \right)^{1/2}, \tag{25}$$

where x_{pj} denotes the jth feature value of pattern $\mathbf{x}_p$, max_j and min_j are the maximum and the minimum values computed over all the patterns along the jth direction. Similarity between $\mathbf{x}_p$ and $\mathbf{x}_q$, represented as S_{pq}, can be defined as

$$S_{pq} = e^{-\alpha L_{pq}}; \tag{26}$$

where α is a positive constant. A possible value of $\alpha = \frac{-ln 0.5}{\widehat{L}}$, where $\widehat{L}$ is the mean distance among patterns computed over the entire data set. Hence α is determined by the given data and can be calculated automatically.

Entropy [33] of a pattern $\mathbf{x}_p$ with respect to all other patterns is calculated as

$$E_p = -\sum_{\mathbf{x}_q \in \Upsilon}^{\mathbf{x}_p \neq \mathbf{x}_q} \left(S_{pq} log_2 S_{pq} + (1 - S_{pq}) log_2 (1 - S_{pq}) \right). \tag{27}$$

Here Υ is a set of all patterns. Entropy of overall data set is defined by

$$E = \sum_{\mathbf{x}_p \in \Upsilon} E_p = -\sum_{\mathbf{x}_p \in \Upsilon} \sum_{\mathbf{x}_q \in \Upsilon}^{p \neq q} \left(S_{pq} log_2 S_{pq} + (1 - S_{pq}) log_2 (1 - S_{pq}) \right). \tag{28}$$

It is to be noted that, entropy is less for stable configuration of patterns (data has well formed clusters), and is more for disordered configuration, i.e., data is uniformly distributed in the feature space.

6.3 *Parameter Details*

Experiments are conducted on three hyperspectral data sets, namely, Indian Pine, KSC and Botswana. Details about the data sets are given in Sect. 6.1. As already mentioned in the previous section, the clustering oriented KPCA based method first perform DBSCAN clustering technique on pixels to choose N representative patterns and then perform KPCA based transformation on the data set to reduce the dimensionality.

DBSCAN clustering algorithm uses two parameters, namely, minimum distance with respect to a point for which neighborhood is calculated (denoted as *Eps*) and the minimum number of points in an *Eps*-neighborhood of that point (denoted by *MinPts*). Ester et al. [29] suggested to use *MinPts* equal to 4 and used a method which considers the variation of the number of points with respect to their 4th nearest neighbor distance to calculate the value of *Eps*. Although higher values for *MinPts* have also been tested, it did not produce better results. The value of *Eps* is taken to be the location of the first valley of this graph. In the clustering oriented KPCA based strategy, *MinPts* and *Eps* are calculated in accordance to Ester et al. [29]. For Indian Pine data set, *Eps* value is 110, which is the 4th nearest neighbor distance of the first valley of the graph described at Ester et al. [29] with *MinPts* equal to 4. There are about 19 clusters of pixels and few isolated pixels which do not belong to any cluster. It is better to discard the isolated pixels and not consider them in formation of representative patterns, because KPCA is susceptible to noise. Generally, the principle for selecting representative patterns from each cluster is discussed in the proposed method section. But the percentage of total patterns which are selected for representative patterns, is needed to determine. Here, 2–12% of total patterns are selected for representative patterns for calculating kernel matrix of KPCA and the performance of the clustering oriented KPCA based method in terms of overall accuracy for 18 number of extracted features for Indian Pine data is depicted in Table 4. From the table, it is observed that 8–10% data patterns are sufficient for calculating kernel matrix. Similar observations are also found for the other data sets. So, 10% data from each cluster are selected for making representative patterns. So in the set of representative patterns, a small cluster has less number of pixels and vice verse. For example, the number of representative patterns for Indian Pine data is about 850.

To assess the performance of the above mentioned methods, classification of pixels is performed using transformed features. After completing the feature extraction, fuzzy *k*-NN based classification (in theory, any good classification algorithm can be used) is performed using the transformed features in 10-fold cross validation manner. 10-fold cross validation is a well-known technique for choosing training and testing data for classification. In this method, the whole data set is randomly partitioned into 10 blocks. Each time one block of data is treated as a testing data, and the remaining 9 blocks are training data. The whole process is repeated 10 times with different training and test data sets and the average overall accuracy is calculated. There may be overlapping of information between neighboring pixels of the hyperspectral

Table 4 Performance of the clustering oriented KPCA based method in terms of OA for 18 number of extracted features with different number of small representative samples for calculating kernel matrix of KPCA for Indian Pine data

N (%)	2	5	8	10	12
OA (%)	65.57	79.45	86.36	87.69	87.58

images. Fuzzy k-NN, rather than other classification techniques, is used to take care of the fuzziness present in the hyperspectral images.

The desired number of transformed features is not known apriori, because it varies with data set. In the present investigation, experiments are carried out for different number of features ranging from 4 to 30 with a step size of 2. Overall classification accuracy (OA), kappa coefficient (κ), class separability (S) and entropy (E) are calculated for the transformed set of features to assess the effectiveness of the feature extraction methods.

6.4 *Analysis of Results*

The cumulative eigenvalues of PCA, KPCA and clustering oriented KPCA based methods are depicted in Table 5 in percentage for Indian Pine data set. The cumulative eigenvalues represent the cumulative variance of the data [22, 34]. It shows that ninety five percent of cumulative variance of PCA is retained by the first six components, while KPCA and clustering oriented KPCA based methods need 14 to 18 components. In PCA most of the information content is retained in the first few features, where as, KPCA and clustering oriented KPCA based methods require more number of components.

The obtained OA and κ for Indian Pine data after applying fuzzy k-NN classifier over the transformed set of features by PCA, segmented PCA (SPCA), kernel PCA (KPCA) and clustering oriented KPCA based methods are given in Table 6. For PCA based method, OA becomes saturated when the number of transformed feature is 10 and after that it is stabilized. For KPCA and clustering oriented KPCA based methods, OA saturated at 18 and 16 number of features, respectively. It is due to

Table 5 Percentage of cumulative eigenvalues of principal components of PCA, KPCA and clustering oriented KPCA based methods for Indian Pine data

No. of PCs	PCA	KPCA	Clustering oriented KPCA
	(Cum.%)	(Cum.%)	(Cum.%)
2	72.32	57.74	63.18
4	85.89	68.11	73.74
6	96.69	76.41	81.54
8	98.37	83.37	88.62
10	99.06	87.16	91.97
12	99.23	89.78	94.24
14	99.33	91.71	95.86
16	99.37	93.48	96.92
18	99.42	94.82	97.60
20	99.46	95.84	98.15

Table 6 Overall accuracy and kappa coefficients of PCA, SPCA, KPCA and clustering oriented KPCA based methods for different number of extracted features for Indian Pine data

No. of features	PCA		SPCA		KPCA		Clustering oriented KPCA	
	OA(%)	κ	OA(%)	κ	OA(%)	κ	OA(%)	κ
4	64.91	0.5952	52.19	0.4502	60.95	0.5516	**61.15**	**0.5539**
6	75.36	0.7167	66.24	0.6106	73.49	0.6948	**74.86**	**0.7102**
8	82.84	0.8031	74.31	0.7044	78.62	0.7541	**80.18**	**0.7713**
10	83.87	0.8144	78.92	0.7578	81.36	0.7859	**82.58**	**0.7997**
12	83.96	0.8154	82.31	0.7769	83.39	0.8092	**84.27**	**0.8187**
14	84.01	0.8159	83.68	0.8128	84.21	0.8180	**86.49**	**0.8436**
16	83.84	0.8140	84.78	0.8245	85.14	0.8287	**87.61**	**0.8559**
18	83.93	0.8149	85.16	0.8288	85.56	0.8332	**87.73**	**0.8573**
20	83.82	0.8137	85.02	0.8273	85.59	0.8336	**87.66**	**0.8565**
22	84.01	0.8159	84.98	0.8269	85.78	0.8356	**87.49**	**0.8546**
24	83.71	0.8124	85.13	0.8286	85.53	0.8328	**87.58**	**0.8556**
26	83.54	0.8102	85.01	0.8272	85.64	0.8341	**87.82**	**0.8584**
28	82.56	0.7995	84.91	0.8261	85.51	0.8324	**87.46**	**0.8542**
30	83.78	0.8132	85.10	0.8282	85.43	0.8316	**87.38**	**0.8533**

the fact that the number of principal components for PCA, KPCA and clustering oriented KPCA methods, for containing most of the variance of data, are 10, 18 and 16, respectively (shown in Table 5). It is noticed from Table 6 that Kernel PCA based methods (i.e., KPCA and clustering oriented KPCA) give better results than PCA and segmented PCA based methods. From Table 6, it is also observed that clustering oriented KPCA method achieves better results in terms of OA and κ for different number of transformed features. The reason behind this finding is that all the four methods transform the original set of features into a new set of features considering the maximum variance of data. Moreover, KPCA based methods incorporate the non linearity in transformation. The clustering oriented KPCA method gives better results than KPCA, because the representative patterns, for calculating kernel matrix for KPCA, are not selected randomly (like KPCA). The DBSCAN clustering technique is used to select the representative patterns so that it properly represents all the clusters of the data set, as well as, discard noisy pattern.

Figure 5 depicts the variation of average *OA* (in percentage) with number of features for all the methods used in the experiment. The graph corroborates to our earlier findings. For Indian Pine data, ground truth image with 16 classes is shown in Fig. 6, where different colors are used to distinguish the pixels among classes. Figure 7a–d shows the pictorial representation of the classified image with the best subset of features extracted using PCA, SPCA, KPCA and clustering oriented KPCA based techniques, correspondingly. A view of the classified images show that the clustering oriented KPCA based technique transforms a better set of features for classification

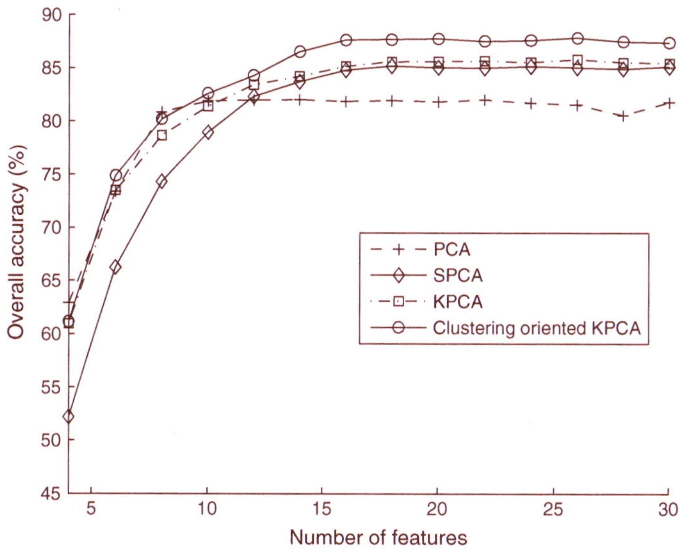

Fig. 5 Comparison of the performance of PCA, SPCA, KPCA and clustering oriented KPCA based methods in terms of overall accuracy with respect to the number of features used for Indian Pine data

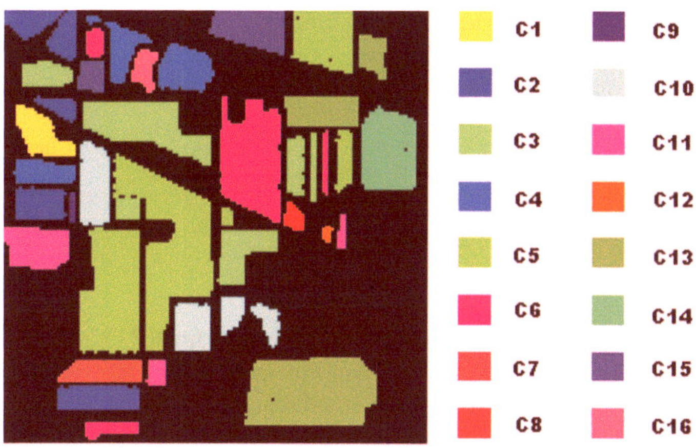

Fig. 6 Ground truth image of Indian Pine data

of given hyperspectral images as compared to other methods. It is clearly observed that the classified Indian Pine image with transformed feature set using clustering oriented KPCA based method has very less misclassified pixels compared to other methods. Table 10 contains the optimum value of OA, κ, S and E for all three hyperspectral data sets for all four methods. From this table, it is noticed that clustering oriented KPCA based method gives less value of S and E, which is better with respect to the other three methods used in our experiments. It shows that the clustering oriented KPCA method transforms better subset of features which gives well separated classes as well as stable configuration of patterns compared to other methods.

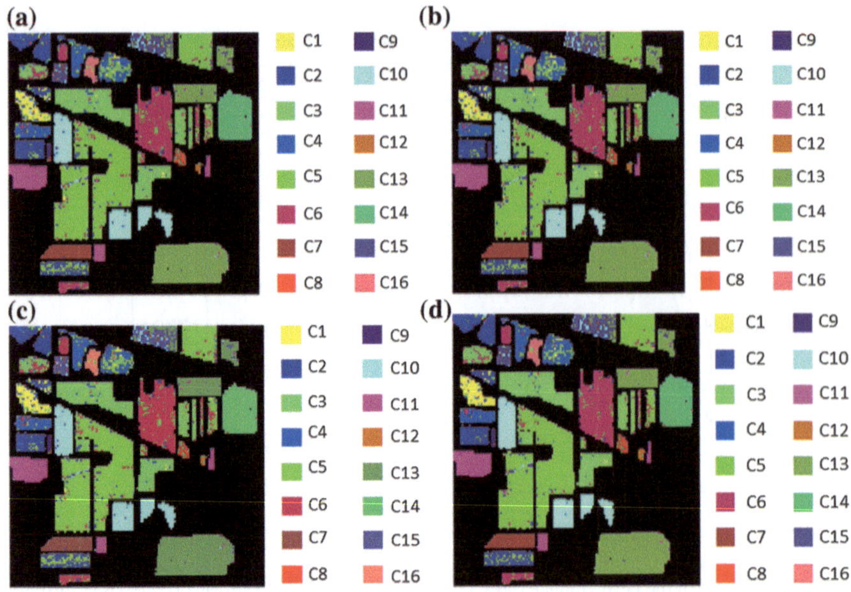

Fig. 7 Classified images of Indian Pine data with extracted feature set using (a) PCA, (b) SPCA, (c) KPCA, and (d) clustering oriented KPCA based methods

Table 7 CPU time for PCA, SPCA, KPCA and clustering oriented KPCA based methods using Indian Pine data

	PCA	SPCA	KPCA	Clustering oriented KPCA
CPU time (s)	15.29	48.24	79.12	61.31

For comparing the computational costs, using an Intel(R) Core(TM) i7 2600 CPU @ 3.40-GHz processor and an Indian Pine image with 185 features of 145×145 pixels, clustering oriented KPCA method required about 61.31 s. Programs are developed in C. Table 7 gives a simple quantitative analysis of the computational cost of each method for Indian Pine data. The clustering oriented KPCA method takes much less time than KPCA, where all the patterns are used for kernel matrix, but it takes little more time than PCA based methods (i.e., PCA and SPCA).

Overall accuracy (OA) and kappa coefficient (κ) for KSC and Botswana data sets are put in Tables 8 and 9, respectively. From the table, it is observed that clustering oriented KPCA based method is producing better results than the other methods for both the data sets. A variation of OA for these methods with the number of transformed features are depicted graphically in Figs. 8 and 9, respectively, for KSC and Botswana data. Results for these data sets corroborate to our earlier findings. It is also observed that KPCA based transformation (KPCA and clustering oriented KPCA) are found to be better than PCA based methods (PCA and Segmented PCA).

Table 8 Overall accuracy and kappa coefficients of PCA, SPCA, KPCA and clustering oriented KPCA based methods for different number of extracted features for KSC data

No. of features	PCA		SPCA		KPCA		Clustering oriented KPCA	
	OA(%)	κ	OA(%)	κ	OA(%)	κ	OA(%)	κ
4	69.44	0.6483	59.07	0.5311	65.84	0.6054	**67.21**	**0.6213**
6	78.37	0.7513	73.41	0.6943	77.31	0.7390	**78.19**	**0.7493**
8	86.21	0.8407	79.87	0.7703	85.37	0.8311	**85.81**	**0.8359**
10	88.50	0.8658	85.31	0.8304	87.21	0.8514	**88.47**	**0.8655**
12	88.72	0.8682	89.01	0.8714	88.79	0.8690	**89.21**	**0.8736**
14	88.84	0.8695	90.10	0.8835	89.38	0.8755	**89.98**	**0.8822**
16	88.89	0.8701	89.92	0.8815	90.61	0.8898	**90.72**	**0.8910**
18	88.69	0.8679	90.03	0.8827	90.72	0.8910	**91.92**	**0.9042**
20	88.42	0.8649	90.21	0.8847	90.89	0.8929	**91.93**	**0.9044**
22	88.51	0.8659	90.16	0.8841	90.81	0.8920	**91.87**	**0.9036**
24	88.10	0.8614	89.77	0.8798	90.64	0.8901	**91.89**	**0.9039**
26	88.27	0.8633	90.01	0.8825	90.32	0.8859	**91.75**	**0.9023**
28	88.48	0.8656	89.98	0.8822	90.57	0.8893	**91.87**	**0.9036**
30	88.46	0.8654	90.08	0.8833	90.61	0.8898	**91.82**	**0.9031**

If higher order statistics of a hyperspectral data set are considered with variance of data, then the methods give better results than others.

Class separability and entropy values are also calculated for both the KSC and Botswana data sets. Results of these data sets provide similar findings with the results of Indian Pine data. Table 10 incorporates the optimum values (for all the three data sets) in terms of OA, κ, S and E. The optimum value of all the methods are achieved in different number of extracted features which are also depicted in this table. The different numbers of extracted features for different methods (for optimum results) are in between 14 and 22, because different methods follow different extraction principles. The best results are marked in bold. This table also confirms the fact that clustering oriented KPCA based feature extraction algorithm gives better transformed set of features for classification than the other methods used in our experiment.

It also has been noticed that richness of the information of hyperspectral data is not fully handled using only variance of the data (by PCA method), it needs variance as well as higher order statistics of the data (like KPCA based methods). The KPCA based methods can extract more information from the hyperspectral data than the conventional PCA. Also, a proper choice of representative patterns for kernel matrix calculation, like clustering oriented KPCA based methods, produces better subset of features.

Table 9 Overall accuracy and kappa coefficients of PCA, SPCA, KPCA and clustering oriented KPCA based methods for different number of extracted features for Botswana data

No. of features	PCA		SPCA		KPCA		Clustering oriented KPCA	
	OA(%)	κ	OA(%)	κ	OA(%)	κ	OA(%)	κ
4	69.82	0.6552	58.31	0.5228	66.12	0.6085	**68.02**	**0.6302**
6	78.51	0.7529	72.42	0.6834	76.53	0.7244	**77.23**	**0.7321**
8	86.94	0.8486	79.98	0.7715	83.32	0.8084	**85.21**	**0.8293**
10	88.38	0.8645	86.31	0.8416	86.17	0.8439	**88.19**	**0.8624**
12	89.30	0.8746	87.83	0.8533	89.92	0.8815	**90.71**	**0.8909**
14	89.32	0.8748	89.39	0.8756	90.67	0.8904	**91.32**	**0.8976**
16	89.26	0.8740	90.43	0.8872	91.23	0.8966	**91.78**	**0.9026**
18	89.33	0.8749	90.72	0.8908	91.25	0.8988	**92.68**	**0.9125**
20	89.17	0.8730	90.74	0.8910	91.19	0.8961	**92.89**	**0.9151**
22	89.22	0.8735	90.63	0.8894	91.31	0.8975	**92.71**	**0.9130**
24	89.23	0.8736	90.68	0.8899	91.07	0.8968	**92.34**	**0.9089**
26	89.08	0.8719	90.21	0.8847	91.24	0.8987	**92.21**	**0.9075**
28	89.14	0.8726	90.19	0.8845	91.17	0.8979	**92.46**	**0.9103**
30	89.21	0.8734	90.52	0.8881	91.21	0.8983	**92.61**	**0.9119**

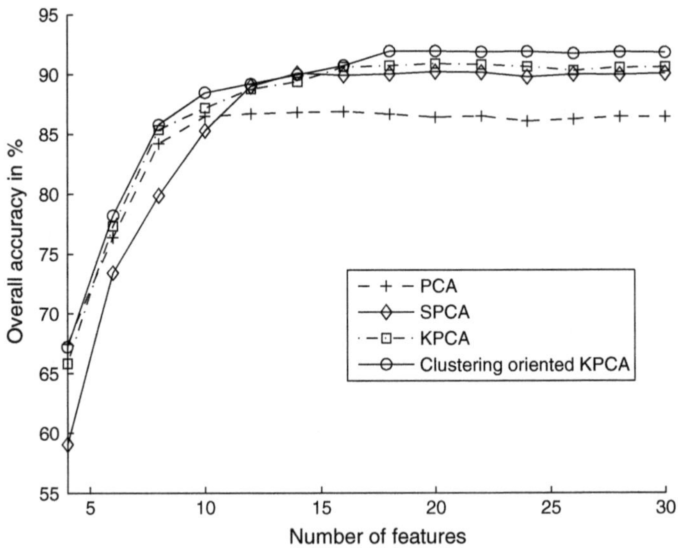

Fig. 8 Comparison of the performance of PCA, SPCA, KPCA and clustering oriented KPCA based methods in terms of overall accuracy with respect to the number of features used for KSC data

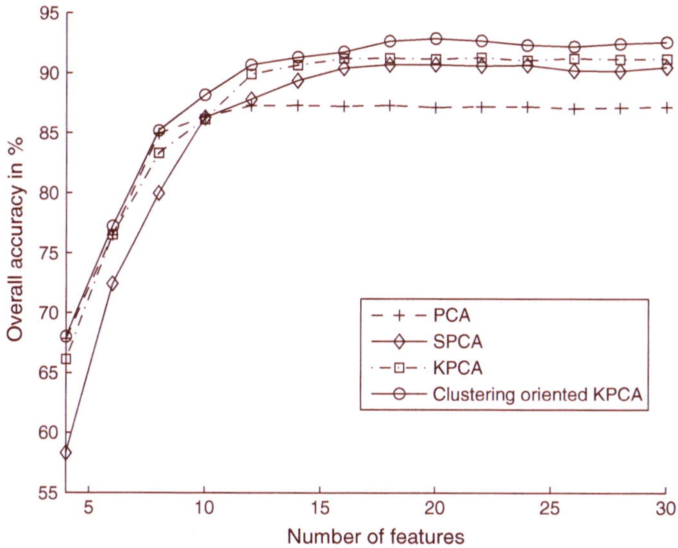

Fig. 9 Comparison of the performance of PCA, SPCA, KPCA and clustering oriented KPCA based methods in terms of overall accuracy with respect to the number of features used for Botswana data

Table 10 Comparison of feature extraction methods for hyperspectral data sets

Data set used	Method	Selected feature no.	Evaluation criterion			
			E	*S*	*OA*	κ
Indian Pine D = 185	PCA	14	0.6013	0.2659	84.01	0.8159
	SPCA	18	0.5929	0.2607	85.16	0.8288
	KPCA	22	0.5815	0.2559	85.78	0.8356
	Clustering oriented KPCA	18	**0.5567**	**0.2413**	**87.82**	**0.8584**
KSC D = 176	PCA	16	0.5637	0.1307	88.89	0.8701
	SPCA	20	0.5529	0.1279	90.21	0.8847
	KPCA	20	0.5496	0.1241	90.89	0.8929
	Clustering oriented KPCA	20	**0.5403**	**0.1193**	**91.93**	**0.9044**
Botswana D = 145	PCA	16	0.4734	0.1002	89.33	0.8749
	SPCA	20	0.4561	0.0913	90.74	0.8910
	KPCA	22	0.4493	0.0896	91.25	0.8988
	Clustering oriented KPCA	20	**0.4376**	**0.0809**	**92.89**	**0.9151**

7 Conclusions

PCA and KPCA based feature extraction techniques for hyperspectral images in unsupervised manner has been presented in this chapter, which transform the original data to a lower dimensional space. PCA is a linear transformation, whereas KPCA is non linear in nature and advantageous to attain the higher order statistics of data. In clustering oriented KPCA, the DBSCAN clustering technique is used to select proper training patterns for calculating kernel matrix for KPCA. To measure the effectiveness of these methods, four evaluation measures (namely, overall accuracy, kappa coefficient, class separability and entropy value) have been used. It is observed from the results that clustering oriented KPCA technique has a significant improvement, and a more consistent and steady behavior for different hyperspectral image data sets (Indian Pine, KSC and Botswana data) with respect to the other methods, i.e., PCA, SPCA and KPCA based methods in terms of all four evaluation measures.

It can be concluded from the above mentioned experimental results that clustering oriented KPCA based method gives better performance with respect to other methods, because the technique considers variance of the data set as well as other higher order statistics by using kernel PCA based transformation. The method also takes necessary steps for choosing the representative patterns as well as avoid noisy patterns for calculating kernel matrix of KPCA, which is a proper representation of the original data set by using DBSCAN clustering algorithm.

References

1. Varshney, P.K., Arora, M.K.: Advanced Image Processing Techniques for Remotely Sensed Hyperspectral Data, 2nd edn. Springer, Berlin (2004)
2. Landgrebe, D.: Hyperspectral image data analysis. IEEE Signal Processing Magazine, pp. 17–28, 2002
3. Manolakis, D., Marden, D., Shaw, G.A.: Hyperspectral image processing for automatic target detection applications. Lincoln Lab. J. **14**(1), 79–116 (2003)
4. Shippert, P.: Introduction to hyperspectral image analysis. Online Journal of Space Communication, 2003
5. Shippert, P.: Why use hyperspectral imagery? Photogrammetric Engineering and Remote Sensing, pp. 377–380, April 2004
6. Jain, A.K., Duin, R.P.W., Mao, J.: Statistical pattern recognition: a review. IEEE Transactions on Pattern Analysis and Macine Intelligence **22**(1), 4–37 (2000)
7. Bishop, C.M.: Neural Networks for Pattern Recognition, 1st edn. Oxford University Press, New Delhi (1995)
8. Devijver, P.A., Kittler, J.: Pattern Recognition: A Statistical Approach, 1st edn. Prentice-Hall International, New Delhi (1982)
9. Ghosh, A., Datta, A., Ghosh, S.: Self-adaptive differential evolution for feature selection in hyperspectral image data. Appl. Soft Comput. **13**(4), 1969–1977 (2013)
10. Jia, X., Kuo, B.-C., Crawford, M.M.: Feature mining for hyperspectral image classification. Proc. IEEE **101**(3), 676–697 (2013)
11. Datta, A., Ghosh, S., Ghosh, A.: Band elimination of hyperspectral imagery using partitioned band image correlation and capacitory discrimination. Int. J. Remote Sens. **35**(2), 554–577 (2014)

12. Fukunaga, K.: Introduction to Statistical Pattern Recognition, 2nd edn. Acacdemic Press, San Diego (1990)
13. Datta, A., Ghosh, S., Ghosh, A.: Combination of clustering and ranking techniques for unsupervised band selection of hyperspectral images. IEEE J. Sel. Top. Appl. Earth Observations Remote Sens. **8**(6), 2814–2823 (2015)
14. Jia, S., Ji, Z., Shen, L.: Unsupervised band selection for hyperspectral imagery classification without manual band removal. IEEE J. Sel. Top. Appl. Earth Observations Remote Sens. **5**(2), 531–543 (2012)
15. Datta, A., Ghosh, S., Ghosh, A.: Wrapper based feature selection in hyperspectral image data using self-adaptive differential evolution. In: *Proceedings of the International Conference on Image Information Processing*, pp. 1–6 (2011)
16. Datta, A., Ghosh, S., Ghosh, A.: Clustering based band selection for hyperspectral images. In: *Proceedings of the International Conference on Communications, Devices and Intelligent Systems*, pp. 101–104 (2012)
17. Mojaradi, B., Abrishami-Moghaddam, H., Zoej, M.J.V., Duin, R.P.W.: Dimensionality reduction of hyperspectral data via spectral feature extraction. IEEE Trans. Geosci. Remote Sens. **47**(7), 2091–2105 (2009)
18. Datta, A., Ghosh, S., Ghosh, A.: Band elimination of hyperspectral imagery using correlation of partitioned band image. In *Proceedings of the International Conference on Advances in Computing, Communications and Informatics*, pp. 412–417 (2013)
19. Jain, A., Zongker, D.: Feature selection: Evaluation, application, and small sample performance. IEEE Trans. Pattern Anal. Mach. Intell. **19**, 153–189 (1997)
20. Jia, X., Richards, J.A.: Segmented principal components transformation for efficient hyperspectral remote-sensing image display and classification. IEEE Trans. Geosci. Remote Sens. **37**, 538–542 (1999)
21. Datta, A., Ghosh, S., Ghosh, A.: Supervised band extraction of hyperspectral images using partitioned maximum margin criterion. IEEE Geosci. Remote Sens. Lett. **14**(1), 82–86 (2017)
22. Fauvel, M., Chanussot, J., Benediktsson, J.A.: Kernel principal component analysis for the classification of hyperspectral remote sensing data over urban areas. J. Adv. Sig. Process. **2009**, 1–14 (2009)
23. Datta, A., Ghosh, S., Ghosh, A.: Maximum margin criterion based band extraction of hyperspectral imagery. In *Proceedings of the Fourth International Conference on Emerging Applications of Information Technology*, pp. 300–304 (2014)
24. Kuo, B.-C., Landgrebe, D.A.: Nonparametric weighted feature extraction for classification. IEEE Trans. Geosci. Remote Sens. **42**, 1096–1105 (2004)
25. Schölkopf, B., Smola, A., Müller, K.-R.: Nonlinear component analysis as a kernel eigenvalue problem. Neural Comput. **10**(5), 1299–1319 (1998)
26. Datta, A., Ghosh, S., Ghosh, A.: Unsupervised band extraction for hyperspectral images using clustering and kernel principal component analysis. Int. J. Remote Sens. **38**(3), 850–873 (2017)
27. Rodarmel, C., Shan, J.: Principal component analysis for hyperspectral image classification. Surveying Land Inf. Syst. **62**(2), 115–122 (2002)
28. Richards, J.A., Jia, X.: Remote Sensing Digital Image Analysis: An Introduction, 1st edn. Springer, New York (1999)
29. Ester, M., Kriegel, H.P., Sander, J., Xu, X.: A density based algorithm for discovering clusters in large spatial databases with noise. In: *Proceedings of the 2nd International Conference on Knowledge Discovery and Data Mining (KDD)*, pp. 226–231 (1996)
30. Jimenez, L.O., Landgrebe, D.A.: Hyperspectral data analysis and supervised feature reduction via projection pursuit. IEEE Trans. Geosci. Remote Sens. **37**, 2653–2667 (1999)
31. Ham, J., Chen, Y., Crawford, M.M., Ghosh, J.: Investigation of the random forest framework for classification of hyperspectral data. IEEE Trans. Geosci. Remote Sens. **43**(3), 492–501 (2005)
32. Congalton, R.G., Green, K.: Assessing the Accuracy of Remotely Sensed Data, 2nd edn. CRC Press, London (2009)

33. Yao, J., Dash, M., Tan, S.T., Liu, H.: Entropy-based fuzzy clustering and fuzzy modeling. Fuzzy Sets Syst. **113**, 381–388 (2000)
34. Licciardi, G., Marpu, P.R., Chanussot, J., Benediktsson, J.A.: Linear versus nonlinear PCA for the classification of hyperspectral data based on the extended morphological profiles. IEEE Geosci. Remote Sens. Lett. **9**(3), 447–451 (2012)

Principal Component Analysis in the Presence of Missing Data

Marco Geraci and Alessio Farcomeni

Abstract The aim of this chapter is to provide an overview of recent developments in principal component analysis (PCA) methods when the data are incomplete. Missing data bring uncertainty into the analysis and their treatment requires statistical approaches that are tailored to cope with specific missing data processes (i.e., ignorable and nonignorable mechanisms). Since the publication of the classic textbook by Jolliffe, which includes a short, same-titled section on the missing data problem in PCA, there have been a few methodological contributions that hinge upon a probabilistic approach to PCA. In this chapter, we unify methods for ignorable and nonignorable missing data in a general likelihood framework. We also provide real data examples to illustrate the application of these methods using the R language and environment for statistical computing and graphics.

1 Introduction

Missing values occur frequently in all fields of research, including longitudinal studies [1, 18, 24], bioinformatics and gene expression [3, 5, 32], electrophysiology [31], meteorology and satellite imagery [29, 40, 41], oceanology [10, 15], and, more in general, signal processing.

It is well known that including in the analysis only complete cases, i.e. cases that have been observed for all variables in the model, may have undesirable consequences. Firstly, the results of complete case analyses can be biased. Secondly, the cumulative effect of missing data in several variables often leads to the

M. Geraci (✉)
University of South Carolina, 915 Greene Street, Columbia, SC 29209, USA
e-mail: geraci@mailbox.sc.edu

A. Farcomeni
Sapienza - University of Rome, Piazzale Aldo Moro 5, Rome, Italy
e-mail: alessio.farcomeni@uniroma1.it

G. R. Naik (ed.), *Advances in Principal Component Analysis*,
https://doi.org/10.1007/978-981-10-6704-4_3

exclusion of a substantial proportion of the original sample, resulting in a serious loss of precision of the estimates and of power in detecting associations between variables. This clearly has tremendous consequences on the validity of the conclusions drawn in these studies.

The literature on statistical methods that deal with missing data is rich and diverse, with seminal contributions dating back to the early 1970s concurrently with advances in computer technology and programming. Several of the methods available today have their roots in the works by Orchard and Woodbury [33], and by Little and Rubin [26, 39], who have systemized concepts and principles of the treatment of missing data within a likelihood framework. This is the framework to which we refer in Sect. 2 of this chapter and which we adopt in Sect. 3 when we discuss the missing data problem in principal component analysis (PCA).

Many multivariate statistical analyses, including PCA, start from a reduction of the data to the first two moments of the joint distribution. Suppose $(\mathbf{y}_1, \mathbf{y}_2, \ldots, \mathbf{y}_n)$ is sample of size n from some probability distribution F and each $\mathbf{y}_i = (y_{i1}, y_{i2}, \ldots, y_{ip})^\top$, $i = 1, 2, \ldots, n$, is a p-dimensional variate with $p \times 1$ mean $\boldsymbol{\mu}$ and $p \times p$ variance-covariance matrix $\boldsymbol{\Sigma}$. The question of how to calculate

$$s_{jk} = \sum_i^n (y_{ij} - \bar{y}_j)(y_{ik} - \bar{y}_k), \tag{1}$$

where

$$\bar{y}_j = \frac{1}{n} \sum_i y_{ij}, \tag{2}$$

from incomplete data is of central importance [27].

For example, consider the data matrix

$$\mathbf{Y} = \begin{bmatrix} 0.07 & 0.68 & 1.91 \\ 0.60 & 1.72 & 0.98 \\ — & 0.64 & — \\ 1.17 & -0.73 & -0.36 \\ -2.80 & -3.12 & -0.85 \\ 0.21 & — & 0.85 \\ -0.01 & 0.97 & — \\ 3.57 & 0.73 & -1.92 \\ 1.67 & -0.22 & -2.04 \\ -3.44 & -2.51 & — \end{bmatrix},$$

where — denotes a missing value, and suppose that the goal is to estimate s_{jk}, $j, k = 1, 2, 3$, as in (1). The analyst would face some challenges.

- Should the column averages $\bar{y}_j$'s be computed using all available information? In this case, the loss of information would be 10% (i.e., 1 observation out of 10) for $\bar{y}_1$ and $\bar{y}_2$, and 30% for $\bar{y}_3$.

- Should the $\bar{y}_j$'s be computed using only complete observations, i.e. observations for which the values of *all* three variables are available? In this case, the loss would be 40% for all estimates.
- Similarly, should the s_{jk}'s be computed using *pairwise* available cases (with a 20% loss for s_{12}, 30% for s_{13}, and 40% for s_{23}) or should incomplete observations (i.e., observations with a missing value on any of the three variables) be removed first?
- Moreover, what is the value of the lost information? Does the latter just reduce efficiency and thus increase the uncertainty of the inference or does it also radically change the conclusions from such inference because of bias?

Common sense would probably tell the analyst that the smaller the loss of information, the better. Thus, using pairwise complete observations, the estimated variance-covariance matrix of $\mathbf{Y}$ would be

$$\tilde{\mathbf{S}} = \begin{bmatrix} 4.61 & 2.87 & -1.08 \\ 2.87 & 2.69 & 0.92 \\ -1.08 & 0.92 & 2.29 \end{bmatrix}.$$

One could then compare this matrix to the one obtained using complete observations, that is

$$\hat{\mathbf{S}} = \begin{bmatrix} 4.41 & 2.37 & -1.19 \\ 2.37 & 2.83 & 0.92 \\ -1.19 & 0.92 & 2.49 \end{bmatrix}.$$

While apparently there are little numerical differences between $\tilde{\mathbf{S}}$ and $\hat{\mathbf{S}}$, the former, unfortunately, is not positive semi-definite. Although lack of positive semi-definiteness may not be seen necessarily as a problem in some contexts [28], in others this may pose unacceptable consequences. Additionally, evaluating potential for bias should always be a priority for the analyst.

In the next section, we examine the instances in which missing data have potentially serious implications on the results of the analysis and on their subsequent interpretation, and when, in contrast, the missing data problem is more benign. Given the vast literature on this topic, we only introduce basic definitions and traditional statistical approaches to the missing data problem. In Sect. 3, we linger on the application of some of these methods in the context of PCA and we show practical examples supported by R [36] code snippets.

2 Missing Data Mechanisms

The commonly adopted ontology of missing data [27] distinguishes among three cases: missing completely at random (MCAR), missing at random (MAR), and missing not at random (MNAR). These are called *missing data mechanisms*.

In our set up, let $\mathbf{Y}$ denote an n by p matrix of continuous measurements obtained from n units on p possibly correlated random variables $Y_1, Y_2, \ldots, Y_p$. Let also $\mathbf{M}$ denote an n by p matrix with row vectors $\mathbf{m}_i$, $i = 1, 2, \ldots, n$, whose jth entry is given by the binary indicator

$$m_{ij} = \begin{cases} 1 & \text{if the } ij\text{th entry of } \mathbf{Y} \text{ is missing,} \\ 0 & \text{otherwise.} \end{cases}$$

The matrix $\mathbf{M}$ describes the *pattern* of missing data. In a likelihood framework, each element of $\mathbf{M}$ is assumed to be a random variable with marginal distribution given by a Bernoulli with probability π_{ij}, that is, $m_{ij} \sim \text{Bin}(1, \pi_{ij})$. All statements regarding how missing data are generated arise from assumptions on the joint distribution of $\mathbf{Y}$ and $\mathbf{M}$, and these assumptions, in turn, determine which methods are most appropriate to deal with the missing data. Before we discuss MCAR, MAR, and MNAR assumptions in detail, let us introduce some additional notation. Suppose that the ith row of $\mathbf{Y}$ contains $s_i \geq 0$ missing values. Then, $\mathbf{y}_i$ is partitioned into the $s_i \times 1$ vector $\mathbf{z}_i$ and the $(p - s_i) \times 1$ vector $\mathbf{x}_i$. That is, $\mathbf{x}_i$ is the observed part of $\mathbf{y}_i$ while $\mathbf{z}_i$ is its unobserved part, which would have been recorded if at all possible. Finally, let $\mathbf{I}_n$ denote the identity matrix of order n.

The MCAR mechanism assumes that $\mathbf{y}_i$ and $\mathbf{m}_i$ are marginally independent. In other words, if in addition we assume independence among the m_{ij}'s, this is equivalent to tossing a coin whose probability of heads equals π_{ij}, and to deleting the jth entry of $\mathbf{y}_i$ if heads comes up. This is the strongest of the three assumptions, seldom tenable in practice.

Under the MAR mechanism, missingness may depend on the observed data but not on the data that are missing. This means that the joint probability $\boldsymbol{\pi}_i = (\pi_{i1}, \pi_{i2}, \ldots, \pi_{ip})^\top$ has some relationship with $\mathbf{x}_i$, but not with $\mathbf{z}_i$. If in addition the parameters of such relationship are distinct from the parameters involved in the data generating process (i.e., the distribution of $\mathbf{y}_i$), then the missing data are said to be *ignorable* and the analysis can be performed directly on the basis of the observed data. At first sight, the difference between MAR and MCAR may seem of little practical relevance, given that statistical models are often estimated conditionally on the observed data. This is not true. A better understanding of the difference between MAR and MCAR is provided by Heitjan and Basu [14], who gave examples where MCAR and MAR estimates are substantially different.

Finally, the MNAR mechanism is the most difficult situation to deal with since the missing data are not ignorable. The difficulty is as much practical as it is theoretical. Since MNAR models cannot be completely general, their identifiability requires explicit modeling assumptions and some simplifications [1, 4].

Just to reiterate the main concepts, the missing data mechanism is related to the question "*why values are missing?*", while the pattern of missing data is associated with the question "*which values are missing?*". The main reason why missing values should be studied carefully is that if the true missing data mechanism is MNAR but the data are analysed under a MAR assumption, then the estimates can be strongly biased. On the other hand, if MNAR methods are used when the data are actually missing (completely) at random, then a loss of efficiency in the estimators should be expected. Unfortunately, observed data provide only limited information on the nature of the missing data mechanism and, in general, a decision must be made based on the experimental design, including data collection, as well as sensitivity analyses. We refer the reader to [4, 30] for a more theoretical discussion on this issue.

We conclude this section with a brief remark about the missing data pattern. When the missing data occur following regular patterns, this information can be used to simplify model assumptions or, at least, to improve model estimation. For example, a *univariate* pattern is one where groups of items are either entirely observed or entirely missing for each subject. In a *monotone* pattern, a missing item of the variable Y_j for a particular individual implies that all subsequent variables $Y_k, k > j$, are missing for that individual. This occurs in longitudinal experiments when subjects drop-out from the study. In the remainder of this chapter, we generally assume that the missing data follow an *arbitrary* pattern, that is, either a monotone or a non-monotone pattern. For more discussion on this topic, we refer the reader to [27].

2.1 Missing Completely at Random

When the distribution of the missing data indicator does not depend on either the observed or unobserved data, i.e. when

$$\Pr(\mathbf{m}_i|\mathbf{z}_i, \mathbf{x}_i) = \Pr(\mathbf{m}_i), \tag{3}$$

then the missing data are said to be *missing completely at random*.

Lack of measurement is therefore unpredictable (as if tossing a coin, so to speak) and, as such, *not informative*. One could therefore proceed with a *complete case* (CC) analysis without the risk of incurring into estimation bias. As mentioned in our introductory example, there are two possible choices for a CC analysis in the multivariate context: observations can be discarded listwise or component-wise. A listwise approach proceeds by discarding $\mathbf{y}_i$ as soon as $s_i > 0$, that is, if any variable has not been measured, then the entire unit is discarded. A component-wise approach proceeds by using as much information contained in $\mathbf{x}_i$ as possible ("nothing goes to waste"). For instance, if $m_{i1} = m_{i2} = 0$ and $m_{i3} = 1$, then y_{i1} and y_{i2} will contribute to the estimation of s_{12} in a component-wise approach, but not in a listwise approach.

2.2 Missing at Random

When the distribution of the missing data indicator depends only on the observed data, i.e. when

$$\Pr(\mathbf{m}_i|\mathbf{z}_i, \mathbf{x}_i) = \Pr(\mathbf{m}_i|\mathbf{x}_i), \tag{4}$$

then the missing data are said to be *missing at random*. This setting is more general than MCAR's. Indeed, MCAR data are always MAR. However, the converse is false.

The event that a measurement is missing may depend on the measurement itself (e.g., wealthy people may be prone to refuse disclosing their income in a survey). However, conditional on the observed data, this dependence disappears (e.g., the propensity of people to disclose their income is completely explained by the knowledge of their assets).

On the one hand, in many cases a CC analysis under a MAR assumption is perfectly valid. As most statistical procedures are conditional on the observed data, m cannot give any additional information on z. Hence, missing values can be simply ignored. On the other hand, it might be possible to incorporate in the analysis the uncertainty for not knowing z by predicting z from x. This procedure is known as *imputation*. There are several techniques that fall under this label.

A popular technique is *hot deck* single imputation. This involves selecting a number of *donors*, that is, units that are "similar" to the unit with missing values and then predicting the missing values through the average of the donors' observed values. This method is simple and grounded on the fact that units that are similar with respect to the observed values should also be similar with respect to the unobserved ones (provided there is a strong association among variables). There are, however, some difficulties with this technique. Firstly, choosing the number of donors can be sometimes difficult. Secondly, hot deck imputation is not based on a statistical model, hence it lacks theoretical ground and it is difficult to adapt to specific problems. Finally, and most importantly, hot deck single imputation (as any other technique based on a single imputation) fails to take into account the uncertainty brought about by imputation. In other words, a predicted value replacing the missing value is effectively treated as a direct measurement.

To overcome the latter limitation, one can consider a *multiple imputation* approach which consists in repeatedly predicting the missing values. The results based on several predictions can be averaged to produce a final estimate, or they can be evaluated with respect to their sensitivity to specific imputed values.

Imputation (either single or multiple) can be made theoretically sound by drawing imputations from probability models. The latter are used to capture the generating mechanism of the missing values and can often formally take into account the imputation's uncertainty.

2.3 Missing Not at Random

When the distribution of the missing data indicator depends on the unobserved data, after conditioning on the observed data, i.e. when

$$\Pr(\mathbf{m}_i|\mathbf{z}_i, \mathbf{x}_i) \neq \Pr(\mathbf{m}_i|\mathbf{x}_i), \tag{5}$$

then the missing data are said to be *missing not at random*. This setting is the most general of all.

In the MNAR scenario the missingness indicator is assumed to be related with unmeasured predictors and/or the unobserved response, even conditionally on the observed data. MNAR data are also referred to as *informative* since the missing values contain information about the MNAR mechanism itself.

It should be stressed that, if the true mechanism is MNAR, simple CC analyses or naïve imputation methods inevitably produce biased results. The extent and direction of this bias is unpredictable, and even relatively small fractions of missing values might lead to a large bias.

There are different approaches to the treatment of this kind of missing data. Model-based procedures are most commonly adopted. These aim at modeling the joint distribution of the measurement process and the dropout process, by specifying a *missing data model* (MDM). The MDM must take into account the residual dependence between the missingness indicator and the unobserved response. Below, we summarize the three main approaches.

- *Pattern-mixture models*. The joint distribution of $\mathbf{y}_i$ and $\mathbf{m}_i$ is factorized as

$$\Pr(\mathbf{y}_i, \mathbf{m}_i) = \Pr(\mathbf{y}_i|\mathbf{m}_i)\Pr(\mathbf{m}_i).$$

  This approach involves formulating separate submodels $\Pr(\mathbf{y}_i|\mathbf{m}_i)$ for each possible configuration of $\mathbf{m}_i$, or, at least, for each observed configuration. This is appealing for studies where the main objective is to compare the response distribution in subgroups with possibly different missing value patterns. On the other hand, its specification can be cumbersome, while its interpretation at the population level may become difficult.
- *Selection models*. The joint distribution of $\mathbf{y}_i$ and $\mathbf{m}_i$ is factorized as

$$\Pr(\mathbf{y}_i, \mathbf{m}_i) = \Pr(\mathbf{m}_i|\mathbf{y}_i)\Pr(\mathbf{y}_i).$$

  This approach involves an explicit model to handle the distribution of the missing data process given the measurement mechanism. If correctly specified, the model for $\mathbf{y}_i$ is estimated without bias and its interpretation is not compromised.
- *Shared parameter models*. It is assumed that an unobserved variable, say U, contains all the information that would lead to a MAR mechanism. Hence, conditionally on this latent variable, y_i and m_i are independent. This is the so-called *local independence* assumption. The joint distribution of y_i and m_i can be expressed as

$$\Pr(\mathbf{y}_i, \mathbf{m}_i) = \int_u \Pr(\mathbf{y}_i|u)\Pr(\mathbf{m}_i|u)f(u)\,du,$$

  where $f(u)$ denotes the density of U. Of course, parametric assumptions are needed for f. Usually it is assumed that U is a zero-mean Gaussian variable with unknown variance.

3 Methods to Handle Missing Data in Principal Component Analysis

In this section, we discuss selected missing data methods in PCA, some of which are relatively recent at the time of writing. PCA, originally introduced by Karl Pearson [34], is arguably one the most popular multivariate analysis techniques. It is often described as a tool for dimensionality reduction. Some authors consider PCA as a descriptive method, which needs not be based on distributional assumptions, whereas others provide probabilistic justifications in relation to sampling errors (see for example [20] or [21] for alternative interpretations of the PCA model). It is not our purpose to get embroiled in this discussion; here we take a probabilistic view as it is an essential framework for a statistical treatment of the missing data problem.

There are basically two main approaches where sampling comes into play: *fixed* and *random* effects PCA. (The random-effects approach can be formulated in either a frequentist or a Bayesian framework. We focus on the former, while more details on the latter can be found in [21]). In the fixed-effects approach, individuals are of direct interest. Therefore, individual-specific scores are parameters to be estimated. In symbols, the fixed-effects PCA model is given by

$$\mathbf{y}_i = \boldsymbol{\mu} + \mathbf{W}\mathbf{b}_i + \boldsymbol{\varepsilon}_i, \quad i = 1, \ldots, n, \tag{6}$$

where $\boldsymbol{\mu} = (\mu_1, \mu_2, \ldots, \mu_p)^\top$ is the vector with the mean of each variable, $\mathbf{b}_i = (b_{i1}, b_{i2}, \ldots, b_{iq})^\top$ is the ith row vector of fixed scores and $\mathbf{W}$ is a $p \times q$ matrix of unknown loadings with elements w_{jh}, $j = 1, \ldots, p$, $h = 1, \ldots, q$, with $q \leq p$. The error is assumed to be zero-centered Gaussian with homoscedastic variance, $\boldsymbol{\varepsilon}_i \sim N(\mathbf{0}, \psi \mathbf{I}_p)$, where ψ is a positive scalar. Model (6)'s parameter, $\boldsymbol{\theta} = (\boldsymbol{\mu}, \mathbf{W}, \psi, \mathbf{b}_1, \mathbf{b}_2, \ldots, \mathbf{b}_n)$, can be estimated via maximum likelihood (or, equivalently, least squares) estimation (MLE). A downside of this approach is that the dimension of $\boldsymbol{\theta}$ increases with the sample size.

The random-effects specification of the probabilistic representation of PCA [38, 42] is given by the model

$$\mathbf{y}_i = \boldsymbol{\mu} + \mathbf{W}\mathbf{u}_i + \boldsymbol{\varepsilon}_i, \quad i = 1, \ldots, n, \tag{7}$$

where $\mathbf{u}_i = (u_{i1}, u_{i2}, \ldots, u_{iq})^\top$ is the ith row vector of latent scores and $\mathbf{W}$ is, as above, a matrix of unknown loadings. Furthermore, it is assumed that $\mathbf{u}$ is stochastically independent from $\boldsymbol{\varepsilon}$. Conventionally, $\mathbf{u}_i \sim N(\mathbf{0}, \mathbf{I}_q)$. If in addition the error is assumed to be zero-centered Gaussian with covariance matrix $\boldsymbol{\Psi}$, $\boldsymbol{\varepsilon}_i \sim N(\mathbf{0}, \boldsymbol{\Psi})$, we obtain the multivariate normal distribution $\mathbf{y}_i \sim N(\boldsymbol{\mu}, \mathbf{C})$, $\mathbf{C} = \mathbf{W}\mathbf{W}^\top + \boldsymbol{\Psi}$. We also assume that $\boldsymbol{\Psi} = \psi \mathbf{I}_p$, so that the elements of $\mathbf{y}_i$ are conditionally independent, given $\mathbf{u}_i$. The parameter $\boldsymbol{\mu} = (\mu_1, \mu_2, \ldots, \mu_p)^\top$ allows for a location-shift fixed effect.

The marginal log-likelihood of model (7) is thus given by

$$l(\boldsymbol{\theta}; \mathbf{Y}) \equiv \sum_{i=1}^{n} \log \int_{\mathbb{R}^q} f(\mathbf{y}_i, \mathbf{u}_i)\, d\mathbf{u}_i = -\frac{n}{2}\left\{p \log(2\pi) + \log|\mathbf{C}| + \mathrm{tr}\left(\mathbf{C}^{-1}\mathbf{S}\right)\right\}, \tag{8}$$

where $f(\mathbf{y}_i, \mathbf{u}_i) = f(\mathbf{y}_i|\mathbf{u}_i) f(\mathbf{u}_i)$ is the joint density of the response and the random effects and $\mathbf{S} = \frac{1}{n}\sum_{i=1}^{n} = (\mathbf{y}_i - \boldsymbol{\mu})(\mathbf{y}_i - \boldsymbol{\mu})^\top$. Model (7)'s parameter $\boldsymbol{\theta} = (\boldsymbol{\mu}, \mathbf{W}, \psi)$ can be estimated via MLE, while the individual scores can be predicted via the best linear predictor $\mathrm{E}(\mathbf{u}|\mathbf{y})$ [42].

In the sections that follow, we consider alternative approaches to the estimation of $\boldsymbol{\theta}$ when some values in $\mathbf{Y}$ are missing. We discuss methods in relation to different missing data mechanisms and we illustrate their application using the R programming language [36].

3.1 *Missing (Completely) at Random*

3.1.1 Single Imputation

R code 3.1 Summary statistics of the Orange data set.

```
> data(orange, package = "missMDA")
> Y <- orange
> summary(Y)
 Color.intensity Odor.intensity  Attack.intensity     Sweet
 Min.    :4.083  Min.    :4.292  Min.    :3.917    Min.    :4.083
 1st Qu.:4.448   1st Qu.:4.958   1st Qu.:4.833     1st Qu.:4.510
 Median :4.646   Median :5.292   Median :5.292     Median :4.938
 Mean   :5.083   Mean   :5.326   Mean   :5.319     Mean   :4.943
 3rd Qu.:5.948   3rd Qu.:5.938   3rd Qu.:5.375     3rd Qu.:5.479
 Max.   :6.583   Max.   :6.167   Max.   :7.417     Max.   :5.792
 NA's   :2       NA's   :1       NA's   :3         NA's   :4
      Acid            Bitter           Pulp           Typicity
 Min.   :4.125   Min.   :2.833   Min.   :1.292   Min.   :3.417
 1st Qu.:4.375   1st Qu.:3.104   1st Qu.:1.510   1st Qu.:3.958
 Median :5.042   Median :3.583   Median :2.479   Median :4.438
 Mean   :5.065   Mean   :3.536   Mean   :3.312   Mean   :4.462
 3rd Qu.:5.292   3rd Qu.:3.792   3rd Qu.:4.521   3rd Qu.:5.042
 Max.   :6.750   Max.   :4.375   Max.   :7.333   Max.   :5.250
 NA's   :3       NA's   :4       NA's   :2
```

We begin this section by introducing a toy data set which is available in the R package `missMDA` [16]. The data consist of 8 sensory measurements of 12 orange juices (R code 3.1). Seven out of the eight variables are incomplete, although the number of missing values is small and ranges between 1 and 4.

A crude approach to single imputation is mean imputation, whereby y_{ij} is replaced with $\bar{y}_j$ if y_{ij} is missing. This is the default approach used in `FactoMineR` [25] and is shown in the R code 3.2. To simplify the reporting of the results shown further below in Table 1 and Fig. 1, we kept only the first two components (`ncp = 2`). The warning message indicates that the missing values were replaced by the mean of the (observed) values from the corresponding variables. So, for example, the two missing values of color intensity were replaced by 5.083, while all four missing measurements of bitterness by 3.536. The same message suggests using a different imputation approach which we now describe briefly.

R code 3.2 Principal component analysis with mean imputation.

```
> FactoMineR::PCA(Y, ncp = 2, graph = FALSE)
**Results for the Principal Component Analysis (PCA)**
The analysis was performed on 12 individuals, described by 8
variables *The results are available in the following objects:

   name               description
1  "$eig"             "eigenvalues"
2  "$var"             "results for the variables"
3  "$var$coord"       "coord. for the variables"
...
[omitted]

Warning message:
In FactoMineR::PCA(Y, graph = FALSE) :
  Missing values are imputed by the mean of the variable: you
  should use the imputePCA function of the missMDA package
```

The so-called EM-PCA and regularized algorithms, discussed by [21] and available in the R package `missMDA` [16], can be used as a preliminary step to fill the missing values in and then apply a standard PCA to the completed data set. The former algorithm (EM-PCA) consists in iteratively fitting a *fixed effects* PCA, while the

Table 1 Estimates of the first (PC1) and second (PC2) principal axes from the complete case PCA and the single imputation EM-PCA. Bold denotes EM-PCA estimates that differ more than 20% from the corresponding complete case estimate

	Complete case		EM-PCA	
	PC1	PC2	PC1	PC2
Color intensity	0.807	−0.068	0.774	**−0.085**
Odor intensity	0.570	0.693	0.563	0.717
Attack intensity	0.849	−0.247	0.935	−0.212
Sweet	−0.688	−0.428	−0.650	**−0.520**
Acid	0.809	−0.174	0.727	**−0.110**
Bitter	0.458	0.563	0.532	**0.359**
Pulp	−0.632	0.590	−0.533	**0.722**
Typicity	−0.826	0.205	−0.836	0.244

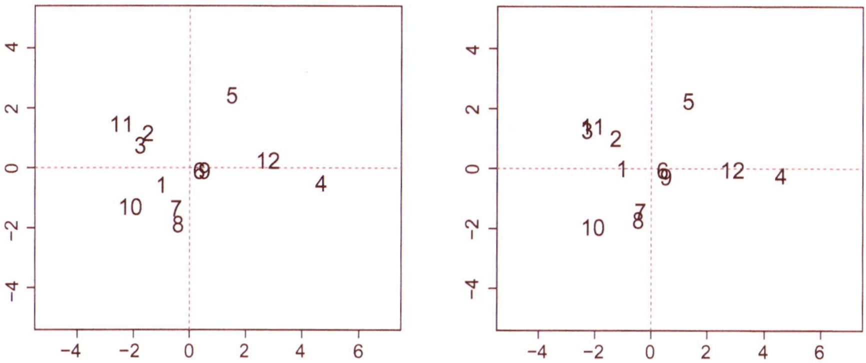

Fig. 1 Individuals maps for the Orange data set obtained from the complete case PCA (left) and the EM-PCA (right)

latter (regularized) makes use of shrinkage to solve overfitting problems. As shown in the R code 3.3, the function `imputePCA` requires the data matrix **Y**, the number of components `ncp` used for predicting the missing values, and the specific imputation algorithm. Note that the number of components used for imputing the missing values (in this example, we used as many as possible) does not necessarily coincide with the number of components kept in the second stage of the analysis.

R code 3.3 Principal component analysis with iterative imputation.

```
> Yhat <- missMDA::imputePCA(Y, ncp = 7, method = "EM")$completeObs
> FactoMineR::PCA(Yhat, ncp = 2, graph = FALSE)
```

The results from the CC analysis and the (single imputation) EM-PCA are shown in Table 1 and Fig. 1. These two analyses produced loadings with the same signs. In contrast, the magnitude of the coefficients differed between the imputation and the CC analysis, with some of the differences being over 20%. As a result, the individual projections gave slightly different maps (Fig. 1).

3.1.2 Multiple Imputation

As mentioned before, single imputation methods treat imputed missing values as fixed (known). This means that the uncertainty related to the missing values is ignored, which generally leads to deflated standard errors. Josse et al. [23] proposed to deal with this issue by first performing a residual bootstrap procedure to obtain B estimates of the PPCA parameters and then generate B data matrices, each completed with samples from the predictive distribution of the missing values conditional on the observed values and the corresponding bootstrapped parameter set. More formally, one can proceed as follows:

1. obtain an initial estimate $\hat{\boldsymbol{\mu}}$, $\hat{\mathbf{W}}$, $\hat{\mathbf{b}}_i$, $i = 1, \ldots, n$, of the parameters in model (6) (e.g., via EM-PCA estimation). Reconstruct the data $\hat{\mathbf{Y}}$ with the first q dimensions and calculate $\mathbf{R} = \mathbf{Y} - \hat{\mathbf{Y}}$, where the $n \times p$ matrix $\mathbf{R}$ of residuals has missing entries corresponding to those of $\mathbf{Y}$;
2. draw B random samples from the non-missing rows of $\mathbf{R}$. Denote each replicate by $\mathbf{R}_b^*$, $b = 1, \ldots, B$;
3. calculate $\mathbf{Y}_b^* = \hat{\mathbf{Y}} + \mathbf{R}_b^*$, $b = 1, \ldots, B$ and estimate the PPCA parameters $\hat{\boldsymbol{\mu}}_b^*$, $\hat{\mathbf{W}}_b^*$, $\hat{\mathbf{b}}_{b,i}^*$, $i = 1, \ldots, n$, from $\mathbf{Y}_b^*$, $b = 1, \ldots, B$;
4. for $\{i : s_i > 0\}$, calculate $\mathbf{y}_{b,i}^* = \hat{\boldsymbol{\mu}}_b^* + \hat{\mathbf{W}}_b^* \hat{\mathbf{b}}_{b,i}^* + \mathbf{r}^*$, where $\mathbf{r}^*$ is a newly sampled residual from $\mathbf{R}$. Complete the vector $\mathbf{y}_i$ with $\mathbf{y}_{b,i}^*$ to obtain the bth complete data matrix $\mathbf{Y}_b$, $b = 1, \ldots, B$.

There are several ways to obtain bootstrapped residuals. One approach is to draw a sample with replacement from the entries of the matrix $\mathbf{R}$. Another approach, recommended by [21], is to sample the residuals from a zero-centered Gaussian with variance estimated from the non-missing entries of $\mathbf{R}$. Improved results might be obtained with corrected residuals (e.g., leave-one-out residuals).

Once B complete data matrices have been generated, the simplest analytic approach is to carry out a PPCA on each $\mathbf{Y}_b$, and then calculate the average of the B sets of parameters. A multiple imputation PPCA of the Orange data set is given in R code 3.4. Our example is based on $B = 100$ replicates, with $q = 2$. The individuals and variables maps obtained from average scores and loadings are plotted in Fig. 2. As compared to the complete case PCA and the single imputation EM-PCA (Table 1), the multiple imputation PCA produced noticeably different estimates of $\hat{\mathbf{W}}$, especially for the second principal axis (R code 3.4). The uncertainty due to the missing values is also shown in Fig. 2. For example, individual scores 1, 9, and 10 showed more total variability, as given by the area of the ellipses, and more variability

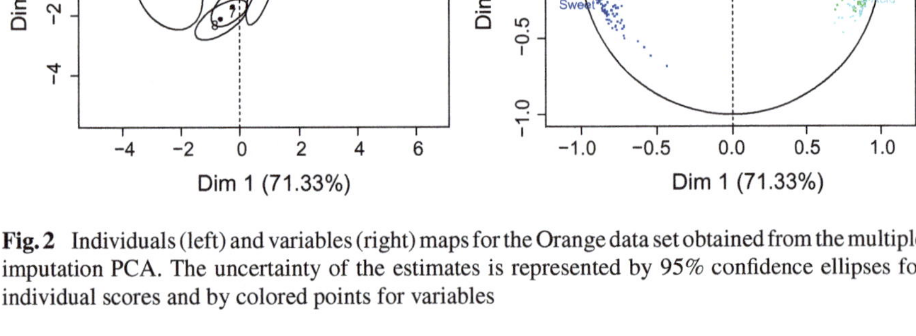

Fig. 2 Individuals (left) and variables (right) maps for the Orange data set obtained from the multiple imputation PCA. The uncertainty of the estimates is represented by 95% confidence ellipses for individual scores and by colored points for variables

along the second axis, as reflected in the eccentricity of the ellipses. The uncertainty was greater for sweetness, color intensity, and bitterness, and, as in the case of the individual scores, it was more prominent in relation to the second axis.

3.1.3 EM Algorithm for PPCA

In their seminal paper, Dempster et al. [6] discussed the use of the EM algorithm in factor analysis. As a particular case of the standard factor analysis model, Tipping and Bishop [42] proposed an EM algorithm to estimate the parameter $\boldsymbol{\theta}$ in (8), where the incomplete part of the data is represented by the latent variables. Let $\mathbf{U}$ be the $n \times q$ matrix of latent scores. Given the complete data log-likelihood

$$l\left(\boldsymbol{\theta}; \mathbf{Y}, \mathbf{U}\right) = \sum_{i}^{n} \log f\left(\mathbf{y}_i | \mathbf{u}_i, \boldsymbol{\theta}\right) + \log f\left(\mathbf{u}_i\right), \tag{9}$$

R code 3.4 Principal component analysis with multiple imputation.

```
> set.seed(190)
> Yhat <- missMDA::MIPCA(Y, ncp = 2, method.mi = "Boot",
+ nboot = 100)
> fit.mi <- lapply(Yhat$res.MI, function(x)
+ FactoMineR::PCA(x, ncp = 2, graph = FALSE))
> tmp <- lapply(fit.mi, function(x) x$var$coord)
> What <- 0
> for(i in 1:100) What <- What + tmp[[i]]/100
> round(What, 3)
                     Dim.1  Dim.2
Color.intensity      0.790  0.299
Odor.intensity       0.587  0.660
Attack.intensity     0.919 -0.163
Sweet               -0.841 -0.250
Acid                 0.874 -0.195
Bitter               0.872  0.301
Pulp                -0.679  0.657
Typicity            -0.830  0.374
> plot(Yhat)
```

the EM approach alternates between an

(i) expectation step (E-step) $Q(\boldsymbol{\theta}|\boldsymbol{\theta}^{(t)}) = \mathrm{E}_{\mathbf{u}|\mathbf{y},\boldsymbol{\theta}^{(t)}}\left\{l\left(\boldsymbol{\theta}; \mathbf{Y}, \mathbf{U}\right)\right\}$; and a

(ii) maximization step (M-step) $\boldsymbol{\theta}^{(t+1)} = \underset{\boldsymbol{\theta}}{\arg\max}\, Q(\boldsymbol{\theta}|\boldsymbol{\theta}^{(t)})$,

where $\boldsymbol{\theta}^{(t)}$ is the estimate of the parameter after t cycles. The expectation in step (i) is taken with respect to the conditional distribution $f\left(\mathbf{u}|\mathbf{y}, \boldsymbol{\theta}^{(t)}\right)$ which, in the

Gaussian PPCA, is normal with parameters depending on $\boldsymbol{\theta}^{(t)}$ only. Closed-form solutions to steps (i) and (ii) are provided by [42].

The EM estimation approach might have little immediate appeal as opposed to the usual diagonalization of the sample covariance matrix. However, it provides a computationally efficient strategy in analysing high-dimensional large datasets [38] and it is particularly enticing in the presence of MAR values. In this case, the incomplete part of the data would become $(\mathbf{z}, \mathbf{u})$ and the expectation in (i) would be taken with respect to the density $f\left(\mathbf{u}, \mathbf{z}|\mathbf{x}, \boldsymbol{\theta}^{(t)}\right)$. See for example [42] for an application of the PPCA EM algorithm in the presence of missing data.

3.2 *Missing Not at Random*

A number of statistical approaches have been developed to cope with nonignorable missing mechanisms and such approaches have been applied in different analytic frameworks. However, the approaches used in PCA have typically focused on assumptions of ignorability [19, 21, 42]. Recently, Geraci and Farcomeni [11] extended Tipping and Bishop's [42] EM approach to the case in which the vector $\mathbf{y}$ is partially observed and the missing data mechanism is nonignorable. Specifically, they proposed an adaptation of Ibrahim et al.'s [17] methods for missing responses in random-effects models with non-monotone patterns of missing data.

Suppose that $\mathbf{y}_i$ contains s_i, $s_i < p$, missing values. The ith contribution to the complete data density of $(\mathbf{y}_i, \mathbf{u}_i, \mathbf{m}_i)$ is given by

$$f\left(\mathbf{y}_i, \mathbf{u}_i, \mathbf{m}_i|\boldsymbol{\theta}, \boldsymbol{\eta}\right) = f\left(\mathbf{y}_i|\mathbf{u}_i, \boldsymbol{\theta}\right) f\left(\mathbf{u}_i\right) f\left(\mathbf{m}_i|\mathbf{y}_i, \boldsymbol{\eta}\right), \qquad i = 1, \ldots, n, \tag{10}$$

where the additional factor $f\left(\mathbf{m}_i|\mathbf{y}_i, \boldsymbol{\eta}\right)$, indexed by the parameter $\boldsymbol{\eta}$, is the MDM, which we assume to be independent from $\mathbf{u}_i$. This assumption simplifies the subsequent steps of the estimation algorithm, although it can be relaxed at the cost of increased computational time (see [17, 18] for a discussion).

Estimation of $\boldsymbol{\theta}$ would in general require marginalizing the log-likelihood based on (10) over the unobserved data, which however leads to a rather intractable integral of dimension $s_i + q$. Instead, the EM algorithm can be applied. The E-step at the $(t+1)$th iteration is defined as follows:

$$Q(\boldsymbol{\lambda}|\boldsymbol{\lambda}^{(t)}) = \mathrm{E}_{\mathbf{z},\mathbf{u}|\mathbf{x},\mathbf{m},\boldsymbol{\lambda}^{(t)}}\left\{l\left(\boldsymbol{\lambda}; \mathbf{Y}, \mathbf{U}, \mathbf{M}\right)\right\}, \tag{11}$$

with $\boldsymbol{\lambda} = (\boldsymbol{\theta}, \boldsymbol{\eta})$ and $l\left(\boldsymbol{\lambda}; \mathbf{Y}, \mathbf{U}, \mathbf{M}\right) = \sum_i^n \log f\left(\mathbf{y}_i|\mathbf{u}_i, \boldsymbol{\theta}\right) + \log f\left(\mathbf{u}_i\right) + \log f\left(\mathbf{m}_i|\mathbf{y}_i, \boldsymbol{\eta}\right)$, and where the expectation is taken with respect to the conditional distribution of $\mathbf{z}_i$ and $\mathbf{u}_i$, given the observed data, evaluated at $\boldsymbol{\lambda}^{(t)}$.

The E-step (11), however, does not yet offer a computational advantage since it is not easy to solve analytically. Therefore, Geraci and Farcomeni [11] applied a Monte Carlo E-step [17]. They considered an adaptive rejection Metropolis sampling (ARMS) algorithm [12] and specified the following MDM

$$f\left(\mathbf{m}_i|\mathbf{y}_i,\boldsymbol{\eta}\right)=\prod_{j=1}^{p}\pi_{ij}^{m_{ij}}\left(1-\pi_{ij}\right)^{1-m_{ij}}, \tag{12}$$

where π_{ij} is the probability that y_{ij} is missing, conditional on the response $\mathbf{y}_i$. The ARMS algorithm is convenient as, basically, no tuning is needed. In addition, it is run in parallel for each row of the data matrix, which therefore greatly speeds up the computation. Moreover, the PPCA Gaussian model provides scope for further reductions in computational time during the calculation of the E-step. All the technical details are given in appendix.

We now show an application of the MNAR approach to accelerometer data obtained from seven-year-old children of the UK Millennium Cohort Study (MCS) [13]. The dataset, previously analyzed by [11], consists of six physical activity outcomes derived from high sampling frequency accelerometer measurements that were collected continuously over seven days. The outcomes, aggregated by day of the week, are: (a) total number of counts ($\times$1000), (b) total number of steps ($\times$1000), (c) proportion of time spent in sedentary behaviour (SB), (d) proportion of time spent in moderate-to-vigorous activity (MVPA), (e) duration (minutes) of sporadic MVPA bouts, and (f) frequency of sporadic bouts. Here, a sporadic bout is defined as a short burst of intense activity that lasts less than ten minutes. An example of an individual weekly physical activity profile is given in Table 2.

Figure 3 shows the heat map of the correlation between physical activity outcomes. As expected, acceleration counts (a) and steps (b) are positively correlated one to each other since they both quantify the amount of movement. As such, they show a negative correlation with sedentary behaviour (c) but positive with all other variables measuring activity. There are also temporal patterns within outcomes, with

Table 2 Example of weekly physical activity profile summary with six outcome variables for one child of the Millennium Cohort Study (Reproduced from [11])

		Mon	Tue	...	Sun
Total activity					
(a)	Counts ($\times$1000)	287.7	564.7	...	305.7
(b)	Steps ($\times$1000)	7.4	13.3	...	6.5
Proportion of time (%) spent					
(c)	In sedentary behaviour	61.4	51.5	...	57.9
(d)	In MVPA[a]	5.0	11.0	...	5.4
Sporadic MVPA[a] bouts					
(e)	Total time (minutes)	37.5	92.75	...	38
(f)	Frequency	91	169	...	96

[a] *Moderate-to-vigorous activity*

stronger correlations between most of the weekdays but weaker between weekdays and weekends.

Let us define $\mathbf{y}_i = \left(y_{i1}^{(Mon)}, \dots, y_{i6}^{(Mon)}, y_{i7}^{(Tue)}, \dots, y_{i12}^{(Tue)}, \dots, y_{i42}^{(Sun)}\right)^\top$ as the ith response of dimensions 42×1 made up of the variables listed in Table 2 on each day of the week. Essentially, repeated measurements for each child are treated as columns, analogously to multivariate methods for time series data such as singular spectrum analysis (SSA) [20]. However, unlike SSA, the temporal correlation is not explicitly modelled.

A large proportion of children (4,042 out of 5,682) did not provide valid observations for all seven days of the week. Overall, 22% of expected child-days ($n \cdot 7$) were missing. Moreover, the percentage of missing values by day of the week was

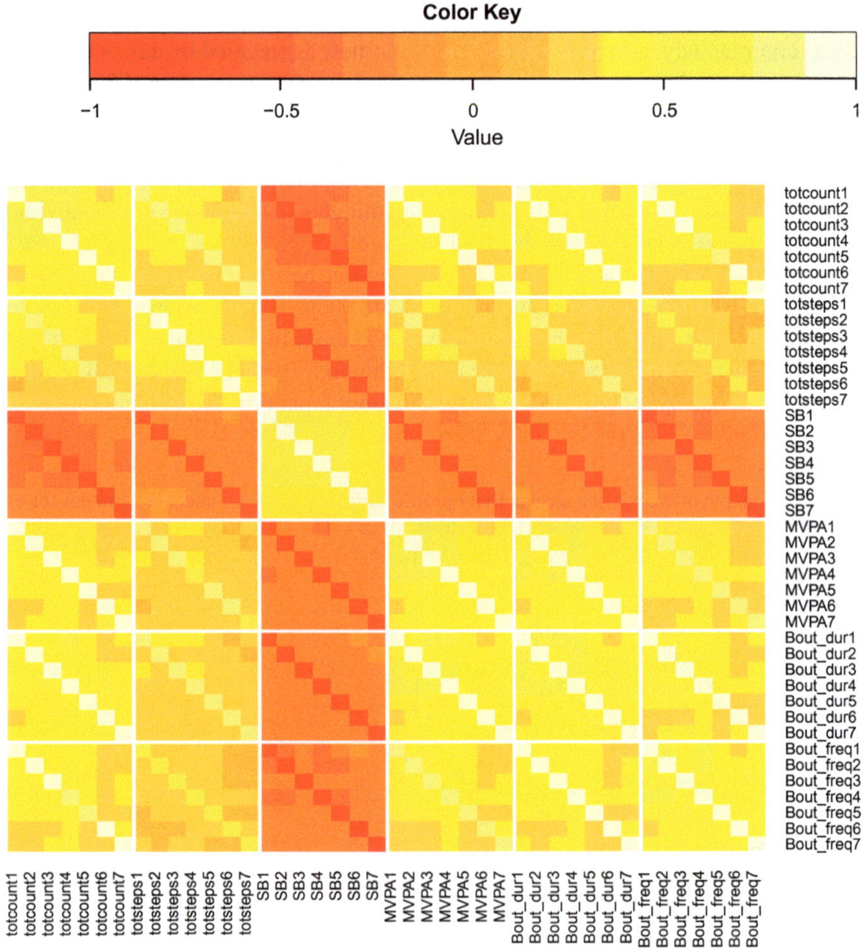

Fig. 3 Heat map of the correlation between physical activity outcomes (Table 2). The numbers in the variable labels denote the days of the week (1, Monday; 2, Tuesday; ...; 7, Sunday)

equal to 17 (Monday), 18 (Tuesday), 16 (Wednesday), 15 (Thursday), 15 (Friday), 32 (Saturday), and 44 (Sunday). In a recent study on predictors of nonresponse using the same data [37], it was found that children who exercised once a week or less according to the MCS questionnaire-based data, were less likely to provide reliable accelerometer measurements for all days of the week. Concerns about the informativeness of the missing data process are therefore justifiable and statistical methods to reduce associated bias [27] are warranted.

Geraci and Farcomeni [11] analyzed the MCS physical activity data in two steps:

1. *Missing data mechanism*. Several models using AIC and BIC were compared. The final model used for the analysis was the logistic regression

$$\text{logit}\left\{\pi_{ij}\right\} = \mathbf{t}_i^\top \boldsymbol{\eta},$$

where $\mathbf{t}_i$ is a 11×1 vector made up of 5 of the variables listed in Table 2 with each variable averaged over weekdays and weekend days, resulting in a 11×1 parameter vector $\boldsymbol{\eta}$, including an intercept. (Duration of MVPA bouts was not included to avoid identifiability issues due to its near-unity correlation with time spent in MVPA.)

2. *Probabilistic principal component*. The number q of principal components to be estimated was fixed to eight. This choice was motivated by the results obtained in a separate complete-case PCA. A generalized cross-validation (GCV) approach [22] gave $q = 11$ as optimal number of components. However, the value of the GCV criterion for $q = 11$ was not substantially different from that for $q = 8$. Moreover, the simplified E-step (15) was carried out as described in appendix using a Monte Carlo sample size $K = 100$. The difference in scale between variables was taken into account through standardization at each step of the EM algorithm, whereby variables were divided by the current standard deviation estimates (however, during the sampling step all the variables were transformed back to their original scales).

We first discuss the results reported by [11] related to the MDM and then those related to the PPCA.

During weekdays, the predicted probability of data being missing was lower with higher volumes of activity as measured by total counts (Table 3). Intuitively, this could be explained by the lower occurrence of non-wear periods (i.e., extended time intervals of 20 min or more during which the accelerometer values are zero) when more activity is recorded. The negative coefficient for sedentary time on the one hand, but positive for steps and MVPA on the other, are perhaps a consequence of higher compliance rates observed between Monday and Friday, hence when children might be less active because they are involved in day-to-day routines.

In contrast, opposite associations were observed during weekend days, when the fraction of missing values tend to be much higher. These results could be interpreted as a consequence of a process by which children are more likely not to follow the study protocol if they do not participate in moderate-to-vigorous activities. This clearly leads to underestimate the volume of physical activity and the proportion of

Table 3 Maximum likelihood estimates and standard errors (SE) from the Millennium Cohort Study data for the missing data mechanism (Reproduced from [11])

	Weekdays		Weekends	
	Estimate	SE	Estimate	SE
Total counts	−4.682	0.162	4.239	0.090
Total steps	0.289	0.026	−0.817	0.024
Time in sedentary behaviour	−0.818	0.036	0.202	0.022
Time in MVPA	2.794	0.105	−1.759	0.054
Frequency of MVPA bouts	0.852	0.035	−1.671	0.032

sedentary time, but only during weekend days. This finding might also explain the association found by [37] between lack of exercise and nonresponse in the MCS data.

The barplot of the estimated loadings in Fig. 4 aids interpretation of the first eight principal components. The first and most important component (35.6%) was driven by the negative correlation between sedentary and active behaviours. In other words, children with higher scores on this dimension tend to have higher levels of moderate-to-vigorous physical activity (MVPA) and lower levels of sedentary behaviour. Therefore, the first component relates to the 'predominant behaviour'. The second component (9.4%) contrasted weekday and weekend activity patterns, while the third (7.2%) contrasted Saturday and Sunday patterns. Note also that, along the third component, higher activity levels on Saturday are paralleled by higher levels of sedentary behaviour on Sunday, and vice versa. Hence, it is reasonable to associate the second and third dimensions with 'weekend behaviours'.

Components four to seven, each accounting between 6.9 and 7.4% of the variability, presented correlations with activities during distinct days of the week. Finally, the eighth component (4.2%) specifically related to sedentary behaviour but was otherwise minimally or not correlated with the other outcomes. It is important to stress that, while the first component establishes a trade-off between sedentary and active behaviours, the eighth dimension determines the relative location of children in terms of sedentariness, independently from their predominant behaviour.

Figure 5 shows the contribution of the first eight eigenvalues relative to the sum of all 42 eigenvalues when ignoring the MDM as compared to that observed in the nonignorable model. There is indication that the weights are redistributed when accounting for the missing data, with a substantial reduction in weight for the first component.

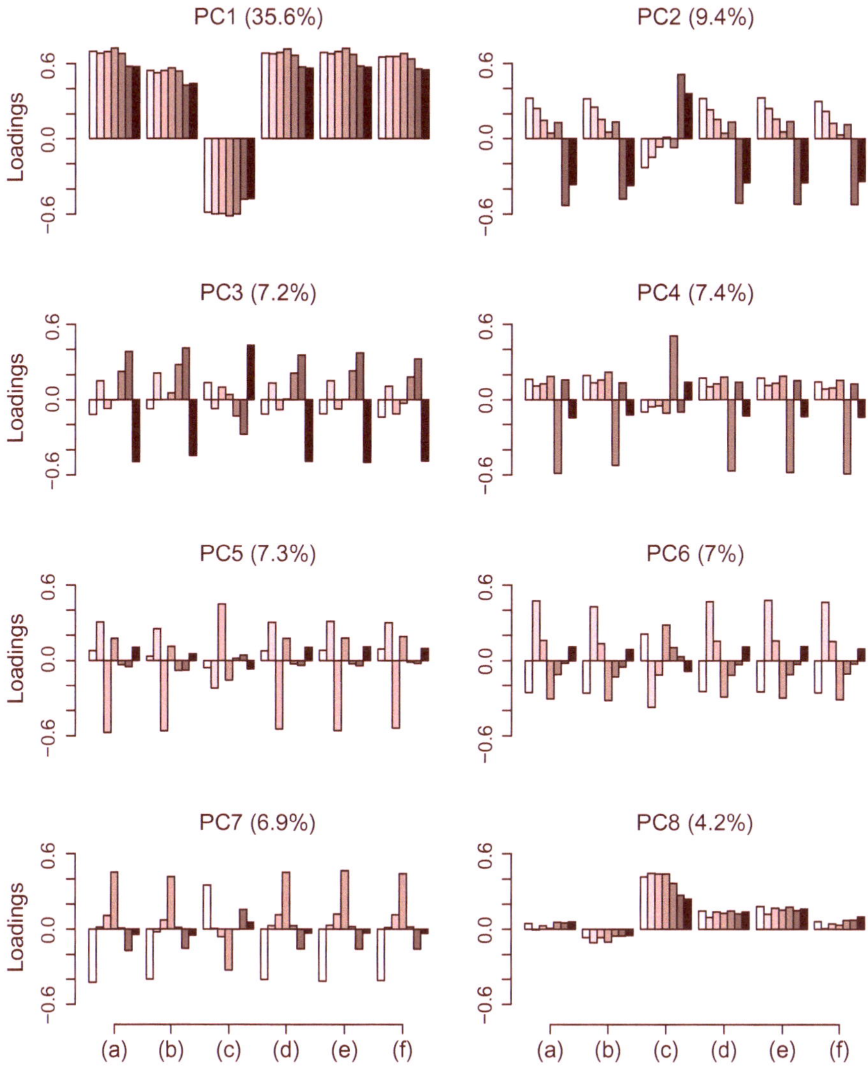

Fig. 4 Results of the nonignorable principal component analysis of physical activity outcomes in the Millennium Cohort Study: barplots of the loadings for the first eight components (PC1-PC8) and proportion (%) of variability explained. Bars for total counts (a) and steps (b), sedentary behaviour (c), moderate-to-vigorous activity (d), duration (e) and frequency (f) of bouts are colour-coded by day of the week starting from Monday (lightest grey) to Sunday (darkest grey) (Reproduced from [11])

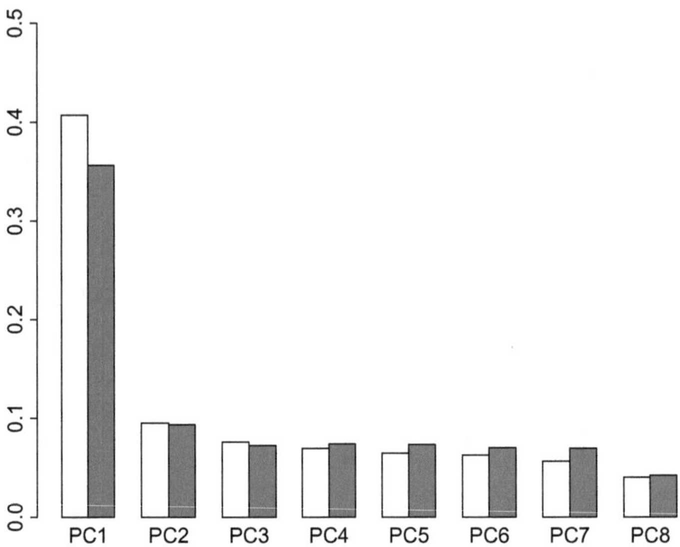

Fig. 5 Proportion of variability explained by the first eight principal components from the Millennium Cohort Study data for the ignorable (white) and nonignorable (grey) models (Reproduced from [11])

4 Conclusion

In principal component analysis, increased uncertainty and potential bias of the estimates due to the presence of missing data can be tackled in a likelihood framework via a probabilistic approach to PCA. The application of missing data methods such as multiple imputation or selection models is facilitated by the introduction of parametric assumptions about the multivariate distribution of the variables of interest. In this chapter, we often entertained assumptions of normality. There exist a number of alternative approaches whereby models are robustified and, thus, made less sensitive to parametric specifications [9, 20]. However, one must not lose sight of the specific nature of PCA and its properties when considering such alternatives. For example, the L_1-norm variant of PCA, which has been advocated for its robustness to outliers, is not rotational invariant [7]. Several robust PCA methods are discussed in [8], and their extension to the case of missing data is ground for further work. Moreover, computational efficiency is also a fundamental aspect when evaluating statistical methods for missing data, especially due to the multidimensional nature of PCA and the increasingly larger size of the datasets that are now available for analysis. This is particularly important for the case of robust methods.

Appendix – EM Algorithm for PPCA with MNAR Values

In this appendix, we provide additional details on the Monte Carlo EM algorithm introduced in Sect. 3.2 and we derive a simplified E-step where the random effects are integrated out from the complete data log-likelihood.

The Monte Carlo E-step requires sampling from $f\left(\mathbf{z}_i, \mathbf{u}_i|\mathbf{x}_i, \mathbf{m}_i, \boldsymbol{\lambda}^{(t)}\right)$. This task can be carried out efficiently via ARMS [12] using the full conditionals

$$f\left(\mathbf{z}_i|\mathbf{x}_i, \mathbf{u}_i, \mathbf{m}_i, \boldsymbol{\lambda}^{(t)}\right) \propto f\left(\mathbf{y}_i|\mathbf{u}_i, \boldsymbol{\lambda}^{(t)}\right) f\left(\mathbf{m}_i|\mathbf{y}_i, \boldsymbol{\lambda}^{(t)}\right), \tag{13}$$

$$f\left(\mathbf{u}_i|\mathbf{x}_i, \mathbf{z}_i, \mathbf{m}_i, \boldsymbol{\lambda}^{(t)}\right) \propto f\left(\mathbf{y}_i|\mathbf{u}_i, \boldsymbol{\lambda}^{(t)}\right) f\left(\mathbf{u}_i\right). \tag{14}$$

An implementation of ARMS is available in the R package `HI` [35].

A sample $\boldsymbol{\xi}_{i1}, \ldots, \boldsymbol{\xi}_{iK}$ for $i = 1, \ldots, n$ is obtained at each EM iteration t, where the $(s_i + q) \times 1$ vector $\boldsymbol{\xi}_{ik} = (\tilde{\mathbf{z}}_{ik}, \tilde{\mathbf{u}}_{ik})$, $k = 1, \ldots, K$, contains 'imputed' values for $\mathbf{z}_i$ and $\mathbf{u}_i$ (with the understanding that $\boldsymbol{\xi}_{ik} = \tilde{\mathbf{u}}_{ik}$ if $s_i = 0$). Here the Monte Carlo sample size K is kept constant throughout. Alternative strategies with varying $K^{(t)}$ that may increase the speed or the accuracy of the EM algorithm can be considered [2, 17]. The E-step (11) is approximated by

$$Q(\boldsymbol{\lambda}|\boldsymbol{\lambda}^{(t)}) = \frac{1}{K}\sum_{i=1}^{n}\sum_{k=1}^{K} l\left(\boldsymbol{\lambda}; \boldsymbol{\xi}_{ik}, \mathbf{x}_i, \mathbf{m}_i\right). \tag{15}$$

The maximization of (15) with respect to $\boldsymbol{\lambda}$ is straightforward. Define $\tilde{\mathbf{y}}_{ik} = (\tilde{\mathbf{z}}_{ik}, \mathbf{x}_i)$ if $s_i > 0$ or $\tilde{\mathbf{y}}_{ik} = \mathbf{y}_i$ if $s_i = 0$, $i = 1, \ldots, n$, $k = 1, \ldots, K$. The maximum likelihood solution of the M-step at the $(t + 1)$th iteration is given by

$$\hat{\boldsymbol{\mu}}^{(t+1)} = \frac{1}{nK}\sum_{i=1}^{n}\sum_{k=1}^{K}\left(\tilde{\mathbf{y}}_{ik} - \hat{\mathbf{W}}^{(t)}\tilde{\mathbf{u}}_{ik}\right), \tag{16}$$

$$\hat{\mathbf{W}}^{(t+1)} = \left\{\sum_{i=1}^{n}\sum_{k=1}^{K}\left(\tilde{\mathbf{y}}_{ik} - \hat{\boldsymbol{\mu}}^{(t+1)}\right)\tilde{\mathbf{u}}_{ik}^{\top}\right\}\left(\sum_{i=1}^{n}\sum_{k=1}^{K}\tilde{\mathbf{u}}_{ik}\tilde{\mathbf{u}}_{ik}^{\top}\right)^{-1}, \tag{17}$$

$$\hat{\psi}^{(t+1)} = \frac{1}{nKp}\sum_{i=1}^{n}\sum_{k=1}^{K}\|\tilde{\mathbf{y}}_{ik} - \hat{\boldsymbol{\mu}}^{(t+1)} - \hat{\mathbf{W}}^{(t+1)}\tilde{\mathbf{u}}_{ik}\|_2^2. \tag{18}$$

Analogously, the MLE of $\boldsymbol{\eta}$ can be easily obtained using standard results for generalized linear models.

Note that the computational burden can be alleviated by first integrating out the random effects in (11) and then sampling from $f\left(\mathbf{z}_i|\mathbf{x}_i, \mathbf{m}_i, \boldsymbol{\lambda}^{(t)}\right)$ during the Monte Carlo E-step. We obtain what we call a simplified E-step

$$
\begin{aligned}
Q_i(\boldsymbol{\lambda}|\boldsymbol{\lambda}^{(t)}) &= \iint \{\log f(\mathbf{y}_i, \mathbf{u}_i|\boldsymbol{\theta}) + \log f(\mathbf{m}_i|\mathbf{y}_i, \boldsymbol{\eta})\} f\left(\mathbf{z}_i, \mathbf{u}_i|\mathbf{x}_i, \mathbf{m}_i, \boldsymbol{\lambda}^{(t)}\right) d\mathbf{u}_i d\mathbf{z}_i \\
&= \iint \{\log f(\mathbf{y}_i|\mathbf{u}_i, \boldsymbol{\theta}) + \log f(\mathbf{u}_i)\} f\left(\mathbf{z}_i, \mathbf{u}_i|\mathbf{x}_i, \mathbf{m}_i, \boldsymbol{\lambda}^{(t)}\right) d\mathbf{u}_i d\mathbf{z}_i \\
&\quad + \int \log f(\mathbf{m}_i|\mathbf{y}_i, \boldsymbol{\eta}) f\left(\mathbf{z}_i|\mathbf{x}_i, \mathbf{m}_i, \boldsymbol{\lambda}^{(t)}\right) d\mathbf{z}_i \\
&= \int \left\{-\frac{p}{2}\log(\psi) - \frac{1}{2\psi}\operatorname{tr}\left(\mathbf{W}^\top\mathbf{W}\mathbf{B}^{(t)}\right) - \frac{1}{2\psi}\|\mathbf{y}_i - \boldsymbol{\mu} - \mathbf{W}\mathbf{v}_i^{(t)}\|_2^2 \right. \\
&\quad \left. + \log f(\mathbf{m}_i|\mathbf{y}_i, \boldsymbol{\eta}) \right\} f\left(\mathbf{z}_i|\mathbf{x}_i, \mathbf{m}_i, \boldsymbol{\lambda}^{(t)}\right) d\mathbf{z}_i \\
&\equiv \mathrm{E}_{\mathbf{z}|\mathbf{x},\mathbf{m},\boldsymbol{\lambda}^{(t)}}\{l(\boldsymbol{\lambda}; \mathbf{y}_i, \mathbf{m}_i)\},
\end{aligned}
\tag{19}
$$

where $\mathbf{v}_i^{(t)} = \mathbf{B}^{(t)}\mathbf{W}^{(t)^\top}\left(\mathbf{y}_i - \boldsymbol{\mu}^{(t)}\right)/\psi^{(t)}$ and $\mathbf{B}^{(t)} = \left\{\mathbf{W}^{(t)^\top}\mathbf{W}^{(t)}/\psi^{(t)} + \mathbf{I}_q\right\}^{-1}$. Note that by assumption $\mathbf{m}_i$ is independent from $\mathbf{u}_i$. The expectation above is now taken with respect to

$$
f\left(\mathbf{z}_i|\mathbf{x}_i, \mathbf{m}_i, \boldsymbol{\lambda}^{(t)}\right) \propto \exp\left\{-\frac{1}{2}\left(\mathbf{y}_i - \boldsymbol{\mu}^{(t)}\right)^\top \mathbf{C}^{(t)^{-1}}\left(\mathbf{y}_i - \boldsymbol{\mu}^{(t)}\right)\right\} f\left(\mathbf{m}_i|\mathbf{y}_i, \boldsymbol{\eta}^{(t)}\right),
$$

$\mathbf{C}^{(t)} = \mathbf{W}^{(t)}\mathbf{W}^{(t)^\top} + \boldsymbol{\psi}^{(t)}$.
Again, we obtain a sample $\tilde{\mathbf{z}}_{ik}$, $i = 1, \ldots, n$, $k = 1, \ldots, K$ and calculate the approximate E-step

$$
Q(\boldsymbol{\lambda}|\boldsymbol{\lambda}^{(t)}) = \frac{1}{K}\sum_{i=1}^{n}\sum_{k=1}^{K} l(\boldsymbol{\lambda}; \tilde{\mathbf{z}}_{ik}, \mathbf{x}_i, \mathbf{m}_i). \tag{20}
$$

The MLE equations of the M-step which follow from maximizing the log-likelihood in (20) are similar to equations (27) and (28) in [42] and they do not require explicit computation of the covariance matrix. We omit them for the sake of brevity.

Finally, we note that, based on the linear predictions

$$
\hat{\mathbf{u}}_{ik} = \left(\hat{\mathbf{W}}^\top\hat{\mathbf{W}} + \hat{\boldsymbol{\psi}}\right)^{-1}\hat{\mathbf{W}}^\top(\tilde{\mathbf{y}}_{ik} - \hat{\boldsymbol{\mu}}), \tag{21}
$$

where $\tilde{\mathbf{y}}_{ik} = (\tilde{\mathbf{z}}_{ik}, \mathbf{x}_i)$ is the complete data vector at convergence, we can calculate the element-wise variances of $\frac{1}{K}\sum_{k=1}^{K}\hat{\mathbf{u}}_{ik}$ over the individuals space as estimates of $\delta_1, \ldots, \delta_q$. The quantity $(p - q) \cdot \hat{\boldsymbol{\psi}}$ provides the portion of the total variability associated with the 'discarded' components.

References

1. Bartolucci, F., Farcomeni, A.: A discrete time event-history approach to informative drop-out in mixed latent Markov models with covariates. Biometrics **71**(1), 80–89 (2015)
2. Booth, J.G., Hobert, J.P.: Maximizing generalized linear mixed model likelihoods with an automated Monte Carlo EM algorithm. Journal of the Royal Statistical Society B **61**(1), 265–285 (1999)
3. Creemers, A., Hens, N., Aerts, M., Molenberghs, G., Verbeke, G., Kenward, M.G.: A sensitivity analysis for shared-parameter models for incomplete longitudinal outcomes. Biometrical Journal **52**(1), 111–125 (2010)
4. de Brevern, A., Hazout, S., Malpertuy, A.: Influence of microarrays experiments missing values on the stability of gene groups by hierarchical clustering. BMC Bioinformatics **5**(1), 114 (2004)
5. de Souto, M.C., Jaskowiak, P.A., Costa, I.G.: Impact of missing data imputation methods on gene expression clustering and classification. BMC Bioinformatics **16**(1), 64 (2015)
6. Dempster, A.P., Laird, N.M., Rubin, D.B.: Maximum likelihood from incomplete data via the EM algorithm. Journal of the Royal Statistical Society B **39**(1), 1–38 (1977)
7. Ding, C., Zhou, D., He, X., Zha, H.: L_1-PCA: rotational invariant L_1-norm principal component analysis for robust subspace factorization. In: Proceedings of the 23rd International Conference on Machine Learning, pp. 281–288. ACM
8. Farcomeni, A., Greco, L.: Robust methods for data reduction. CRC Press, Boca Raton, FL (2015)
9. Geraci, M.: Estimation of regression quantiles in complex surveys with data missing at random: An application to birthweight determinants. Statistical Methods in Medical Research **25**(4), 1393–1421 (2016)
10. Geraci, M., Bottai, M.: Use of auxiliary data in semi-parametric spatial regression with nonignorable missing responses. Statistical Modelling **6**(4), 321–336 (2006)
11. Geraci, M., Farcomeni, A.: Probabilistic principal component analysis to identify profiles of physical activity behaviours in the presence of nonignorable missing data. Journal of the Royal Statistical Society C **65**(1), 51–75 (2016)
12. Gilks, W.R., Wild, P.: Adaptive rejection sampling for Gibbs sampling. Journal of the Royal Statistical Society C **41**(2), 337–348 (1992)
13. Griffiths, L.J., Cortina-Borja, M., Sera, F., Pouliou, T., Geraci, M., Rich, C., Cole, T.J., Law, C., Joshi, H., Ness, A.R., Jebb, S.A., Dezateux, C.: How active are our children? Findings from the Millennium Cohort Study. BMJ Open **3**(8), e002,893 (2013)
14. Heitjan, D.F., Basu, S.: Distinguishing "missing at random" and "missing completely at random". The American Statistician **50**(3), 207–213 (1996)
15. Houseago-Stokes, R.E., Challenor, P.G.: Using PPCA to estimate EOFs in the presence of missing values. Journal of Atmospheric and Oceanic Technology **21**(9), 1471–1480 (2004)
16. Husson, F., Josse, J.: missMDA: Handling missing values with/in multivariate data analysis (principal component methods) (2013). https://CRAN.R-project.org/package=missMDA. R package version 1.7.2
17. Ibrahim, J.G., Chen, M.H., Lipsitz, S.R.: Missing responses in generalised linear mixed models when the missing data mechanism is nonignorable. Biometrika **88**(2), 551–564 (2001)
18. Ibrahim, J.G., Molenberghs, G.: Missing data methods in longitudinal studies: A review. Test **18**(1), 1–43 (2009)
19. Ilin, A., Raiko, T.: Practical approaches to principal component analysis in the presence of missing values. Journal of Machine Learning Research **11**(Jul), 1957–2000 (2010)
20. Jolliffe, I.T.: Principal component analysis, 2nd edn. Springer-Verlag, New York, NY (2002)
21. Josse, J., Husson, F.: Handling missing values in exploratory multivariate data analysis methods. Journal de la Société Française de Statistique **153**(2), 79–99 (2012)
22. Josse, J., Husson, F.: Selecting the number of components in principal component analysis using cross-validation approximations. Computational Statistics and Data Analysis **56**(6), 1869–1879 (2012)

23. Josse, J., Pagès, J., Husson, F.: Multiple imputation in principal component analysis. Advances in Data Analysis and Classification **5**(3), 231–246 (2011)
24. Laird, N.M.: Missing data in longitudinal studies. Statistics in Medicine **7**(1–2), 305–315 (1988)
25. Lê, S., Josse, J., Husson, F.: FactoMineR: A package for multivariate analysis. Journal of Statistical Software **25**(1), 1–18 (2008)
26. Little, R.J.A., Rubin, D.B.: Statistical analysis with missing data. Wiley, New York, NY (1987)
27. Little, R.J.A., Rubin, D.B.: Statistical analysis with missing data, 2nd edn. Wiley, Hoboken, NJ (2002)
28. Mehrotra, D.V.: Robust elementwise estimation of a dispersion matrix. Biometrics **51**(4), 1344–51 (1995)
29. Melgani, F., Mercier, G., Lorenzi, L., Pasolli, E.: Recent methods for reconstructing missing data in multispectral satellite imagery. In: R.S. Anderssen, P. Broadbridge, Y. Fukumoto, K. Kajiwara, T. Takagi, E. Verbitskiy, M. Wakayama (eds.) Applications + Practical Conceptualization + Mathematics = fruitful Innovation: Proceedings of the Forum of Mathematics for Industry 2014, pp. 221–234. Springer Japan, Tokyo (2016)
30. Molenberghs, G., Beunckens, C., Sotto, C., Kenward, M.G.: Every missingness not at random model has a missingness at random counterpart with equal fit. Journal of the Royal Statistical Society B **70**(2), 371–388 (2008)
31. Morelli, M.S., Giannoni, A., Passino, C., Landini, L., Emdin, M., Vanello, N.: A cross-correlational analysis between electroencephalographic and end-tidal carbon dioxide signals: Methodological issues in the presence of missing data and real data results. Sensors (Basel, Switzerland) **16**(11), e1828 (2016)
32. Oh, S., Kang, D.D., Brock, G.N., Tseng, G.C.: Biological impact of missing-value imputation on downstream analyses of gene expression profiles. Bioinformatics **27**(1), 78–86 (2011)
33. Orchard, T., Woodbury, M.A.: A missing information principle: theory and applications. In: Proceedings of the Sixth Berkeley Symposium on Mathematical Statistics and Probability, Volume 1: Theory of Statistics, Sixth Berkeley Symposium on Mathematical Statistics and Probability, pp. 697–715. University of California Press
34. Pearson, K.: LIII. on lines and planes of closest fit to systems of points in space. The London, Edinburgh, and Dublin Philosophical Magazine and Journal of. Science **2**(11), 559–572 (1901)
35. Petris, G., Tardella, L.: HI: Simulation from distributions supported by nested hyperplanes (2013). https://CRAN.R-project.org/package=HI. R package version 0.4
36. R Core Team: R: A Language and Environment for Statistical Computing. R Foundation for Statistical Computing, Vienna, Austria (2016). https://www.R-project.org/
37. Rich, C., Cortina-Borja, M., Dezateux, C., Geraci, M., Sera, F., Calderwood, L., Joshi, H., Griffiths, L.J.: Predictors of non-response in a UK-wide cohort study of children's accelerometer-determined physical activity using postal methods. BMJ Open **3**(3), e002290 (2013)
38. Roweis, S.: EM algorithms for PCA and SPCA. In: M.I. Jordan, M.J. Kearns, S.A. Solla (eds.) Advances in neural information processing systems 10: Proceedings of the 1997 conference, vol. 10, pp. 626–632. MIT Press, Cambridge, MA (1998)
39. Rubin, D.B.: Inference and missing data. Biometrika **63**(3), 581–592 (1976)
40. Sattari, M.T., Rezazadeh-Joudi, A., Kusiak, A.: Assessment of different methods for estimation of missing data in precipitation studies. Hydrology Research (2016). https://doi.org/10.2166/nh.2016.364
41. Schneider, T.: Analysis of incomplete climate data: Estimation of mean values and covariance matrices and imputation of missing values. Journal of Climate **14**(5), 853–871 (2001)
42. Tipping, M.E., Bishop, C.M.: Probabilistic principal component analysis. Journal of the Royal Statistical Society B **61**(3), 611–622 (1999)

Robust PCAs and PCA Using Generalized Mean

Jiyong Oh and Nojun Kwak

Abstract In this chapter, a robust principal component analysis (PCA) is described, which can overcome the problem that PCA is prone to outliers included in training set. Different from the other alternatives which commonly replace L_2-norm by other distance measures, our method alleviates the negative effect of outliers using the characteristic of the generalized mean keeping the use of the Euclidean distance. The optimization problem based on the generalized mean is solved by a novel method. We also present a generalized sample mean, which is a generalization of the sample mean, to estimate a robust mean in the presence of outliers. The proposed method shows better or equivalent performance than the conventional PCAs in various problems such as face reconstruction, clustering, and object categorization.

1 Introduction

Dimensionality reduction [1] is a classical problem in pattern recognition and machine learning societies, and numerous methods have been proposed to reduce dimensionality of data. Principal component analysis (PCA) [2] is one of the most popular unsupervised dimensionality reduction methods which tries to find a subspace where the average reconstruction error of training data is minimized. It is

J. Oh
Daegu-Gyeongbuk Research Center, Electronics and Telecommunications Research Institute, 1, Techno sunwhan-ro 10-gil, Yuga-myeon, Dalseong-gun, Daegu 42994, Korea
e-mail: jiyongoh@etri.re.kr

N. Kwak (✉)
Graduate School of Convergence Science and Technology, Seoul National University, 1, Gwanak-ro,Gwanak-gu, Seoul 08826, Korea
e-mail: nojunk@snu.ac.kr

G. R. Naik (ed.), *Advances in Principal Component Analysis*,
https://doi.org/10.1007/978-981-10-6704-4_4

useful in representation of input data in a low dimensional space and it has been successfully applied to face recognition [3, 4], visual tracking [5], clustering [6, 7], and so on.

When automatically collecting a large data set, outliers may be contained in the collected data since it is very difficult to examine whether each sample of data is outlier or not [8]. It is well known that, in this case, the conventional PCA is sensitive to outliers because it minimizes the reconstruction errors of training data in terms of the mean squared error and a few outliers with large errors dominate the objective function. This problem has been addressed in many studies [8–16], some of which are described in this chapter. Several studies to achieve the purpose commonly utilized L_1-norm instead of L_2-norm in the formulation of optimization problem to improve the robustness of PCA against outliers [10, 13, 14]. In [13], the cost function for optimization was constructed based on L_1-norm and a convex programming was employed to solve the problem. R_1-PCA [10] was presented to obtain a solution with the rotational invariance, which is a fundamental desirable property for learning algorithms [17]. In [14], PCA-L_1 was proposed, which maximizes an L_1 dispersion in the reduced space and an extension of PCA-L_1 using L_p-norm with arbitrary p was also proposed in [15]. Other method utilizing L_p-norm was also presented in [16]. On the other hand, some of robust PCAs were recently developed using information theoretic measures [11, 12]. He et al. [11] proposed MaxEnt-PCA which finds a subspace where Renyi's quadratic entropy [18] is maximized. The Renyi's entropy was estimated by a non-parametric Parzen window technique. In [12], HQ-PCA was developed based on the maximum correntropy criterion [19].

After describing the above methods, we then introduce a new robust PCA method based on the power mean or the generalized mean [20], which can become the arithmetic, geometric, and harmonic means depending on the value of its parameter. The proposed method, *PCA-GM*, is a generalization of the conventional PCA by replacing the arithmetic mean with the generalized mean. The proposed method can effectively prevent outliers from dominating objective function by controlling the parameter in the generalized mean. Moreover, it is rotational invariant because it still uses the Euclidean distance as the distance measure between data samples. In doing so, we also introduce a generalized sample mean, which is an enhancement of the conventional algebraic sample mean against outliers to address the problem that the sample mean is easily affected by outliers. It is used in PCA-GM instead of the sample mean. The optimization problems based on the generalized mean are efficiently solved using a mathematical property of the generalized mean. Different from our original work [21], we present MATLAB codes of the generalized sample mean and PCAGM via this chapter.[1] Recently, Candés et al. proposed a robust PCA [22], which is sometimes referred to as RPCA in the literature, where data matrix is tried to be represented as a sum of a low rank matrix, which corresponds to reconstructions of data, and a sparse matrix, which corresponds to reconstruction errors different from the methods mentioned above. It can model pixel-wise noise effectively using the sparse matrix, thus it has been known that RPCA is useful in the applications

[1] The MATLAB codes can be downloaded in http://mipal.snu.ac.kr/index.php/Jiyong_Oh.

such as background modeling from surveillance video and removing shadows and specularities from face images [22] by using each element in the reconstruction error vector (the column of the sparse matrix). On the other hand, in this chapter, an entire sample is considered as an outlier if it has a large norm of the reconstruction error vector.

The remainder of this chapter is organized as follows. Section 2 briefly introduces PCA and the state-of-the-art robust PCAs. The proposed method is described in Sect. 3. It is demonstrated in Sect. 4 that the proposed method gives better performances in face reconstruction and clustering problems than other variants of PCA. Finally, Sect. 5 concludes this paper.

2 PCA and Robust PCAs

Let us consider a training set of N n-dimensional samples $\{\mathbf{x}_i\}_{i=1}^N$. Assuming that the samples have zero-mean, PCA is to find an orthonormal projection matrix $\mathbf{W} \in \mathbb{R}^{n \times m}$ ($m \ll n$) by which the projected samples $\{\mathbf{y}_i = \mathbf{W}^T\mathbf{x}_i\}_{i=1}^N$ have the maximum variance in the reduce space. It is formulated as the following:

$$\mathbf{W}_{PCA} = \arg\max_{\mathbf{W}} tr(\mathbf{W}^T\mathbf{S}\mathbf{W}),$$

where $\mathbf{S} = \frac{1}{N}\sum_{i=1}^N \mathbf{x}_i\mathbf{x}_i^T$ is a sample covariance matrix and $tr(\mathbf{A})$ is the trace of a square matrix $\mathbf{A}$. The projection matrix $\mathbf{W}_{PCA}$ can be also found from the viewpoint of projection errors, i.e., it minimizes the average of the squared projection errors or reconstruction errors. Mathematically, it is represented as the optimization problem minimizing the following cost function:

$$J_{L_2}(\mathbf{W}) = \frac{1}{N}\sum_{i=1}^N ||\mathbf{x}_i - \mathbf{W}\mathbf{W}^T\mathbf{x}_i||_2^2,$$

where $||\mathbf{x}||_2$ is the L_2-norm of a vector $\mathbf{x}$. The two optimization problems are equivalent and easily solved by obtaining the m eigenvectors associated with the m largest eigenvalues of $\mathbf{S}$. Although PCA is simple and powerful, it is prone to outliers [8, 13] because $J_{L_2}(\mathbf{W})$ is based on the mean squared reconstruction error. To learn a subspace robust to outliers, Ke and Kanade [13] proposed to minimize an L_1-norm based objective function as follows:

$$J_{L_1}(\mathbf{W}) = \frac{1}{N}\sum_{i=1}^N ||\mathbf{x}_i - \mathbf{W}\mathbf{W}^T\mathbf{x}_i||_1,$$

where $||\mathbf{x}||_1$ is the L_1-norm of a vector $\mathbf{x}$. They also present an iterative method to obtain the solution for minimizing $J_{L_1}(\mathbf{W})$.

Although L_1-PCA minimizing $J_{L_1}(\mathbf{W})$ can relieve the negative effect of outliers, it is not invariant to rotations. In [10], Ding et al. proposed R_1-PCA, which is rotational invariant, at the same time is robust to outliers. It is to minimize the following objective function:

$$J_{R_1}(\mathbf{W}) = \sum_{i=1}^{N} \rho\left(\sqrt{\mathbf{x}_i^T\mathbf{x}_i - \mathbf{x}_i^T\mathbf{W}\mathbf{W}^T\mathbf{x}_i}\right),$$

where $\rho(\cdot)$ is a generic loss function and Cauchy function or Huber's M-estimator [23] was used for $\rho(\cdot)$ in [10]. The Huber's M-estimator $\rho_H(s)$ is defined as

$$\rho_H(s) = \begin{cases} s^2 & \text{if } |s| \leq c, \\ 2c|s| - c^2 & \text{otherwise} \end{cases} \tag{1}$$

where c is the cutoff parameter that controls the regularization effect of weights in a weighted covariance matrix. Note that $\rho_H(s)$ becomes a quadratic or a linear function of $|s|$ depending on the value of s. The solution for minimizing $J_{R_1}(\mathbf{W})$ was obtained by performing a subspace iteration algorithm [24].

On the other hand, PCA-L_1 was developed in [14] motivated by the duality between maximizing variance and minimizing reconstruction error. It maximizes an L_1 dispersion among the projected samples, $\sum_{i=1}^{N} ||\mathbf{W}^T\mathbf{x}_i||_1$. A novel and efficient method for maximizing the L_1 dispersion was also presented in [14]. The method allows PCA-L_1 to be performed by much less computational effort than R_1-PCA.

HQ-PCA is formulated based on the maximum correntropy criterion in terms of information theoretic learning. Without the zero-mean assumption, which is necessary in other variants of PCA, HQ-PCA maximizes the correntropy estimated between a set of training samples $\{\mathbf{x}_i\}_{i=1}^N$ and the set of their reconstructed samples $\{\mathbf{W}\mathbf{y}_i + \mathbf{m}\}_{i=1}^N$, where $\mathbf{m}$ is a data mean. Mathematically, HQ-PCA tries to maximize the following objective function:

$$\arg\max_{\mathbf{W},\mathbf{m}} \sum_{i=1}^{N} g\left(\sqrt{\overline{\mathbf{x}}_i^T\overline{\mathbf{x}}_i - \overline{\mathbf{x}}_i^T\mathbf{W}\mathbf{W}^T\overline{\mathbf{x}}_i}\right), \tag{2}$$

where $g(x) = exp(-x^2/2\sigma^2)$ is the Gaussian kernel and $\overline{\mathbf{x}}_i = \mathbf{x}_i - \mathbf{m}$. Note that HQ-PCA finds a data mean as well as a projection matrix. Using the Welsch M-estimator $\rho_W(x) = 1 - g(x)$, HQ-PCA is regarded as a robust M-estimator formulation because it is equivalent to finding $\mathbf{W}_H$ and $\mathbf{m}_H$ that minimize the following objective function:

$$J_{HQ}(\mathbf{W}, \mathbf{m}) = \sum_{i=1}^{N} \rho_W\left(\sqrt{\overline{\mathbf{x}}_i^T\overline{\mathbf{x}}_i - \overline{\mathbf{x}}_i^T\mathbf{W}\mathbf{W}^T\overline{\mathbf{x}}_i}\right). \tag{3}$$

In [12], the optimization problem in (2) was effectively solved in the half-quadratic optimization framework, which is often used to address nonlinear optimization problems in information theoretic learning.

3 Robust Principal Component Analysis Based on Generalized Mean

3.1 Generalized Mean

For a $p \neq 0$, the generalized mean or power mean $\mathcal{M}_p$ of $\{a_i > 0,\ i = 1, \ldots, N\}$ [20] is defined as

$$\mathcal{M}_p\{a_1, \ldots, a_N\} = \left(\frac{1}{N}\sum_{i=1}^{N} a_i^p\right)^{1/p}.$$

Figure 1 [25] shows that $\mathcal{M}_p\{1, 2, \ldots, 10\}$ varies continuously as p changes from -10 to 10. The arithmetic mean, the geometric mean, and the harmonic mean are special cases of the generalized mean when $p = 1$, $p \to 0$, and $p = -1$, respectively. Furthermore, the maximum and the minimum values of the numbers can also be approximated from the generalized mean by making $p \to \infty$ and $p \to -\infty$, respectively. Note that as p decreases (increases), the generalized mean is more affected by the smaller (larger) numbers than the larger (smaller) ones, i.e., controlling p makes it possible to adjust the contribution of each number to the generalized mean. This characteristic is useful in the situation where data samples should be

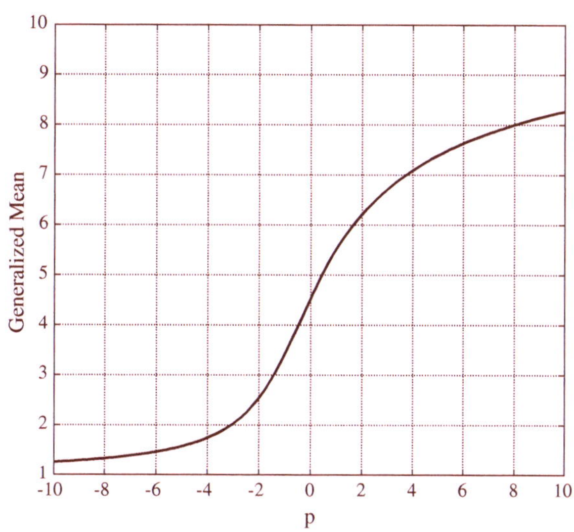

Fig. 1 The generalized mean of $\{1, \ldots, 10\}$ for various values of p

differently handled according to their importance, for example, when outliers are contained in the training set.

In [25], it was shown that the generalized mean of a set of positive numbers can be expressed by a nonnegative linear combination of the elements in the set as the following:

$$\left(\frac{1}{K}\sum_{i=1}^{K} a_i^p\right)^{1/p} = c_1 a_1 + \cdots + c_K a_K. \tag{4}$$

Each c_i in this equation can be obtained by differentiating this equation with respect to c_i.

$$c_i = \left(\frac{1}{K}\sum_{i=1}^{K} a_i^p\right)^{\frac{1}{p}-1} \frac{a_i^{p-1}}{K}, \tag{5}$$

where $i = 1, \ldots, K$. In this chapter, it is further simplified as the following:

$$\sum_{i=1}^{K} a_i^p = b_1 a_1 + \cdots + b_K a_K$$
$$b_i = a_i^{p-1}, \quad i = 1, \ldots, K. \tag{6}$$

Note that each weight b_i has the same value of 1 if $p = 1$, where the generalized mean becomes the arithmetic mean. It is also noted that, if p is less than one, the weight b_i increases as a_i decreases. This means that, when $p < 1$, the generalized mean is more influenced by the small numbers in $\{a_i\}_{i=1}^{K}$, and the extent of the influence increases as p decreases. This equation plays an important role in solving the optimization problems using the generalized mean.

3.2 Generalized Sample Mean

Most conventional PCAs commonly assume that training samples have zero-mean. To satisfy this assumption, all of the samples are subtracted by the sample mean, i.e., $\mathbf{x}_i - \mathbf{m}_S$ for $i = 1, \ldots, N$, where $\mathbf{m}_S = \frac{1}{N}\sum_{i=1}^{N} \mathbf{x}_i$. The conventional sample mean can be considered as the center of the samples in the sense of the least square, i.e.,

$$\mathbf{m}_S = \arg\min_{\mathbf{m}} \frac{1}{N}\sum_{i=1}^{N} ||\mathbf{x}_i - \mathbf{m}||_2^2. \tag{7}$$

In (7), a small number of outliers in the training samples dominate the objective function because the objective function in (7) is constructed based on the squared distances. To obtain a robust sample mean in the presence of outliers, a new optimization problem is formulated by replacing the arithmetic mean in (7) with the generalized mean as

$$\mathbf{m}_G = \arg\min_{\mathbf{m}} \left(\frac{1}{N} \sum_{i=1}^{N} \left(||\mathbf{x}_i - \mathbf{m}||_2^2 \right)^p \right)^{1/p}.$$

This problem is equivalent to (7) if $p = 1$. As mentioned in the previous subsection, the contribution of a large number to the objective function decreases as p decreases. Thus, the negative effect of outliers can be alleviated if $p < 1$. From now on, we will call $\mathbf{m}_G$ as the *generalized sample mean*. Using the fact that x^p with $p > 0$ is a monotonic increasing function of x for $x > 0$, this problem can be converted to

$$\mathbf{m}_G = \arg\min_{\mathbf{m}} \sum_{i=1}^{N} \left(||\mathbf{x}_i - \mathbf{m}||_2^2 \right)^p. \tag{8}$$

Although the minimization in (8) should be changed into the maximization when $p < 0$, we only consider positive values of p in this paper.

The necessary condition for $\mathbf{m}_G$ to be a local minimum is that the gradient of the objective function in (8) with respect to $\mathbf{m}$ is equal to zero, i.e.,

$$\frac{\partial}{\partial \mathbf{m}} \sum_{i=1}^{N} \left(||\mathbf{x}_i - \mathbf{m}||_2^2 \right)^p = 0.$$

However, it is hard to find a closed-form solution of the above equation. Although any gradient-based iterative algorithms can be applied to obtain $\mathbf{m}_G$, they usually have slow convergence speed. Alternatively, we develop a novel method based on (6), which is more efficient than gradient-based iterative methods. Our method for solving the problem in (8) is an iterative one, similar to the expectation-maximization algorithm [26].

In the derivation, we decompose (8) into the form of (6) and consider the weight b_i in (6) as a constant. Then, (8) can be approximated by a quadratic function of $||\mathbf{x}_i - \mathbf{m}||_2$ which can easily be optimized. The details are as follows. Let us denote the value of $\mathbf{m}$ after t iterations as $\mathbf{m}^{(t)}$. The first step of the update rule is, for $\mathbf{m}$ close to a fixed $\mathbf{m}^{(t)}$, to represent the objective function in (8) as a linear combination of $||\mathbf{x}_i - \mathbf{m}^{(t)}||_2^2$ using (6), i.e.,

$$\sum_{i=1}^{N} \left(||\mathbf{x}_i - \mathbf{m}||_2^2 \right)^p \approx \sum_{i=1}^{N} \alpha_i^{(t)} ||\mathbf{x}_i - \mathbf{m}||_2^2,$$

where

$$\alpha_i^{(t)} = \left(||\mathbf{x}_i - \mathbf{m}^{(t)}||_2^2 \right)^{p-1}. \tag{9}$$

Here, the approximation becomes exact when $\mathbf{m} = \mathbf{m}^{(t)}$. Note that the objective function near $\mathbf{m}^{(t)}$ can be approximated as a quadratic function of $\mathbf{m}$ without computing the Hessian matrix of the objective function. The next step is to find $\mathbf{m}^{(t+1)}$ that minimizes the approximated function based on the computed $\alpha_i^{(t)}$, i.e.,

$$\frac{\partial}{\partial \mathbf{m}} \sum_{i=1}^{N} \alpha_i^{(t)} ||\mathbf{x}_i - \mathbf{m}||_2^2 = 0.$$

The solution of this equation is just the weighted average of the samples as follows:

Algorithm 1 Generalized sample mean

Input: $\{\mathbf{x}_1, \ldots, \mathbf{x}_N\}$, $p > 0$.
$t \longleftarrow 0$.
$\mathbf{m}^{(t)} \longleftarrow \mathbf{m}_S$.
repeat
  Approximation: For fixed $\mathbf{m}^{(t)}$, compute $\alpha_1^{(t)}, \ldots, \alpha_N^{(t)}$ according to (9).
  Minimization: Using the computed $\alpha_1^{(t)}, \ldots, \alpha_N^{(t)}$, update $\mathbf{m}^{(t+1)}$ according to (10).
  $t \longleftarrow t + 1$.
until A stop criterion is satisfied
Output: $\mathbf{m}_G = \mathbf{m}^{(t)}$.

$$\mathbf{m}^{(t+1)} = \frac{1}{\sum_{j=1}^{N} \alpha_j^{(t)}} \sum_{i=1}^{N} \alpha_i^{(t)} \mathbf{x}_i. \tag{10}$$

This update rule with the two steps is repeated until a convergence condition is satisfied. This procedure is summarized in Algorithm 1 and the corresponding MATLAB code can be found in Appendix 1. Note that a weighted average is computed at each iteration in Algorithm 1. Thus, it can be said that Algorithm 1 is a special case of the mean shift algorithm [27]. It is also noted that the number of initial points is only one, which is set to $\mathbf{m}_S$. Since non-convex optimization methods depend on initial points, they are generally conducted several times started from different initial points and the solution is selected as the one providing the best performance. However, we have empirically found that Algorithm 1 started from $\mathbf{m}_S$ converges to a local optimum point that is enough robust to outliers.

To demonstrate the robustness of the generalized sample mean obtained by Algorithm 1, we randomly generated 100 samples from a two-dimensional Gaussian distribution with the mean $\mathbf{m}_i = 0$ and covariance matrix $\Sigma_i = diag\,[0.5, 0.5]$ for inliers and also generated 10 samples from another two-dimensional Gaussian distribution with the mean $\mathbf{m}_o = [5, 5]^T$ and covariance matrix $\ _o = diag\,[0.3, 0.3]$ for outliers. Using the generated samples, the sample mean was computed and two generalized sample means were also obtained by Algorithm 1 with $p = 0.1$ and $p = 0.2$, respectively. Figure 2 shows the arithmetic sample mean and the two generalized sample means together with the generated samples. It is obvious that the generalized

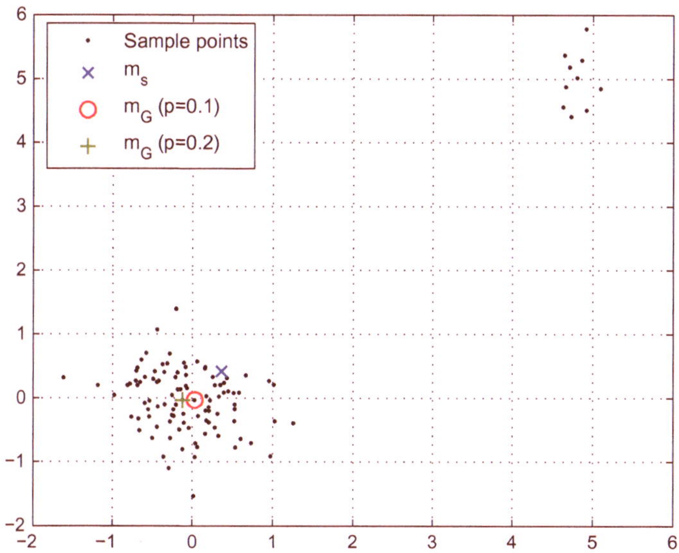

Fig. 2 2D toy example with 100 inliers and 10 outliers. The arithmetic mean ($\mathbf{m}_S$) and the generalized sample mean ($\mathbf{m}_G$) are marked

sample means are located close to the mean of the inliers, $[0, 0]^T$, whereas the arithmetic sample mean is much more biased by the ten outliers. This illustrates that the generalized sample mean with an appropriate value of p is more robust to outliers than the arithmetic sample mean.

3.3 Principal Component Analysis Using Generalized Mean

For a projected sample $\mathbf{W}^T\mathbf{x}$, the squared reconstruction error $e(\mathbf{W})$ can be computed as

$$e(\mathbf{W}) = \widetilde{\mathbf{x}}^T\widetilde{\mathbf{x}} - \widetilde{\mathbf{x}}^T\mathbf{W}\mathbf{W}^T\widetilde{\mathbf{x}},$$

where $\widetilde{\mathbf{x}} = \mathbf{x} - \mathbf{m}$. We use the generalized sample mean $\mathbf{m}_G$ for $\mathbf{m}$. To prevent outliers corresponding to large $e(\mathbf{W})$ from dominating the objective function, we propose to minimize the following objective function:

$$J_G(\mathbf{W}) = \left(\frac{1}{N}\sum_{i=1}^{N}[e_i(\mathbf{W})]^p\right)^{1/p}, \qquad (11)$$

where $e_i(\mathbf{W}) = \widetilde{\mathbf{x}}_i{}^T\widetilde{\mathbf{x}}_i - \widetilde{\mathbf{x}}_i{}^T\mathbf{W}\mathbf{W}^T\widetilde{\mathbf{x}}_i$ is the squared reconstruction error of $\mathbf{x}_i$ with respect to $\mathbf{W}$. Note that $J_G(\mathbf{W})$ is formulated by replacing the arithmetic mean in

$J_{L_2}(\mathbf{W})$ with the generalized mean keeping the use of the Euclidean distance and it is equivalent to $J_{L_2}(\mathbf{W})$ if $p = 1$. The negative effect raised by outliers is suppressed in the same way as in (8). Also, the solution that minimizes $J_G(\mathbf{W})$ is rotational invariant because each $e_i(\mathbf{W})$ is measured based on the Euclidean distance. To obtain $\mathbf{W}_G$, we develop an iterative optimization method similar to Algorithm 1.

Like the optimization problem for $\mathbf{m}_G$ in the previous subsection, under the assumption that $p > 0$, the optimization problem based on (11) is firstly converted as the following:

Algorithm 2 PCA-GM

1: **Input:** $\{\mathbf{x}_1, \ldots, \mathbf{x}_N\}$, $\mathbf{m}_G, m, p$.
2: $t \longleftarrow 0$.
3: $\mathbf{W}^{(t)} \longleftarrow \mathbf{W}_{PCA} \in \mathbb{R}^{n \times m}$.
4: **repeat**
5: **Approximation**: For fixed $\mathbf{W}^{(t)}$, compute $\beta_1^{(t)}, \ldots, \beta_N^{(t)}$ using (13).
6: **Minimization**: Using the computed $\beta_1^{(t)}, \ldots, \beta_N^{(t)}$, find $\mathbf{W}^{(t+1)}$ by solving the eigenvalue problem in (14).
7: $t \longleftarrow t + 1$.
8: **until** A stop criterion is satisfied
9: **Output:** $\mathbf{W}_G = \mathbf{W}^{(t)}$.

$$\begin{aligned} \mathbf{W}_G &= \underset{\mathbf{W}^T\mathbf{W}=\mathbf{I}}{\arg\min} \left(\frac{1}{N} \sum_{i=1}^{N} [e_i(\mathbf{W})]^p \right)^{1/p} \\ &= \underset{\mathbf{W}^T\mathbf{W}=\mathbf{I}}{\arg\min} \sum_{i=1}^{N} [e_i(\mathbf{W})]^p , \end{aligned} \tag{12}$$

Next, let us denote $\mathbf{W}^{(t)}$ as the value of $\mathbf{W} \in \mathbb{R}^{n \times m}$ after the t-th iteration. Near a fixed $\mathbf{W}^{(t)}$, the converted objective function in (12) can be approximated as a quadratic function of $\mathbf{W}$ according to (6) as

$$\sum_{k=1}^{N} [e_i(\mathbf{W})]^p \approx \sum_{i=1}^{N} \beta_i^{(t)} e_i(\mathbf{W}),$$

where

$$\beta_i^{(t)} = \left[e_i(\mathbf{W}^{(t)}) \right]^{p-1}. \tag{13}$$

Here, the approximation becomes exact if $\mathbf{W} = \mathbf{W}^{(t)}$. After calculating each $\beta_i^{(t)}$, $\mathbf{W}^{(t+1)}$ can be computed by minimizing the approximated function as

$$\mathbf{W}^{(t+1)} = \arg\min_{\mathbf{W}} \sum_{i=1}^{N} \beta_i^{(t)} e_i(\mathbf{W})$$
$$= \arg\max_{\mathbf{W}} tr\left(\mathbf{W}^T \mathbf{S}_\beta^{(t)} \mathbf{W}\right), \tag{14}$$

where $\mathbf{S}_\beta^{(t)} = \sum_{i=1}^N \beta_i^{(t)} \tilde{\mathbf{x}}_i \tilde{\mathbf{x}}_i^T$. Note that $\mathbf{S}_\beta^{(t)}$ is a weighted covariance matrix and the columns of $\mathbf{W}^{(t+1)}$ are the m orthonormal eigenvectors associated with the largest m eigenvalues of $\mathbf{S}_\beta^{(t)}$. These two steps are repeated until a convergence criterion is satisfied. Algorithm 2 summarizes this iterative procedure of computing $\mathbf{W}_G$ and the MATLAB code of PCA-GM can be found in Appendix 2. Unfortunately, the update rule in Algorithm 2 does not guarantee that $J_G\left(\mathbf{W}^{(t+1)}\right) < J_G\left(\mathbf{W}^{(t)}\right)$. Nonetheless, the experimental results show that $\mathbf{W}_G$ obtained by the algorithm is good enough.

To help understanding of Algorithm 2, we made another toy example as shown Fig. 3a where 110 two dimensional samples are plotted. Among the samples, 100 samples are regarded as inliers and the others are regarded as outliers. The samples were generated as the following rule:

$$x_i \sim N(0, 1),$$
$$y_i = x_i + \varepsilon_i,$$

where the random noise ε_i is sampled from $N(0, 0.5^2)$ for inliers and $N(0, 3^2)$ for outliers, respectively. Figure 3b shows the objective function of PCA-GM in (11) with $p = 0.3$ for the samples as shown in Fig. 3a. We can see from Fig. 3b that the conventional PCA is based toward the ten outliers because its objective function is minimized around $\mathbf{W} = [\cos 60° \;\; \sin 60°]^T$. However, PCA-GM is robust to the outliers because its objective function is minimized at $\mathbf{W} = [\cos 48.9° \;\; \sin 48.9°]^T$, which is close to the solution without the outliers $\mathbf{W}^* = [\cos 45° \;\; \sin 45°]^T$. Given an initial projection vector $\mathbf{W}^{(0)} = [\cos 30° \;\; \sin 30°]^T$, the approximation step in Algorithm 2 gives a quadratic function corresponding to the red dashed line in Fig. 3b. In the second step, the next iteration $\mathbf{W}^{(1)}$ is determined as $[\cos 32.1° \;\; \sin 32.1°]^T$ by minimizing the approximate function. Interestingly, it can be said that the approximated function plays a similar role of an upper bound of the objective function around $\mathbf{W}^{(0)}$ in this update rule. It is also noted that the approximated function at the local optimal point $\mathbf{W} = [\cos 48.9° \;\; \sin 48.9°]^T$ has its minimum as the same location, which is denoted as the magenta dashed dotted line in Fig. 3b. This means that Algorithm 2 converges to the local minimum point of the objective function for the problem shown in Fig. 3a.

To figure out how different robust PCAs alleviate the negative effect of outliers, we compare the contribution of each sample to objective functions for each method in Table 1. Also, the contributions of the methods with respect to the reconstruction error are plotted in Fig. 4. For R_1-PCA and HQ-PCA, the Huber's and Welsch M-estimators are employed in (1) and (3), respectively and the contribution in PCA-GM is computed based on (4) and (5). It is clear that the contribution of PCA increases

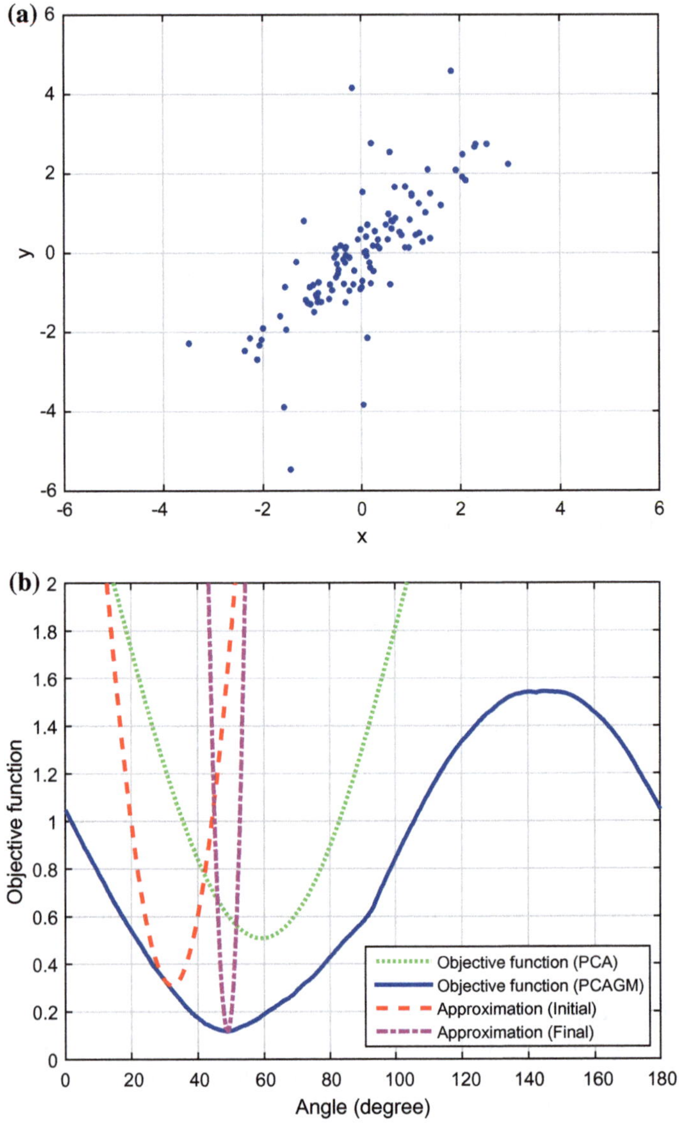

Fig. 3 **a** A toy example to illustrate Algorithm 2. **b** The values of the objective functions $J_{L_2}(\mathbf{W})$, $J_G(\mathbf{W})$ are plotted. The quadratic approximations of $J_G(\mathbf{W})$ at the initial point $\mathbf{W}^{(0)} = [\cos 30^\circ \ \ \sin 30^\circ]^T$ and the final point $\mathbf{W}^{(t)} = [\cos 48.9^\circ \ \ \sin 48.9^\circ]^T$ are also plotted. This figure is best viewed in colors

Table 1 Comparison of different versions of PCAs

Method	Contribution of each sample $\mathbf{x}_i$ to the objective function
PCA	$\mathbf{x}_i^T\mathbf{x}_i - \mathbf{x}_i^T\mathbf{W}\mathbf{W}^T\mathbf{x}_i = \|\|\mathbf{x}_i - \mathbf{W}\mathbf{W}^T\mathbf{x}_i\|\|_2^2$
L_1-PCA	$\|\|\mathbf{x}_i - \mathbf{W}\mathbf{W}^T\mathbf{x}_i\|\|_1$
R_1-PCA	$\rho_H\left(\|\|\mathbf{x}_i - \mathbf{W}\mathbf{W}^T\mathbf{x}_i\|\|_2\right)$
HQ-PCA	$\rho_W\left(\|\|\mathbf{x}_i - \mathbf{W}\mathbf{W}^T\mathbf{x}_i\|\|_2\right)$
PCA-GM	$\left(\|\|\mathbf{x}_i - \mathbf{W}\mathbf{W}^T\mathbf{x}_i\|\|_2^2\right)^p$

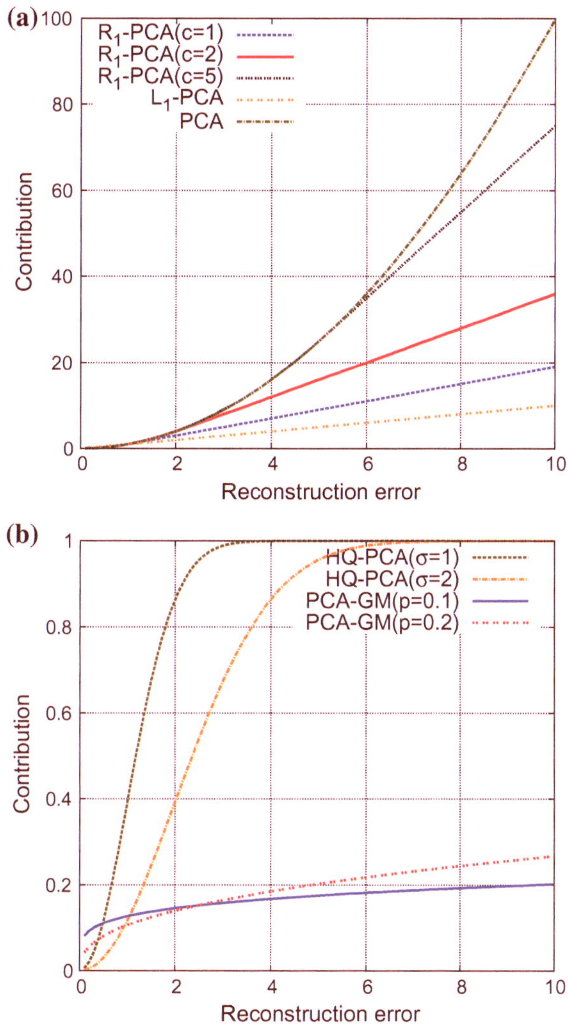

Fig. 4 Contribution of a reconstruction error to objective function **a** PCA, L_1-PCA, and R_1-PCA **b** HQ-PCA and PCA-GM

quadratically, so that the contributions corresponding to outliers with large reconstruction errors become very large. It is noticeable that PCA-GM indirectly modifies the contribution of each sample by minimizing the generalized mean of the squared L_2 reconstruction errors whereas the other methods directly uses the L_1-norm and other distance measures in the formulation of their optimization problems.

In practice, when $e_i(\mathbf{W}^{(t)})$ is zero or very small for any i, $\left[e_i(\mathbf{W}^{(t)})\right]^{p-1}$ is numerically unstable if $p < 1$, and Algorithm 2 can not proceed anymore. This problem can also occur in Algorithm 1. It can be overcome by adding a small constant δ into each $e_i(\mathbf{W})$ as

$$e_i(\mathbf{W})' = \widetilde{\mathbf{x}}_i^T \widetilde{\mathbf{x}}_i - \widetilde{\mathbf{x}}_i^T \mathbf{W}\mathbf{W}^T \widetilde{\mathbf{x}}_i + \delta, \tag{15}$$

where δ should be small enough that the modified objective function is not affected too much. This perturbation also changes $\mathbf{S}_\beta^{(t)}$ in (14) into $\widehat{\mathbf{S}}_\beta^{(t)}$ as

$$\widehat{\mathbf{S}}_\beta^{(t)} = \sum_{i=1}^{N} \beta_i^{(t)} \left(\widetilde{\mathbf{x}}_i \widetilde{\mathbf{x}}_i^T + \frac{\delta}{n} \right),$$

where n is the original dimensionality of data.

4 Experiments

To evaluate the proposed method, we considered face reconstruction, digit clustering, and object categorization problems, the first two of which were addressed in [14] and [12], respectively. The proposed method was compared with PCA, PCA-L_1, R_1-PCA, and HQ-PCA. Except the conventional PCA, they have the parameters to be predetermined and we determined the values of the parameters according to the recommendations in [10, 14], and [12]. Also, in PCA-GM, the generalized sample mean was used instead of the sample mean, and the perturbation parameter δ in (15) was set to 0.01 times the minimum of $e_i(\mathbf{W}_{PCA})$ for $i = 1, \ldots, N$. For the iterative algorithms as R_1-PCA, HQ-PCA, PCA-GM, the number of iterations was limited to 100.

4.1 Face Reconstruction

We collected 800 facial images from the subset 'fa' of the Color FERET database [28] for the face reconstruction problem. Each face image was normalized to a size of 40 × 50 pixels using the eye coordinates, which were obtained in the database. We simulated two types of outliers. For the first type of outliers, some of the facial images were randomly selected, and each of the selected images was occluded by a rectangular area, each pixel in which was randomly set to 0 (black) or 255 (white).

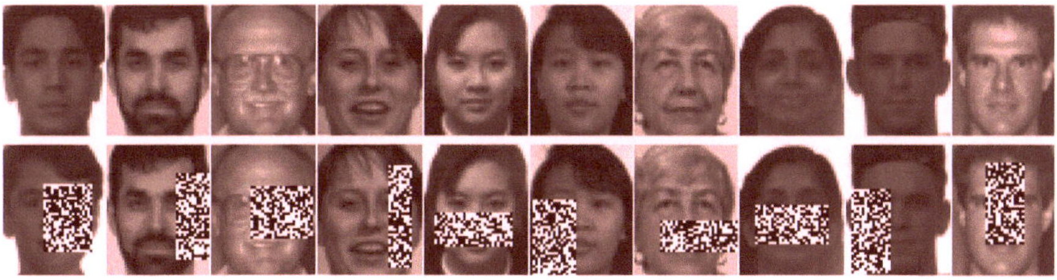

Fig. 5 Examples of original face images (upper row) and the corresponding images (lower row) occluded by rectangular noise

The size and location of the rectangular area were randomly determined. Figure 5 shows examples of original normalized faces in the upper row and their corresponding faces occluded by the rectangular noise in the lower row. To evaluate the proposed method with different noise levels, we selected 80, 160, and 240 images from the 800 facial images and occluded them by rectangular noise, so that we made three training sets including 80, 160, and 240 occluded images. For the second type of outliers, other three training sets were constructed by adding 80, 160, and 240 dummy images (outlier) with the same size to the original 800 face images (inlier), so that the numbers of inliers and outliers in the three training sets are (800,80), (800,160), and (800,240). Each pixel in the dummy images was also randomly set to 0 or 255.

After applying different versions of PCA to the training sets with the various numbers of extracted features m from 5 to 100, we compared the average reconstruction errors as in [14] defined as

$$\frac{1}{N}\sum_{i=1}^{N} || \left(\mathbf{x}_i^{ori} - \mathbf{m}\right) - \mathbf{W}\mathbf{W}^T \left(\mathbf{x}_i - \mathbf{m}\right) ||_2, \tag{16}$$

where $\mathbf{x}_i^{ori}$ and $\mathbf{x}_i$ are the i-th original unoccluded image and the corresponding training image, respectively, N is the number of the face images, and $\mathbf{m}$ is the mean of the original normalized faces. For the training sets related to the second type of outliers, the dummy images were excluded when measuring the average reconstruction errors, and $\mathbf{x}_i^{ori}$ and $\mathbf{x}_i$ were identical. Note that $\mathbf{W}$ is the projection matrix obtained from PCA, PCA-L_1, R_1-PCA, HQ-PCA, and PCA-GM for the various values of m. Moreover, PCA-GM was performed using 0.1, 0.2, 0.3, and 0.4 for the value of p to figure out the effect of it.

Figures 6 and 7 show the average reconstruction errors measured as in (16) for the training sets constructed to simulate two types of outliers when $5 \leq m \leq 100$. As shown in the figures, PCA-GM and HQ-PCA generally gave better performances than PCA, PCA-L_1, and R_1-PCA regardless of the types of outliers and the level of noise, and they yielded competitive results to each other. When the number of the occluded images is 240, which corresponds to Fig. 6c, HQ-PCA provided lower average reconstruction errors than PCA-GM for $m \leq 40$ while PCA-GM with $p = 0.1$

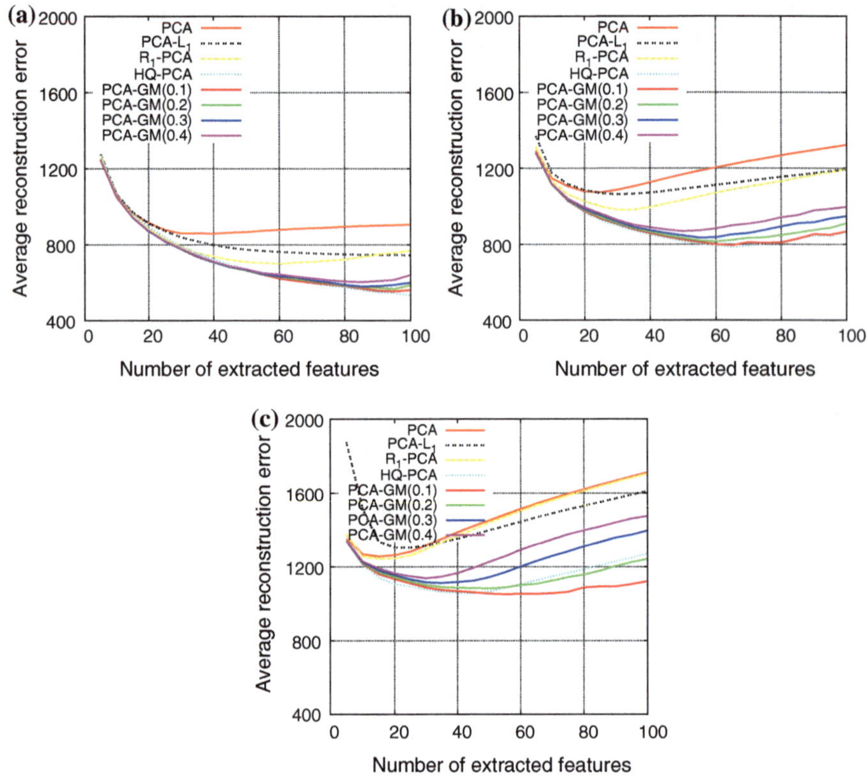

Fig. 6 Average reconstruction errors of different PCA methods for the data sets where the numbers of inliers and outliers (occlusion) are **a** (720,80), **b** (640,160), and **c** (560,240). These plots are best viewed in colors

and $p = 0.2$ gave better performances than HQ-PCA for $m \geq 60$. When the number of the dummy images is 80, which corresponds to Fig. 7a, the lower reconstruction errors could be obtained by PCA-GM rather than HQ-PCA when $m \leq 60$ while HQ-PCA preformed better than PCA-GM for $80 \leq m < 100$.

The effectiveness of the proposed method can also be found by visualizing projection matrices in terms of the *Eigenfaces* [4]. Figure 8 shows the first ten of Eigenfaces obtained by different PCA methods when $m = 40$ and the number of outliers is 240 for both types of outliers. We can see that the Eigenfaces of HQ-PCA and PCA-GM are less contaminated from the outliers than PCA, PCA-L_1, and R_1-PCA. Also, it can be seen from the figure that PCA-L_1 yielded projection matrix with different property. This may be a reason of the fact that PCA-L_1 provided relatively large reconstruction errors when m is small as shown in Figs. 6c and 7c.

The effect of p in PCA-GM was as expected. For the occlusion noise as shown in Fig. 6, the lower values of p gave better performances and the performance differences are more distinct as m and noise level increase. For the dummy images as

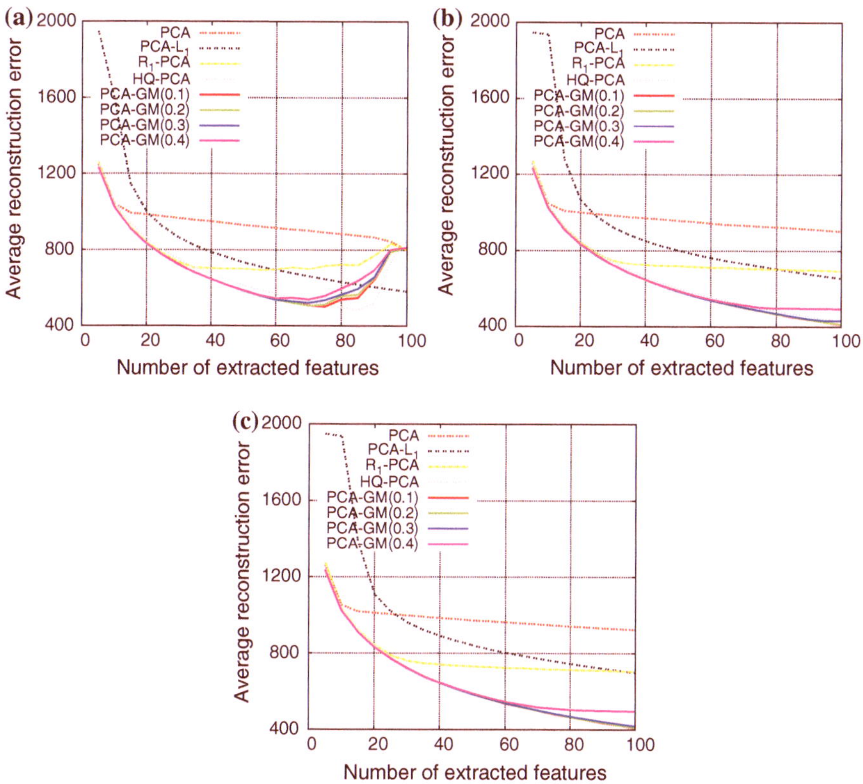

Fig. 7 Average reconstruction errors of different PCA methods for the data sets where the numbers of inliers and outliers (dummy images) are **a** (800,80), **b** (800,160), and **c** (800,240). These plots are best viewed in colors

shown in Fig. 7, PCA-GM showed almost similar performances for all the values of p except 0.4 when $m \geq 70$. These results agree with the fact that the generalized mean of a set of positive numbers depends on small numbers more and more as p gets smaller.

4.2 Clustering

The clustering problem was dealt with using a subset of the MNIST handwritten digit database,[2] which includes a training set of 60,000 examples and a test set of 10,000 examples. We randomly gathered 100 examples per the digits '3', '8', and '9' from the first 10,000 examples in the training set. To simulate outliers, we also

[2]http://yann.lecun.com/exdb/mnist/.

(a)

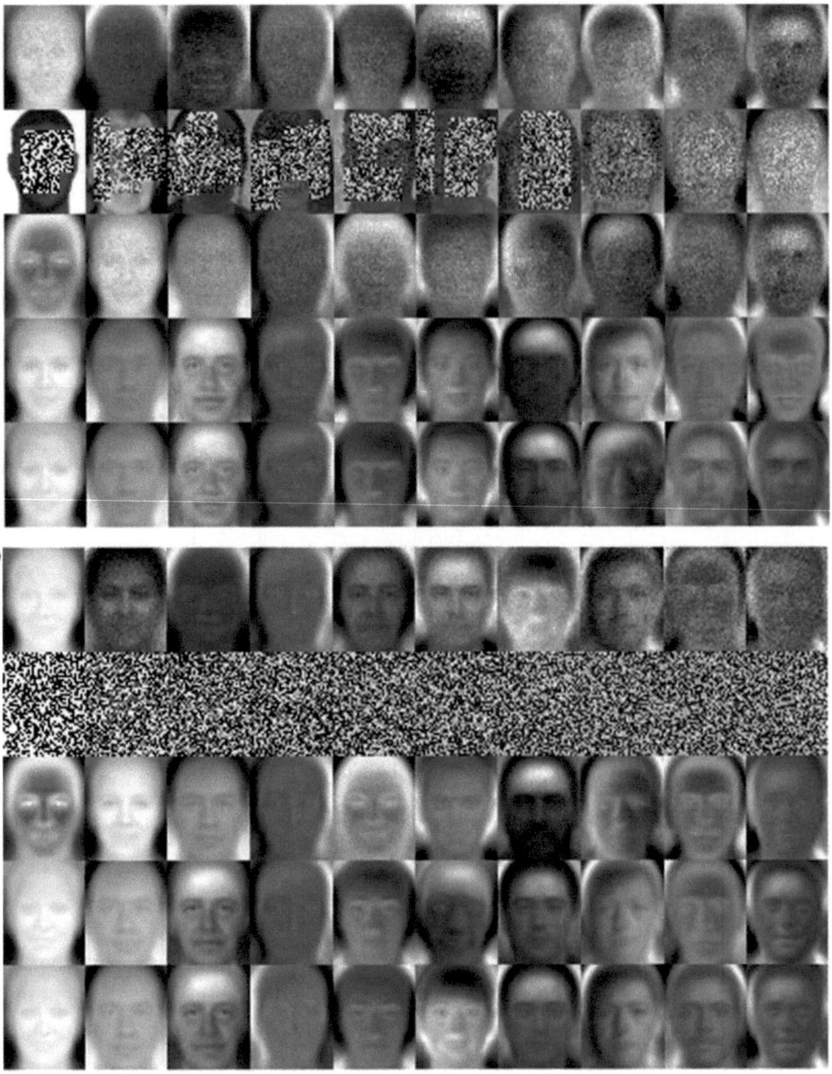

(b)

Fig. 8 Eigenfaces obtained by PCA, PCA-L_1, R_1-PCA, HQ-PCA, and PCA-GM with $p = 0.1$ in order of row. **a** Occlusion noise. **b** Dummy image noise

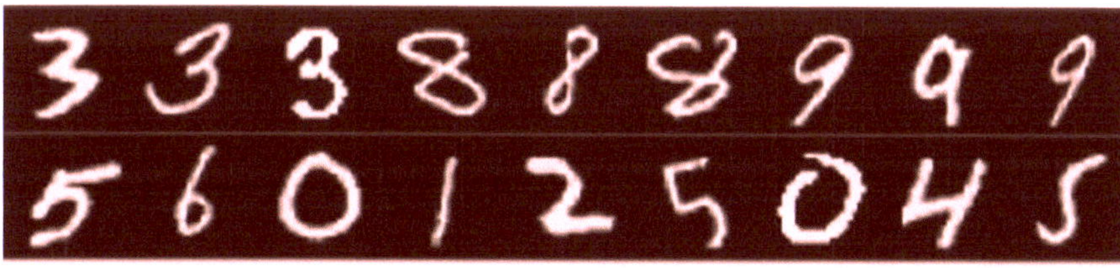

Fig. 9 Examples of MNIST handwritten digit images used as inliers (first row; 3, 8, 9) and outliers (second row; other digits)

Table 2 Clustering accuracy (%) of the digit images corresponding to '3', '8', and '9' in the reduced spaces which are obtained from the training set containing the other digit images as outliers

m	PCA	PCA-L_1	R_1-PCA	HQ-PCA	PCA-GM
50	70.00	69.00	69.67	70.00	70.00
100	70.00	72.00	70.00	69.33	69.67
150	70.67	74.00	70.67	70.00	70.00
200	70.33	73.67	70.33	73.67	**75.00**
250	70.33	73.67	73.67	74.00	74.00
300	70.33	**75.00**	**75.00**	**73.67**	73.67

randomly gathered 60 examples corresponding to the other digits from the same 10,000 examples. Thus, our training set for the clustering problem consists of 300 inliers and 60 outliers, which were normalized to unit norm. Figure 9 shows nine images of the inliers in the upper row and nine images of the outliers in the lower row.

After obtaining projection matrices by applying various versions of PCAs to the training set, K-means clustering with $K = 3$ was performed using the projected inlier examples. For the initial means of the K-means clustering, we selected the two examples with the largest distance and then selected another example which had the largest sum of the distances from the previously selected two examples. The clustering accuracy was computed based on the class labels assigned to the examples in the database. Table 2 shows the clustering accuracy for various numbers of extracted features. As the previous experiments, we conducted PCA-GM using the settings of $p \in \{0.1, 0.2, 0.3, 0.4\}$. The best performance was achieved when $p = 0.3$ which is reported in Table 2. Considering the clustering accuracy without the dimensionality reduction was 70%, PCA-GM improved the clustering accuracy by 5%. Different from the results of the face reconstruction problem in the previous subsection, R_1-PCA and PCA-L_1 gave similar highest clustering accuracy as PCA-GM, while HQ-PCA performed pooly than PCA-GM. However, R_1-PCA and PCA-L_1 provided the highest accuracy when $m = 300$ whereas PCA-GM yielded the best performance when $m = 200$.

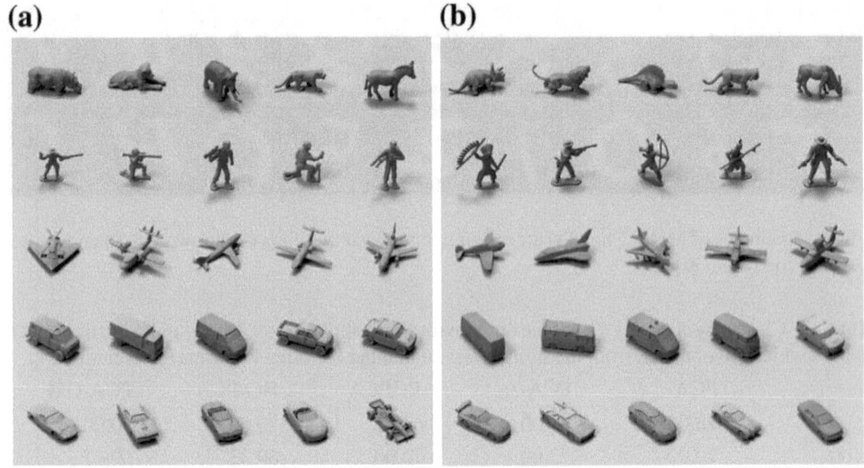

Fig. 10 Images of objects in **a** training and **b** test sets of NORB data set

4.3 *Object Categorization*

We evaluated the proposed method by performing object categorization on Small NORB data set [29]. The NORB data set consists of images of 50 different objects belonging to 5 categories each of which contains 10 objects. For each category, the images of 5 objects shown in Fig. 10a belong to its training set and those of 5 objects shown in Fig. 10b belong to its test set. The Small NORB data set is a subset of the NORB data set comprising 24,300 images for training and 24,300 images for testing, which are normalized with the size of 96×96 pixels on uniform background. Each object in the data set was captured under 18 azimuths, 9 elevations, and 6 light conditions. To evaluate the proposed method for different numbers of training samples, we uniformly sampled the three image capture variables to construct three training sets with 3375, 12150, and 24300 samples. We also resized the images in the data set to 48×48 and 64×64 pixels for computational efficiency. Consequently, we have six training sets with different number of samples (N) and dimensionality of input samples (n).

Although there are various approaches to categorize an arbitrary sample $\mathbf{z}$ corresponding to an image of an object, we performed the categorization by the *nearest-to-subspace*, i.e., $\mathbf{z}$ is determined to belong to the category minimizing the distance from $\mathbf{z}$ to the subspace spanned by the training samples in the category. For the distance from $\mathbf{z}$ to the subspace of the i-th category, we employed the squared residual error of $\mathbf{z}$ to the subspace computed as $\tilde{\mathbf{z}}_i^T\tilde{\mathbf{z}}_i - \tilde{\mathbf{z}}_i^T\mathbf{W}_i\mathbf{W}_i^T\tilde{\mathbf{z}}_i$, where $\tilde{\mathbf{z}}_i = \mathbf{z} - \mathbf{m}_i$ and $\mathbf{W}_i$ is the orthonormal basis of the subspace, which corresponds to the projection matrix and can be obtained by one of the PCA methods aforementioned. Also, $\mathbf{m}_i$ is the mean of the training samples in the i-th category. We used the sample mean $\mathbf{m}_S$ for $\mathbf{m}_i$ in PCA, PCA-L_1, and R_1-PCA while we used $\mathbf{m}_H$ and $\mathbf{m}_G$ instead of $\mathbf{m}_S$

in HQ-PCA and PCA-GM, respectively. For the purpose of comparison, the categorization accuracy was evaluated varying the dimensionality of subspaces (m) from 5 to 50.

Figure 11 shows the the categorization accuracy measured on the 24300 test images in Small NORB data set. It is necessary to note that artificial outliers were not used in this experiments different from the previous ones. We can see that PCA-GM with an appropriate value of p is competitive with the conventional PCA when $N = 3375$ and the proposed method provides higher categorization accuracies than PCA as N increases. Especially when $N = 24300$, the proposed method achieves the best performance for all the cases of m. This trend appears in both cases of $n = 48 \times 48$ and $n = 64 \times 64$. However, the other variants of PCA did not gave higher accuracies than the conventional PCA for most cases. In particular, HQ-PCA, which showed competitive performance in the face reconstruction experiments, resulted in the lowest categorization accuracy. This means that the proposed method can be an effective alternative to PCA in object categorization using the nearest-to-subspace when training data is enough.

Together with the categorization accuracy, we measured number of iterations in PCA-GM and running time of the proposed method to obtain projection matrices from the above six training sets of the Small NORM data set. Table 3 shows the average numbers of iterations performed in the proposed method. From this table, we can find that PCA-GM converges in less than 50 iterations on average. Also, the average number of iterations decreases as the value of p increases from 0.1 to 0.9. This may have been resulted from the fact that the objective function of PCA-GM has many fluctuations when the value of p is close to zero whereas it is similar to one of the conventional PCA, which is quadratic, when the value of p is close to one. The overall running time of the proposed method described in Algorithm 2 varies depending on the number of iterations needed until a stop criterion is satisfied. Thus, we divided the overall running times by the average numbers of iterations performed in computing five projection matrices with respect to five categories for every combination of m, n, and N, which are summarized in Table 4 in the setting of $p = 0.1$. From the other values of p, we could see the similar tendencies. The running times were measured on a 3.4 GHz Intel Xeon workstation with 12 cores using MATLAB. Each iteration in the algorithm consists of two processes, the approximation and the minimization. Compared to the approximation, the minimization requires much more computations. It corresponds to the weighted eigenvalue decomposition, which was implemented by applying the singular value decomposition (SVD) to the weighted data matrix instead of computing the weighted covariance matrix and applying the eigenvalue decomposition to it for efficiency. Thus, the running times reported in Table 4 can be regarded as the running time of the SVD approximately. Considering the average numbers of iterations shown in Table 3, it can be said that the proposed method is feasible enough until $N = 25000$ and $n = 5000$ roughly.

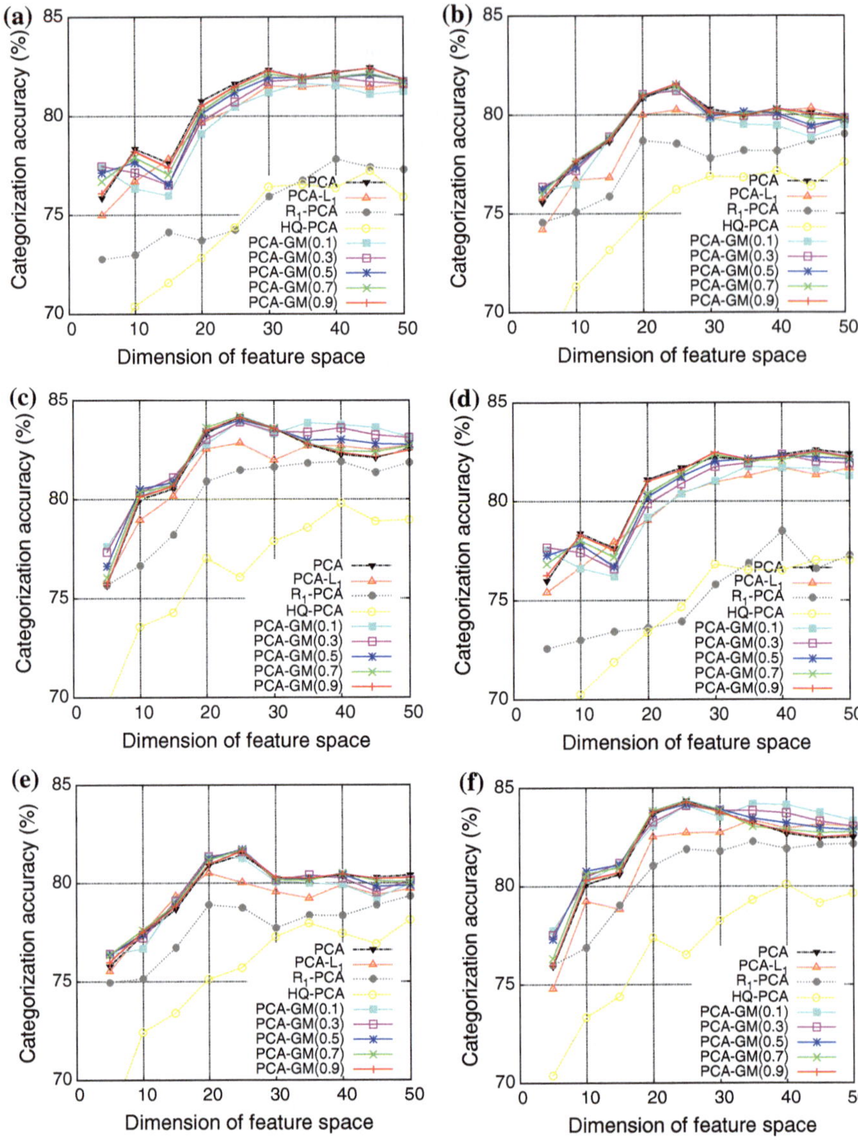

Fig. 11 Categorization accuracy of different PCA methods for the training sets with different sizes of training images (n) and different numbers of training samples (N); **a** (48 × 48, 3375), **b** (48 × 48, 12150), **c** (48 × 48, 24300), **d** (64 × 64, 3375), **e** (64 × 64, 12150), **c** (64 × 64, 24300). These plots are best viewed in colors

Table 3 Average numbers of iterations needed in PCA-GM on Small NORB data set

$p = 0.1$	$p = 0.3$	$p = 0.5$	$p = 0.7$	$p = 0.9$
22.89	23.43	19.27	14.80	8.42

Table 4 Average running time in seconds per each iteration in PCA-GM with $p = 0.1$ on Small NORB data set

m	$n = 48 \times 48$			$n = 64 \times 64$		
	$N = 3375$	$N = 12150$	$N = 24300$	$N = 3375$	$N = 12150$	$N = 24300$
5	1.05	22.55	40.28	1.62	47.37	146.31
10	0.95	25.45	42.90	1.59	51.36	150.91
15	0.98	25.69	37.78	1.59	52.55	134.89
20	0.94	23.74	39.14	1.49	46.61	149.55
25	0.89	17.99	35.50	1.44	36.83	130.75
30	0.88	19.60	38.86	1.42	42.74	98.59
35	0.89	18.34	39.40	1.46	43.60	136.14
40	0.84	20.23	32.62	1.42	42.51	106.12
45	0.85	20.21	33.70	1.38	39.35	126.85
50	0.84	18.35	32.03	1.43	41.43	111.99

5 Conclusion and Discussion

We proposed a robust PCA using the generalized mean to mitigate the negative effect of outliers belonging to the training set. Considering the fact that the sample mean is prone to the outliers, a generalized sample mean was proposed based on the generalized mean as an alternative to the sample mean in the framework of the proposed method. The efficient iterative methods were also developed to solve the optimization problems formulated using the generalized mean. Experiments on the face reconstruction, clustering, and object categorization problems demonstrated that the proposed method performs better than or equal to the other robust PCAs depending on the problems tackled. We expect that the proposed methods can be used in various applications. For example, a trimmed average, which is one of the robust first-order statistics, was used in a scalable robust PCA method [30]. We think that the generalized sample mean can be an effective alternative to the trimmed average.

Acknowledgements This research was supported by Basic Science Research Programs through the National Research Foundation of Korea (NRF) funded by the Ministry of Education (NRF-2013R1A1A2007461, NRF-2013R1A1A1006599).

Appendix 1

This MATLAB code is an implementation of Algorithm 1, which provides the generalized sample mean (`generalizedSampMean`) from the following input arguments.

- dataSamps: a two-dimensional matrix where each column vector corresponds to each data sample.
- p: the intrinsic parameter of the generalized mean.

```
function generalizedSampMean = ...
    GeneralizedSampleMean(dataSamps,p)

nMaxIter = 50;
thresholdRatio = 0.01;

meanSamps = mean(dataSamps);
meanSampsIter = meanSamps;

nSamps = size(dataSamps,1);
meanMat = repmat(meanSamps,[nSamps,1]);
dataSampsZM = dataSamps - meanMat;
objFunc = sum(diag(dataSampsZM*dataSampsZM').^(p));

index = 0;
flag = 0;
while flag == 0
    index = index + 1;

    meanSamps_Before = meanSampsIter;
    objFunc_Before = objFunc;

    meanMat = repmat(meanSampsIter,[nSamps,1]);
    dataSampsZM = dataSamps - meanMat;
    alphas = diag(dataSampsZM*dataSampsZM').^(p-1);
    meanSampsIter = (dataSamps' * alphas / sum(alphas))';

    meanMat = repmat(meanSampsIter,[nSamps,1]);
    dataSampsZM = dataSamps - meanMat;
    objFunc = sum(diag(dataSampsZM*dataSampsZM').^(p));

    diffMeanVec = meanSampsIter - meanSamps_Before;
    diffMeanVecNorm = sqrt(diffMeanVec*diffMeanVec');

    if index ≥ nMaxIter
        flag = 1;
    elseif ...
        diffMeanVecNorm/sqrt(meanSampsIter*meanSampsIter')*100 ..
        < thresholdRatio
        flag = 1;
    elseif objFunc ≥ objFunc_Before
        flag = 1;
        meanSampsIter = meanSamps_Before;
    end
end

generalizedSampMean = meanSampsIter;

end
```

```
function generalizedSampMean = ...
    GeneralizedSampleMean(dataSamps,p)
```

```

nMaxIter = 50;
thresholdRatio = 0.01;

meanSamps = mean(dataSamps);
meanSampsIter = meanSamps;

nSamps = size(dataSamps,1);
meanMat = repmat(meanSamps,[nSamps,1]);
dataSampsZM = dataSamps - meanMat;
objFunc = sum(diag(dataSampsZM*dataSampsZM').^(p));

index = 0;
flag = 0;
while flag == 0
    index = index + 1;

    meanSamps_Before = meanSampsIter;
    objFunc_Before = objFunc;

    meanMat = repmat(meanSampsIter,[nSamps,1]);
    dataSampsZM = dataSamps - meanMat;
    alphas = diag(dataSampsZM*dataSampsZM').^(p-1);
    meanSampsIter = (dataSamps' * alphas / sum(alphas))';

    meanMat = repmat(meanSampsIter,[nSamps,1]);
    dataSampsZM = dataSamps - meanMat;
    objFunc = sum(diag(dataSampsZM*dataSampsZM').^(p));

    diffMeanVec = meanSampsIter - meanSamps_Before;
    diffMeanVecNorm = sqrt(diffMeanVec*diffMeanVec');

    if index ≥ nMaxIter
        flag = 1;
    elseif ...
        diffMeanVecNorm/sqrt(meanSampsIter*meanSampsIter')*100 ...
        < thresholdRatio
        flag = 1;
    elseif objFunc ≥ objFunc_Before
        flag = 1;
        meanSampsIter = meanSamps_Before;
    end
end

generalizedSampMean = meanSampsIter;

end
```

Appendix 2

This MATLAB code is an implementation of Algorithm 2, which provides the projection matrix (`W`) from the following input arguments.

- `dataSamps`: a two-dimensional matrix where each column vector corresponds to each data sample.

- `generalizedSampMean`: the generalized sample mean computed from Algorithm 1.
- `nFeatsPCA`: the dimension of the resuting subspace.
- `p`: the intrinsic parameter of the generalized mean.

```
function W = PCAGM(trainData,generalizedMean,nFeatsPCA,p)

nMaxIter = 100;
threDiff = 10^(-5);
ratio = 0.01;

W = PCA(trainData,nFeatsPCA);

nTrain = size(trainData,1);
meanMat = repmat(generalizedMean,[nTrain,1]);
trainDataZM = trainData - meanMat;

residue = ComptResidualErrors(trainDataZM,W);
objFunc = sum(residue.^(p));

minResidue = min(residue);
eps = ratio * minResidue;

flag = 0;
index = 0;
while flag == 0
    index = index + 1;

    W_Before = W;
    objFunc_Before = objFunc;

    residue = residue + eps*ones(size(residue));
    alphas = residue.^(p-1);

    A = sqrt(diag(alphas));
    S = trainDataZM' * A;
    [eigVectors,¬,¬] = svd(S,0);
    W = eigVectors(:,1:nFeatsPCA);

    residue = ComptResidualErrors(trainDataZM,W);
    objFunc = sum(residue.^(p));

    if objFunc_Before-objFunc < threDiff
        flag = 1;
    elseif objFunc ≥ objFunc_Before
        flag = 1;
        W = W_Before;
    elseif index ≥ nMaxIter
        flag = 1;
    end

end

end

function errors = ComptResidualErrors(trainDataZM,W)

trainDataProj = trainDataZM * W;
```

```
R = trainDataZM*trainDataZM' - trainDataProj*trainDataProj';
errors = diag(R);

end

function W_PCA = PCA(trainData,nFeatsPCA)

[nTrain,nVar] = size(trainData);
meanTrain = mean(trainData);
meanMat = repmat(meanTrain,[nTrain,1]);
if nTrain < nVar
    S = (trainData-meanMat)*(trainData-meanMat)'/nTrain;
    U = diagonal(S);
    W = U(:,1:nFeatsPCA);
    W_PCA = (trainData-meanMat)'*W;
    for k=1:nFeatsPCA
        W_PCA(:,k) = W_PCA(:,k) / sqrt(W_PCA(:,k)'*W_PCA(:,k));
    end
else
    S = (trainData-meanMat)'*(trainData-meanMat)/nTrain;
    W = diagonal(S);
    W_PCA = W(:,1:nFeatsPCA);
end

end

function [Vectors,Values] = diagonal(M)
% the input argument M should be an n by n square matrix.
    [Vectors,D] = svd(M);
    d_1D = diag(D);
    Values = d_1D.^2;
end
```

References

1. Jain, A., Duin, R., Jianchang, M.: Statistical pattern recognition: a review. IEEE Trans. Pattern Anal. Machine Intell. **22**(1), 4–37 (2000)
2. Jolliffe, I.: Principal Component Analysis, 2nd edn. Springer, Berlin (2002)
3. Belhumeur, P., Hespanha, J., Kriegman, D.: Eigenfaces versus fisherfaces: recognition using class specific linear projection. IEEE Trans. Pattern Anal. Machine Intell. **19**(7), 711–720 (1997)
4. Turk, M., Pentland, A.: Eigenfaces for recognition. J. Cognitive Neurosci. **3**(1), 71–86 (1991)
5. Ross, D.A., Lim, J., Lin, R.S., Yang, M.H.: Incremental learning for robust visual tracking. Int. J. Comput. Vision **77**(1–3), 125–141 (2008)
6. Ding, C., He, X.: K-means Clustering via Principal Component Analysis. In: Proceedings of the 21st International Conference on Machine Learning, ICML '04 (2004)
7. Yeung, K.Y., Ruzzo, W.L.: Principal component analysis for clustering gene expression data. Bioinformatics **17**(9), 763–774 (2001)
8. de la Torre, F., Black, M.J.: A framework for robust subspace learning. Int. J. Comput. Vision **54**(1–3), 117–142 (2003)
9. Brooks, J., Dulá, J., Boone, E.: A pure L_1-norm principal component analysis. Comput. Statist. Data Anal. **61**, 83–98 (2013)

10. Ding, C., Zhou, D., He, X., Zha, H.: R_1-PCA: Rotational Invariant L_1-norm Principal Component Analysis for Robust Subspace Factorization. In: Proceedings of the 23rd International Conference on Machine Learning, ICML '06, pp. 281–288 (2006)
11. He, R., Hu, B., Yuan, X., Zheng, W.S.: Principal component analysis based on non-parametric maximum entropy. Neurocomputing **73**(10–12), 1840–1852 (2010)
12. He, R., Hu, B.G., Zheng, W.S., Kong, X.W.: Robust principal component analysis based on maximum correntropy criterion. IEEE Trans. Image Process. **20**(6), 1485–1494 (2011)
13. Ke, Q., Kanade, T.: Robust L_1 Norm Factorization in the Presence of Outliers and Missing Data by Alternative Convex Programming. In: IEEE Computer Society Conference on Computer Vision and Pattern Recognition, 2005, vol. 1, pp. 739–746 (2005)
14. Kwak, N.: Principal component analysis based on L1-norm maximization. IEEE Trans. Pattern Anal. Machine Intell. **30**(9), 1672–1680 (2008)
15. Kwak, N.: Principal component analysis by l_p-norm maximization. IEEE Trans. Cyber. **44**(5), 594–609 (2014)
16. Liang, Z., Xia, S., Zhou, Y., Zhang, L., Li, Y.: Feature extraction based on L_p-norm generalized principal component analysis. Pattern Recognit. Lett. **34**(9), 1037–1045 (2013)
17. Ng, A.Y.: Feature selection, L_1 vs. L_2 regularization, and rotational invariance. In: Proceedings of the 21st International Conference on Machine Learning, ICML '04 (2004)
18. Cover, T.M., Thomas, J.A.: Elements of Information Theory. Wiley, New York (1981)
19. Liu, W., Pokharel, P.P., Principe, J.C.: Correntropy: properties and applications in non-gaussian signal processing. IEEE Trans. Signal Process. **55**(11), 5286–5298 (2007)
20. Bullen, P.: Handbook of Means and Their Inequalities, 2nd edn. Kluwer Academic Publisher, Dordrecht (2003)
21. Oh, J., Kwak, N.: Generalized mean for robust principal component analysis. Pattern Recognit. **54**, 116–127 (2016)
22. Candès, E.J., Li, X., Ma, Y., Wright, J.: Robust Principal Component Analysis? J. ACM **58**(3), 11:1–11:37 (2011)
23. Huber, P.J.: Robust Statistics, 2nd edn. Wiley, New York (2009)
24. Golub, G.H., Loan, C.F.V.: Matrix Computations, 3rd edn. Johns Hopkins, Baltimore (1996)
25. Oh, J., Kwak, N., Lee, M., Choi, C.H.: Generalized mean for feature extraction in one-class classification problems. Pattern Recognit. **46**(12), 3328–3340 (2013)
26. Bishop, C.M.: Pattern Recongntion and Machine Learning. Springer, Berlin (2006)
27. Comaniciu, D., Meer, P.: Mean shift: a robust approach toward feature space analysis. IEEE Trans. Pattern Anal. Machine Intell. **24**(5), 603–619 (2002)
28. Phillips, P., Moon, H., Rizvi, S., Rauss, P.: The FERET evaluation methodology for face-recognition algorithms. IEEE Trans. Pattern Anal. Machine Intell. **22**(10), 1090–1104 (2000)
29. LeCun, Y., Huang, F.J., Bottou, L.: Learning Methods for Generic Object Recognition with Invariance to Pose and Lighting. In: Proceedings of the 2004 IEEE Computer Society Conference on Computer Vision and Pattern Recognition, vol. 2, pp. II–97–104 (2004)
30. Hauberg, S., Feragen, A., Black, M.: Grassmann Averages for Scalable Robust PCA. In: Proceedings of the 2014 IEEE Conference on Computer Vision and Pattern Recognition, pp. 3810–3817 (2014)

Principal Component Analysis Techniques for Visualization of Volumetric Data

Salaheddin Alakkari and John Dingliana

Abstract We investigate the use of Principal Component Analysis (PCA) for the visualization of 3D volumetric data. For static volume datasets, we assume, as input training samples, a set of images rendered from spherically distributed viewing positions, using a state-of-the-art volume rendering technique. We compute a high-dimensional eigenspace, that we can then use to synthesize arbitrary views of the dataset with minimal computation at run-time. Visual quality is improved by subdividing the training samples using two techniques: cell-based decomposition into equally sized spatial partitions and a more generalized variant, which we referred to as band-based PCA. The latter approach is further extended for the compression of time-varying volume data directly. This is achieved by taking, as input, full 3D volumes comprised by the time-steps of the time-varying sequence and generating an eigenspace of volumes. Results indicate that, in both cases, PCA can be used for effective compression with minimal loss of perceptual quality, and could benefit applications such as client-server visualization systems.

1 Introduction

Volumetric data comprises three-dimensional (3D) information in the form of a discretely sampled regular grid. In some cases this is obtained as a 3D stack of 2D images acquired by imaging technologies such as Magnetic Resonance Imaging (MRI), or directly generated by simulation techniques such as used in computational fluid dynamics. In recent years, there has been a trend and a demand to visualize such datasets interactively, so that viewers may peruse the dataset from different viewpoints or based on different viewing parameters. Graphics Processing Units (GPUs), which are becoming integral components in personal computers, have made it

S. Alakkari · J. Dingliana (✉)
Graphics Vision and Visualisation Group, School of Computer Science
and Statistics, Trinity College, Dublin, Ireland
e-mail: John.Dingliana@scss.tcd.ie

S. Alakkari
e-mail: alakkars@tcd.ie

G. R. Naik (ed.), *Advances in Principal Component Analysis*,
https://doi.org/10.1007/978-981-10-6704-4_5

possible to generate high-fidelity interactive 3D visualizations of such data. However, the complexity of volumetric datasets in science and medicine has continued to increase to the point that, often, the dataset cannot fit in GPU memory or the processing and bandwidth overheads are too high that many of the advanced rendering techniques cannot be applied in real-time without considerable reduction of the data. At the same time, the use of portable computing devices is becoming ubiquitous, leading to a demand for visualization techniques suitable for such platforms, which are more constrained than traditional desktop graphical workstations. For instance, this has motivated the development of a number of *client-server* techniques, where a limited front-end client delivers the visualization whilst the bulk of the computational load or memory usage is devolved to a remote high-performance server or, indeed, a distributed source such as the cloud.

In this paper, we investigate the feasibility of using Principal Component Analysis (PCA) to improve the efficiency of visualizing 3D volumetric data. In particular, we are motivated by facilitating increased capacity to visualize such data without requiring dedicated high-end computing facilities such as supercomputers. Our proposed techniques are intended to benefit visualizations on standard workstations and eventually even on minimal client devices such as mobile tablets.

Our primary contribution is a prototype approach that uses PCA to generate a high-dimensional eigenspace capturing a view-independent representation of any 3D volumetric dataset. Arbitrary views of the volume can then be reconstructed *on-demand*, at real-time rates. The efficiency of the eigenspace is improved by two adaptive decomposition mechanisms. The approach is further generalized for the compression of large highly-complex time-varying volume datasets. Experimental results indicate that our PCA-based solution can be used to generate high-quality images of 3D volumetric datasets, whilst reducing computational complexity and data bandwidth.

2 Related Work

Volume rendering is an area of computer graphics that deals with the digital presentation of 3D volumetric data. Due to the ubiquity of such data (often referred to as *voxel* datasets), many rendering techniques have been developed over the past three decades, ranging from relatively simple slice-based techniques, that essentially blend 2D images, to highly complex 3D global illumination models. For instance, *volume ray-casting* [12], is a popular technique which has become the de facto gold-standard in interactive volume rendering. Ray-casting has many advantages such as its generality, flexibility and reduced pre-processing requirement, however it is performance intensive, typically requiring a powerful graphical system with 3D texture handling support in order to achieve real-time frame rates. On the other hand, many mobile, portable and web-based graphical systems popular in some visualization domains still have limited support for such hardware features, and thus are limited to simpler rendering techniques such as slice-based rendering.

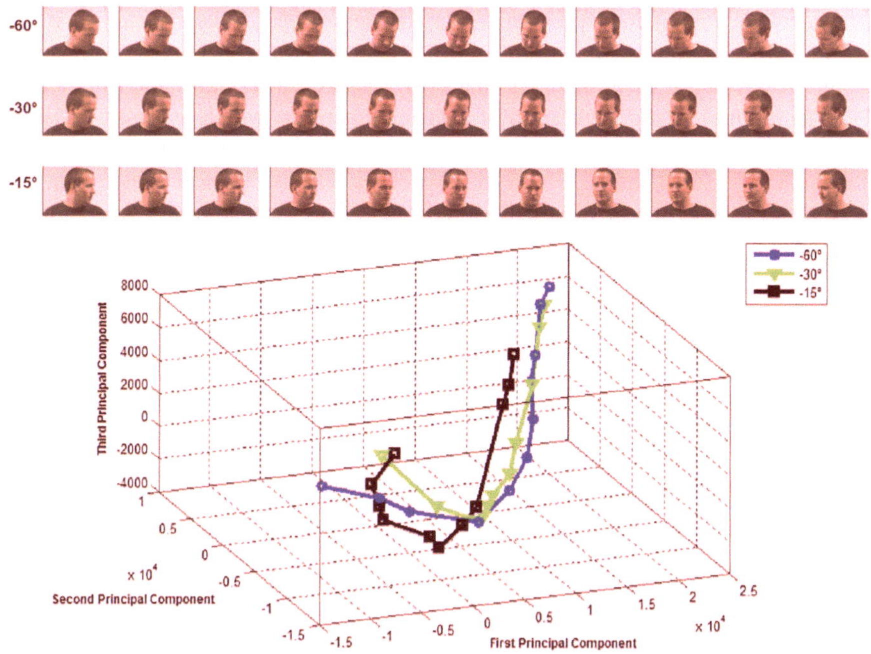

Fig. 1 Face manifolds on the first three principal component for different set of poses. The face images are from the Head Pose Image Database [10]

The use of PCA for analyzing 3D objects has been well reported in the last two decades in Computer Vision and Computer Graphics. Kirby and Sirovich [13] proposed PCA as a means for a holistic representation of the human face in 2D images by extracting a few orthogonal dimensions which form the face-space and were called eigenfaces [28]. Gong et al. [9] were the first to find the relationship between the distribution of samples in the eigenspace, or manifold, and the actual pose in an image of a human face. Figure 1 shows manifold distributions on the first three principal components (also called eigenfaces in this case) for different set of face poses. The use of PCA was extended using *reproducing kernel Hilbert spaces* which non-linearly map the face-space to a much higher dimensional space, Hilbert space [30]. Knittel and Parys [14] employed a PCA-based technique to find initial seeds for vector quantization in image compression.

Nishino et al. [21] proposed a method, called *eigen-texture*, which creates a 3D image from a sample of range images using PCA. They found that partitioning samples into smaller cell-images improved the rendering of surface-based 3D data. Grabner et al. [11] proposed a hardware accelerated technique that uses the multiple eigenspaces method [17] for image-based reconstruction of a 3D polygon mesh model. To our knowledge, PCA has not yet been applied to image based-rendering of volume data, which poses additional challenges as the rendered image typically exposes interior details that need to exhibit consistent occlusion and parallax effects.

There are a number of previous reported uses of PCA-related methods in the visualization literature. For instance, [18] employed PCA for dynamic projections in the visualization of multivariate data. Broersen et al. [2] discussed the use of PCA techniques in the generation of transfer functions, which are used to assign optical properties such as color and opacity to attributes in volume visualization. Takemoto et al. [26] used PCA for feature space reduction to support transfer function design and exploration of volumetric microscopy data. Fout and Ma [7] presented a volume compression method based on transform coding using the Karhunen-Loève Transform (KLT), which is closely related to PCA.

Many remote visualisation techniques have been proposed in the scientific and medical visualization literature. The motivation for these range from facilitating collaborative multi-user systems [15], performance improvements through distributed parallel rendering [8], web-based visualization on browsers [22], remote collaborative analysis by distant experts [24] and to achieve advanced rendering on low-spec client devices [20]. One strategy, in client-server volume rendering is to transmit 3D data on-demand to the client, after compression [20], partitioning [1] or using a progressive rendering [3]. The client in all of these approaches is required to do further processing to render the data. A second alternative, such as employed by [6], is for a high-end server to remotely render the data and transmit only images to the client, which has a much reduced responsibility of simply displaying the pre-rendered image. This strategy, often referred to as *Thin Client* (see Fig. 2), is a popular approach for visualization on portable devices such as a mobile tablets, which may be restricted in terms of computational capacity and GPU components. In between these ends of the spectrum, some image-based approaches pre-compute intermediate 2D images that are post-processed or composited by the client before display [1, 23, 27]. Image-based approaches, in general, have been of interest, for improving the efficiency of volume visualization [4, 5, 19]. At the cost of some additional computational load on the client, such a solution may provide improvements such as reduced latency during interaction and it is in this category that our main contributions lie.

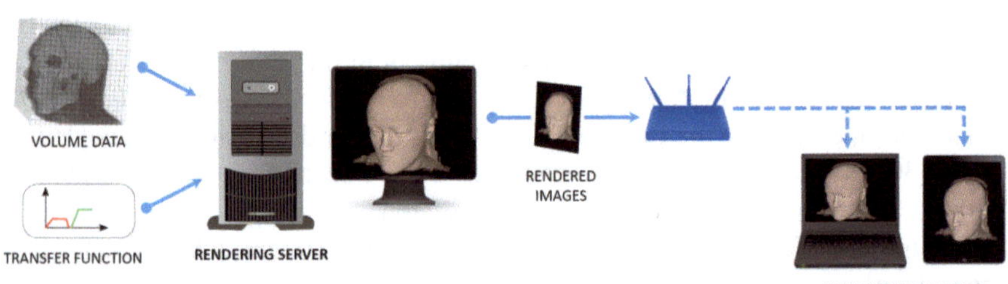

Fig. 2 Framework of a thin-client volume rendering system. Complex rendering operations are performed on a high performance server, then images are streamed in real-time to clients that merely display received images. In contrast our proposed approach is to pre-load data representative of the eigenspace and then at run-time transmit only a small number of scores that are processed by clients to build the image

3 Concepts

In this section we define the essential concepts and general terminology, which will be used in later sections to define our approach for interactive volume visualization using PCA.

The basic approach to PCA is as follows. Given data samples $X = [x_1\, x_2 \ldots x_n] \in \mathbb{R}^{d\times n}$, where each sample is in column vector format, the covariance matrix is defined as

$$C = \frac{1}{n-1}\sum_{i=1}^{n}(x_i - \bar{x})(x_i - \bar{x})^T, \qquad (1)$$

where $\bar{x}$ is the sample mean. After that, we can find the optimal low-dimensional bases that cover most of the data variance by extracting the most significant eigenvectors of the covariance matrix C. Eigenvectors are extracted by solving the following characteristic equation

$$(C - \lambda I)\, v = 0;\; v^T v = 1, \qquad (2)$$

where $v \in \mathbb{R}^d$ is the eigenvector and λ is its corresponding eigenvalue. Eigenvalues describe the variance maintained by the corresponding eigenvectors. Hence, we are interested in the subset of eigenvectors that have the highest eigenvalues $V = [v_1\, v_2 \ldots v_p]$; $p \ll n$. Then we encode a given sample x using its p-dimensional projection values (referred to as *scores*) as follows

$$y = V^T x. \qquad (3)$$

We can then reconstruct the sample as follows

$$x_{reconstructed} = Vy. \qquad (4)$$

Since in the case of $n \ll d$, C will be of rank $n-1$ and hence there are only $n-1$ eigenvectors that can be extracted from (2) and since C is of size $d \times d$, solving (2) becomes computationally expensive. We can find such eigenvectors from the dual eigenspace by computing the $n \times n$ matrix $X^T X$ and then solving the eigenvalue problem

$$\left(X^T X - (n-1)\lambda I\right) v_{dual} = 0 \qquad (5)$$

$$\Rightarrow X^T X v_{dual} = (n-1)\lambda v_{dual};\; v_{dual}^T v_{dual} = 1. \qquad (6)$$

Here, for simplicity, we assumed that the sample mean of X is the zero vector. After extracting the dual eigenvectors, one can note that by multiplying each side of (6) by X, we have

$$X X^T X v_{dual} = (n-1)\lambda X v_{dual}$$

$$\Rightarrow \frac{1}{n-1} X X^T \left(X v_{dual} \right) = \lambda \left(X v_{dual} \right)$$

$$\Rightarrow C \left(X v_{dual} \right) = \lambda \left(X v_{dual} \right)$$

$$\Rightarrow \left(C - \lambda I \right) \left(X v_{dual} \right) = 0$$

which implies that

$$v = X v_{dual}. \tag{7}$$

In order to get the orthonormal eigenvectors, the following formula is used:

$$v_{normalized} = \frac{1}{\left((n-1) \operatorname{Var} \left(v^T X \right) \right)^{\frac{1}{4}}} v.$$

Thus, when $n \ll d$, we only need to extract the dual eigenvectors using (6) and then compute the real eigenvectors using (7). Only the first few eigenvectors $V_p = [v_1 \, v_2 \ldots v_p]$, $p \ll n \ll d$ will be chosen to represent the eigenspace, those with larger eigenvalues. One advantage of PCA is the low computational complexity when it comes to encoding and reconstructing samples.

Definitions and Terminology Used

In this section, we will introduce some important definitions and terminology that are used throughout the rest of this chapter.

Volume data Data that is represented in the form of a discretely sampled 3D field. This is typically stored as a 3D rectilinear grid, with each element of the grid referred to as a volume element or *voxel*. Practical examples include data scanned using Computational Tomography or Magnetic Resonance Imaging.

Transfer function In volume visualization, the transfer function is a mapping of optical properties, such as colour and opacity, to specific voxel values for visualization.

Band A set of attributes. For instance, where data samples are volumes, a band will comprise a set of voxels.

Occurrence The probability of an attribute to belong to a band. This is given by

$$\Pr \left(x \in B_{i=1 \ldots N_{bands}} \right) = \frac{\bar{B}_i}{d},$$

where $\bar{B}_i$ is the cardinality of B_i and d is the total number of attributes.

Cell In the case that bands have equal occurrences, each band is then called a cell.

Sample variance A vector of variances computed for each attribute across the whole set of training samples. This is given by the diagonal elements of the covariance matrix as follows

$$\nu = diag\,(C).$$

Information The amount of variability covered by a band among all training samples. This is defined by

$$info\ (B) = \frac{\text{Var}\,(B)}{\sum_{i=1}^{N_{bands}} \text{Var}\,(B_i)},$$

where $Var\ (B)$ is the pooled variance of band B, which is given by

$$\text{Var}(B) = \frac{\sum_{i \in B} \left(\nu_i + (\bar{x_i} - \bar{x_B})^2\right)}{\bar{B}}.$$

where $\bar{x_i}$ is the mean value of attribute i and $\bar{x_B} = \sum_{i \in B} \bar{x}_i / \bar{B}$.

Information to occurrence ratio The mean amount of information given by an attribute in a band, defined by

$$\varrho\,(B) = \frac{info\,(B)}{\Pr\,(x \in B)}.$$

This ratio is used to detect bands corresponding to background regions. (The term background is used as a generalization of "empty" voxels. In volume data, this may refer to voxels with a zero attribute or that contribute negligible information to the data being represented).

Band-based PCA The case when PCA is applied to each non-background band separately. In this case, the eigenvectors corresponding to each band are called eigenbands.

Cell-based PCA The case when PCA is applied to each non-background cell separately. In this case, the eigenvectors corresponding to each cell are called eigencells.

Eigenvalue The amount of variability covered by an eigenvector.

Explained variance The accumulated ratio of most significant eigenvalues to the total variance of training samples which is given by

$$\Theta = \frac{\sum_{i=1}^{p} \lambda_i}{\sum_{j=1}^{n} \lambda_j}.$$

4 Cell-Based PCA for Volume Data

In this section, we investigate how well PCA is able to learn and preserve visual information in volumetric data. Specifically, we present an approach for image-based rendering of volume data using PCA.

Figure 3 illustrates the steps for reconstructing and rendering a novel view image using three alternative PCA techniques: *Standard PCA* involves applying PCA to

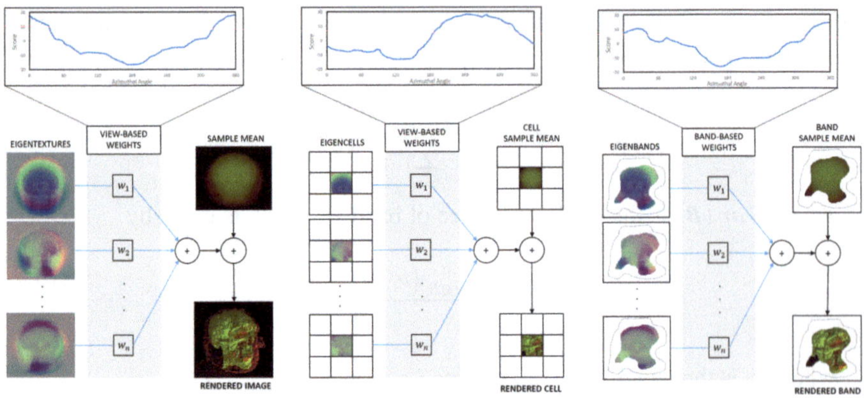

Fig. 3 Overview of our technique for image reconstruction using PCA. Left: Standard PCA; Middle: Cell-based PCA; Right: Band-based PCA

the whole image. In *Cell-based PCA*, the viewport is partitioned into rectangular cells and the eigenspace computed for each cell image individually. This is similar to the approach taken by [21], and we describe in this section how we apply and extend it for volume data visualization. An alternate approach, which we refer to as *Band-based PCA*, is discussed in Sect. 5 and entails partitioning the data more generally into bands of similar attributes that may be randomly distributed in space.

For static volume data, we assume a set of pre-rendered images as training samples. In this case each image is considered a high-dimensional vector and used as input into (1). We compute the eigenspace of the volume dataset by applying PCA to a number of training images from uniformly-spaced viewing positions in a spherical distribution (as illustrated in Fig. 4).

The 3D volumetric dataset is encapsulated using a small number of eigenimages and, at run-time, views are reconstructed using (4), based on the *scores* obtained by projecting sample values into the first p significant eigenvectors (as in (3)). By interpolating between the scores of training samples in the eigenspace, we can further synthesize output samples from novel viewing angles not in the training set.

Note that although a pre-processing stage is required to render the training samples and generate the eigenspace, the advantage of the approach is that run-time performance is independent of the complexity of the dataset or of the rendering technique. Images are effectively reconstructed by computing a weighted sum of the eigenimages, which are much fewer in number than, for instance, the average sampling rate in a ray-caster.

For Standard PCA, the eigenspace is computed for the full-size training images. In the case of Cell-based PCA, we first partition each image into a number of equally-sized cell images. Then, we compute the eigenspace of each cell image individually. It should be noted that the cell-based technique has similar computational complexity and memory footprint as the direct PCA technique as we essentially perform a larger number of much smaller iterations.

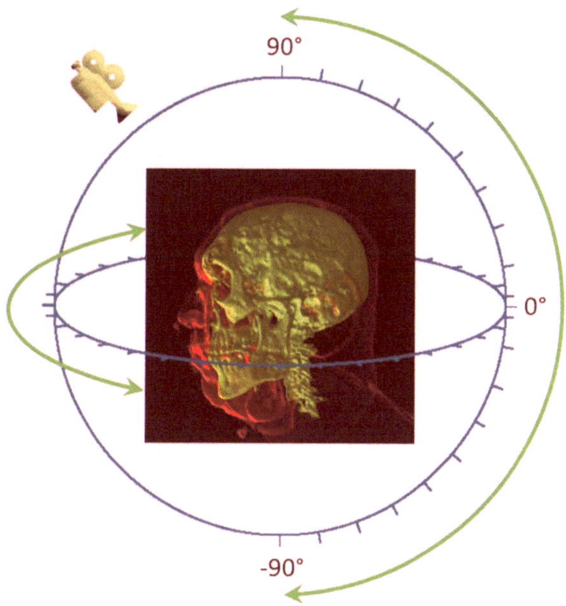

Fig. 4 Volume ray-cast images from spherically distributed sample viewing positions were used as training samples

Results: As a first test, we apply Standard PCA and Cell-based PCA for visualizing the *VisMale Head*[1] from the Visible Human dataset, at a resolution of 300 × 300 pixels. We used 1,500 training images from uniformly-spaced viewing angles (3.6° spacing for the azimuthal angle and 12° spacing for the elevation angle) to generate the input images to compute the eigenspace. The images were rendered using an implementation of a standard GPU volume ray-caster based on the approach by [12] with sampling rate of 1000 samples per ray. We then acquire test samples by applying 0.9° spacing for the azimuthal angle and 30° spacing for the elevation angle leading to a total of 2400 unique views. We used 100 eigenvectors to represent the eigenspace and, for each unique view, we synthesize the corresponding scores (projection values into the first 100 significant eigenvectors).

Figure 5 compares the reconstructed novel view images for both Standard PCA and Cell-based PCA with a ray-cast rendering from the corresponding view. Clearly, the cell-based approach produces much better quality results compared to the somewhat blurry images resulting from the standard technique with the same distribution of training samples, consistent with what was reported in the previous literature [21]. However the cell-based PCA results in subtle discontinuity artefacts at the cell boundaries in the reconstructed images (see Fig. 5d). In terms of complexity, both PCA based methods require only 100 scalar-vector multiplications at run-time, which is computationally much cheaper compared to the operations required in the equivalent ray-cast rendering.

[1] Head dataset is obtained from http://www9.informatik.uni-erlangen.de/External/vollib/.

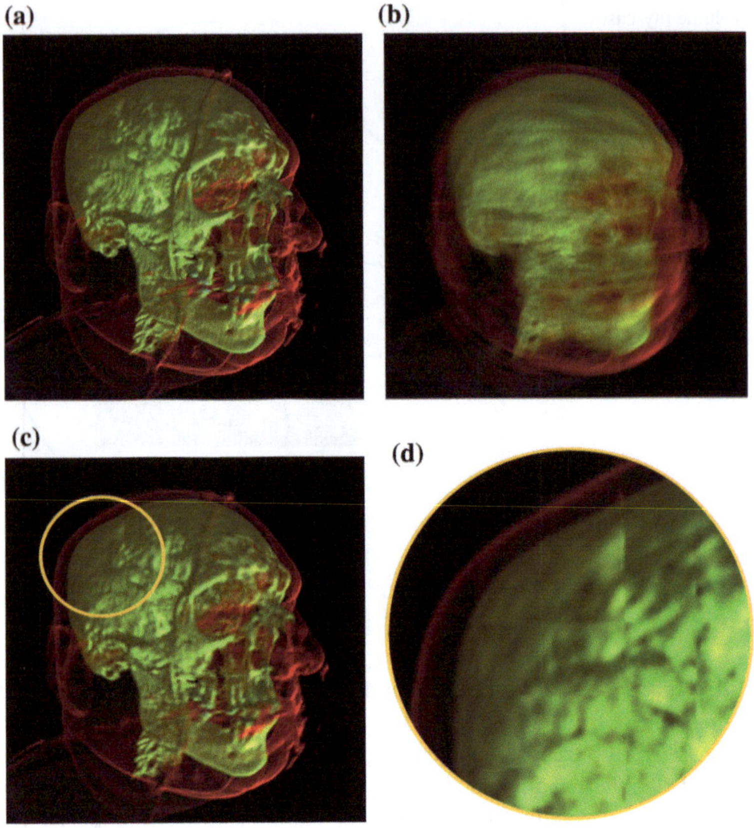

Fig. 5 Novel view of the *Vismale Head* reconstructed at 300 × 300 pixel resolution. **a** Reference image rendered using the volume ray-casting technique. **b** Standard-PCA reconstruction. **c** Cell-PCA reconstruction. **d** Subtle cell-boundary discontinuity artefacts are visible in the cell-based reconstruction when zoomed in. Note that (**d**) has been contrast enhanced to accentuate the error for inspection

Adaptive-cell PCA

A further improvement, is obtained by adaptively varying the number of eigenvectors per cell in order to achieve a more optimal tradeoff between performance and quality. The number required is determined based on total variability explained by the first p eigenvectors. This can be expressed as follows:

$$\Theta = \frac{\sum_{i=1}^{p} \lambda_i}{\sum_{j=1}^{n} \lambda_j} > T, \tag{8}$$

where λ is the eigenvalue, n is the number of first significant eigenvectors, N is the total number of eigenvectors and T is a threshold value, which affects the tradeoff between high variability and low mean number of eigenvectors per cell.

Table 1 Average number of eigenvectors per cell for a given threshold value, which is conservatively chosen for each dataset to ensure a high perceptual similarity

Dataset	VisMale head	Tooth
Volume dimensions	128 × 256 × 256	140 × 120 × 161
Mean number of dimensions per cell	46.27	26.4
T	98%	99.7%
Mean SSIM value	0.9303	0.9934

In our proof-of-concept implementation, the choice of threshold is made manually as desired by the user; in practice a value would be chosen so that it will result in a perceptually adequate result for a particular visualization task. In theory, given a reliable measure of perceptual quality, an iterative automated technique might be employed to select a variance threshold to maximize perceptual quality, whilst minimizing the number of eigenvalues. Given the threshold required, Eq. (8) is used to choose the most significant eigenvectors that have explained variance above the given threshold.

Results: For this study, we apply PCA for visualizing the VisMale *Head* and *Tooth*[2] datasets. The latter is chosen in order to determine how the approach scales to a dataset of different complexity, in this case lower voxel resolution and a structurally simpler object. This time, training samples were acquired using a standard ray-caster of resolution 1080 × 1080 pixels and sampling rate of 1000 samples per ray to achieve an image quality that might be used in a typical real-world application. For each dataset, we computed the eigenspace of each cell using 900 training images from uniformly spaced viewing angles (3.6° spacing for the azimuthal angle and 20° spacing for the elevation angle). Each cell is of size 30 × 30, resulting in a total cell number of 36 × 36 = 1296.

Table 1 shows the threshold and mean number of eigenvectors across all cells for each dataset. Note that the *Tooth* dataset has higher threshold value and lower number of eigenvectors, due to its lower complexity and detail compared to the *Head* dataset. Figure 6 shows two novel views from each dataset reconstructed using the adaptive-cell technique. We found that the adaptive technique leads to visible reduction in artefacts, including at cell boundaries, in comparison to the non-adaptive cell technique. When closely zoomed in some cell-boundary artefacts are still present in some parts of the image, however, at normal viewing resolution, the reconstructed image is almost indistinguishable from the equivalent ray-cast rendering. We analysed the accuracy of the reconstructed image using the structural similarity (SSIM) index [29] in comparison with a ray-cast rendering, which is used as a gold-standard. SSIM scores of 0.9325 and 0.9938 respectively were recorded for the reconstructed Head and Tooth images shown.

---

[2]The Tooth dataset is obtained from http://www9.informatik.uni-erlangen.de/External/vollib/.

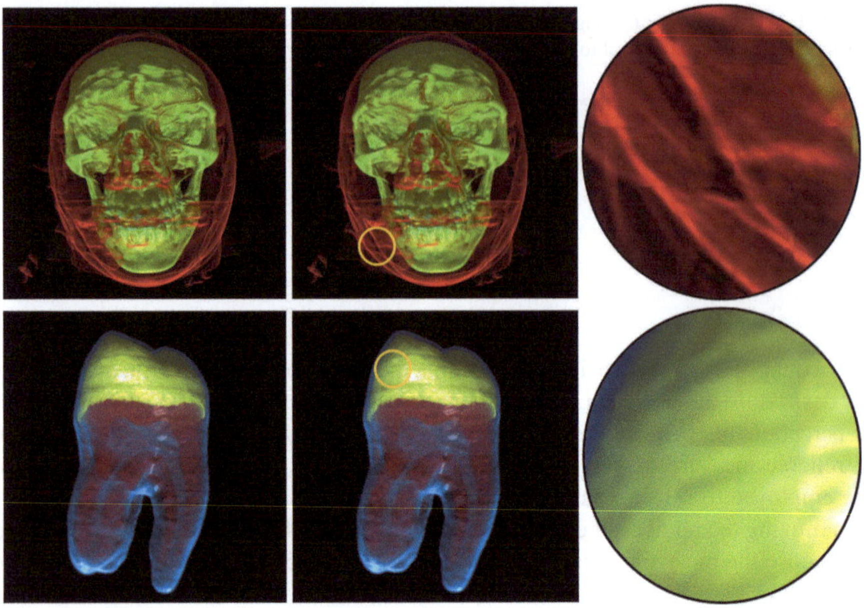

Fig. 6 Novel views of the Head and Tooth dataset visualized at 1080p resolution using the adaptive cell PCA technique. A ray-cast rendering (left) is compared to a reconstructed view (middle). The two PCA images are perceptually indistinguishable from the ray-cast rendering (SSIM 0.9325 and 0.9938 respectively) at normal viewing size but cell-boundary artefacts are visible in some areas when closely zoomed in (right)

5 Band-Based PCA

In the previous section, we partitioned the input training samples into spatial cells of uniform shape and size, with adaptive numbers of eigenvectors per cell. In this section we generalize this further with an approach we refer to as *band-based PCA* (see Fig. 3, right), where attributes are grouped into bands based on their distribution of values and PCA is applied to each non-background band separately. Mapping attributes to different bands can be done in many different ways. We use the following mapping

$$B_i = \left\{ \bar{x}_j \mid \bar{x}_j \in \left(\min(\bar{x}) + \frac{i-1}{N_{bands}} (\max(\bar{x}) - \min(\bar{x})),\ \min(\bar{x}) + \frac{i}{N_{bands}} (\max(\bar{x}) - \min(\bar{x})) \right] \right\},$$

where the range of values of the sample mean is divided into uniform subranges and then each range assigned to a band. We detect background bands (essentially empty voxels or those that have negligible contribution to the final image) by using the information to occurrence ratio as follows

$$\varrho(B) < 0.1$$

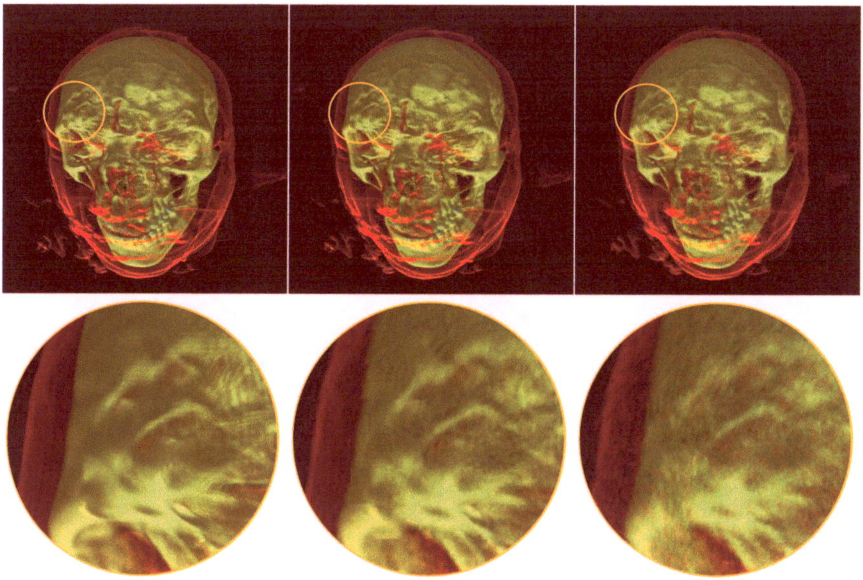

Fig. 7 Comparison of the *Vismale Head* dataset visualized at 1080p resolution with ray-casting (left) and reconstructed using the adaptive cell-based PCA technique (middle) and the band-based PCA (right). When zoomed-in (Bottom Row), cell-boundary artefacts are visible in some areas of the cell-based PCA result and some dithering artefacts are visible in the band-based PCA (right)

In other words bands with information to occurrence ratio less than 0.1 are classified as background. In general cases, we may have more than one layer of background. In such cases, increasing the number of bands will detect such layers more precisely. In contrast to the cell-based approach, the attributes in a band can be randomly distributed across the image, do not have to be contiguous, and the bands need not be spatially uniform. The main premise is that by sub dividing into bands, we provide a more meaningful grouping of attributes and at the same time overcome the cell-boundary artefacts.

Results: Figure 7 shows the Vismale Head rendered at 1080p resolution using volume ray-casting, compared to reconstructions using the cell-based (30×30) and band-based (100 bands) techniques. At normal viewing resolution the reconstructed images are quite similar to a ray-cast image of the original data (SSIM score of 0.9278 and 0.8958 respectively for the cell-based and band-based reconstructions when compared ot the ray-cast rendering). However, when zoomed in closely, we see the aforementioned boundary artefacts in the cell-based PCA. The band-based approach, on the other hand, results in a more subtle dithering-like effect.

Figure 8 shows the effect of changing the number of bands. Increasing from 50 bands (middle) to 100 bands (right) reduces dithered noise artefacts in the zoomed-in view of the data. Increasing the bands further leads to negligible visual differences as observable to the naked eye, so they are not shown here, however the SSIM scores

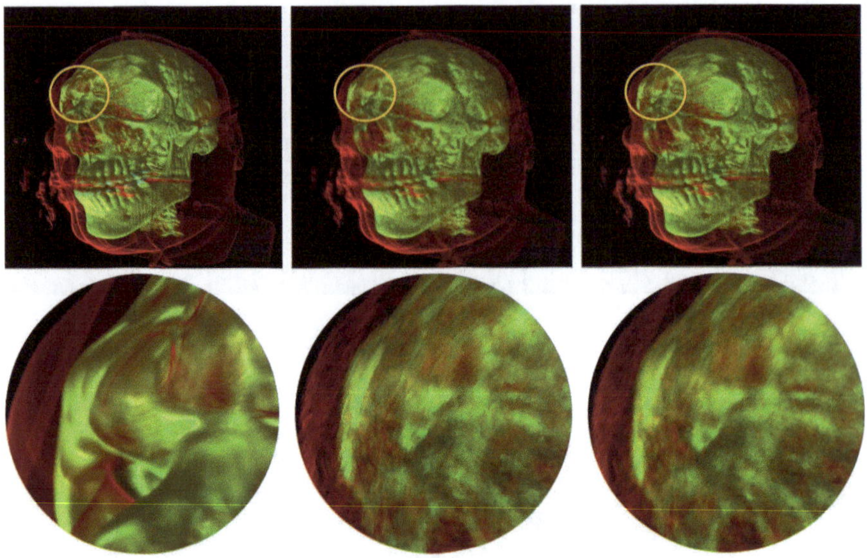

Fig. 8 Sample views of band-based PCA reconstruction Vismale. Left: reference image by ray-casting. Middle: result of using 50 bands (SSIM: 0.8666). Right: 100 bands (SSIM: 0.8779). Inset is a zoomed-in area of each image

indicates an increase in accuracy: 0.866, 0.8779, 0.8905, 0.8911 for 50, 100, 150 and 200 bands respectively.

In order to visually compare with typical volume rendering systems, we used a real-time volume ray-caster to generate the training images in all of our previously presented results. However, a key advantage of our approach is that the run-time cost of generating images is independent of the computational complexity of the rendering process or the dataset. For instance, Fig. 9 shows a proof-of-concept reconstruction of a chest dataset[3] rendered at 1080p (i.e. high definition) resolution using the Exposure Renderer [16], which achieves interactive progressive rendering by exploiting high-end GPUs. In this case we allowed the progressive rendering to converge for 5 seconds for each frame rendered on an Intel PC equipped with a 3.4 GhZ i5-4670 CPU, NVIDIA GeForce GTX 775M GPU and 16 GB RAM. In contrast, once the eigenspace is computed, our approach can be used to efficiently recreate such complex images in high detail at real-time, even on a display device without a powerful GPU (Figs. 10 and 11).

PCA for Time-Varying Volume Data Compression

Up to this point, we used PCA as an image-based rendering technique for static volume datasets; using multiple rendered views as the training set, we used

[3]The chest dataset, ARTIFIX, is obtained from the DICOM Sample Image Library: http://www.osirix-viewer.com/resources/dicom-image-library/.

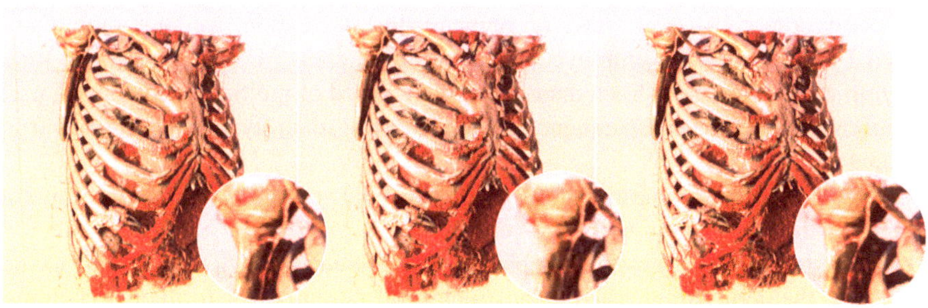

Fig. 9 Sample views of PCA reconstruction of photo-realistic volume renderings. Left: image rendered at 1080p resolution by Exposure Renderer [16]; Middle:Cell-based PCA reconstruction with threshold $\Theta = 99$ and 30×30 cell size (SSIM: 0.9819); Right Band-based PCA Reconstruction with $\Theta = 99$ and 50 bands (SSIM: 0.9991). Inset is a zoomed-in area of each image

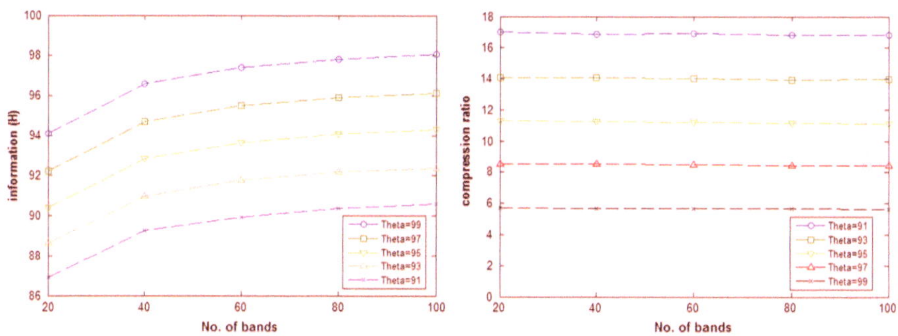

Fig. 10 Preserved information (left) and compression ratio (right) graphs for different number of bands and different theta values

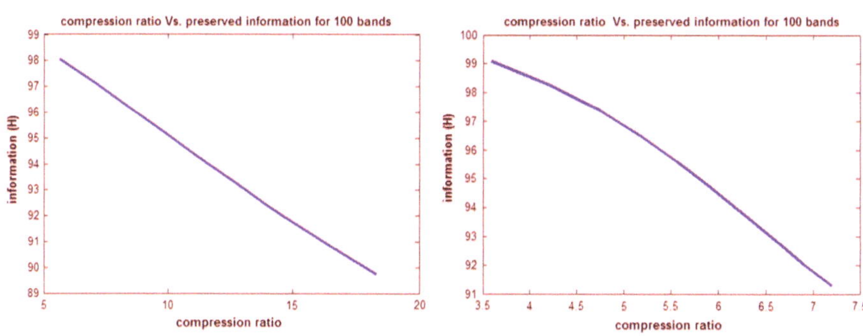

Fig. 11 Preserved information versus compression ratios with 100 bands for the Supernova (left) and Vortex (right)

PCA to reconstruct novel views on demand. In this section, we show how band-based PCA can also be employed to directly compress 3D volume data in time-varying volume datasets. Such datasets are comprised of multiple 3D volumes, each representing a frame in a sequence of time-varying data, typically obtained from simulation processes.

In such a case, we can train PCA using the full 3D volumes from regularly spaced timesteps in order to compute the eigenspace.

Reconstruction of frames takes place as discussed in previous sections, except that, here, the input samples and reconstructed outputs are in the form of 3D volumes rather than view-dependent images as we previously dealt with.

Results: Figure 12 shows rendered example frames from the *Supernova*[4] dataset reconstructed using different Θ values ($\Theta = 95\%$ and $\Theta = 99\%$). Similarly, Fig. 13 shows example frames from the *Turbulent Vortex* dataset.[5] In both cases, there is a visible improvement in reconstruction quality with higher values of Θ.

Figure 14 shows frames from the *Supernova* reconstructed using 20 and 100 bands. We observe that increasing the number of bands leads to a more faithful reconstruction of the original dataset as well as a reduction in the dithered noise artifacts.

Since we now employ PCA to reconstruct full volumes, we need to subsequently apply a transfer function and then a 3D rendering process at run-time to generate the output images. The visibility of noise artifacts in particular is subjective to the choice of transfer function and other viewing parameters, thus the quality of the reconstruction cannot fully be gauged purely on the resulting images. A better indicator of reconstruction quality is obtained by measuring how much information (variability) is preserved in the reconstructed frames as follows

$$H\left(X_{reconstructed}\right) = \Theta \sum_{i \notin Background} info\ (B_i), \tag{9}$$

where Θ is the explained variance defined in the previous section. The preserved information takes into account all non-background voxels and not only those that are visible using a specific transfer function or viewing configuration.

After setting a threshold for Θ and measuring the preserved information, we can then find the compression ratio as follows

$$\zeta(\Theta) = \frac{N \times d}{N_{bands} \times \left(\bar{d}_{band} + N \times \bar{p}\right)},$$

where N is the number of frames (training samples), d is the total number of attributes (dimensionality), $\bar{d}_{band}$ is the mean number of attributes per band and $\bar{p}$ is the mean number of eigenvectors per band.

[4] Supernova dataset obtained from: http://vis.cs.ucdavis.edu/VisFiles/pages/supernova.php.

[5] Turbulent Vortex dataset obtained from Time Varying Volume Data Reporsitory at UC Davis: http://web.cs.ucdavis.edu/~ma/ITR/tvdr.html.

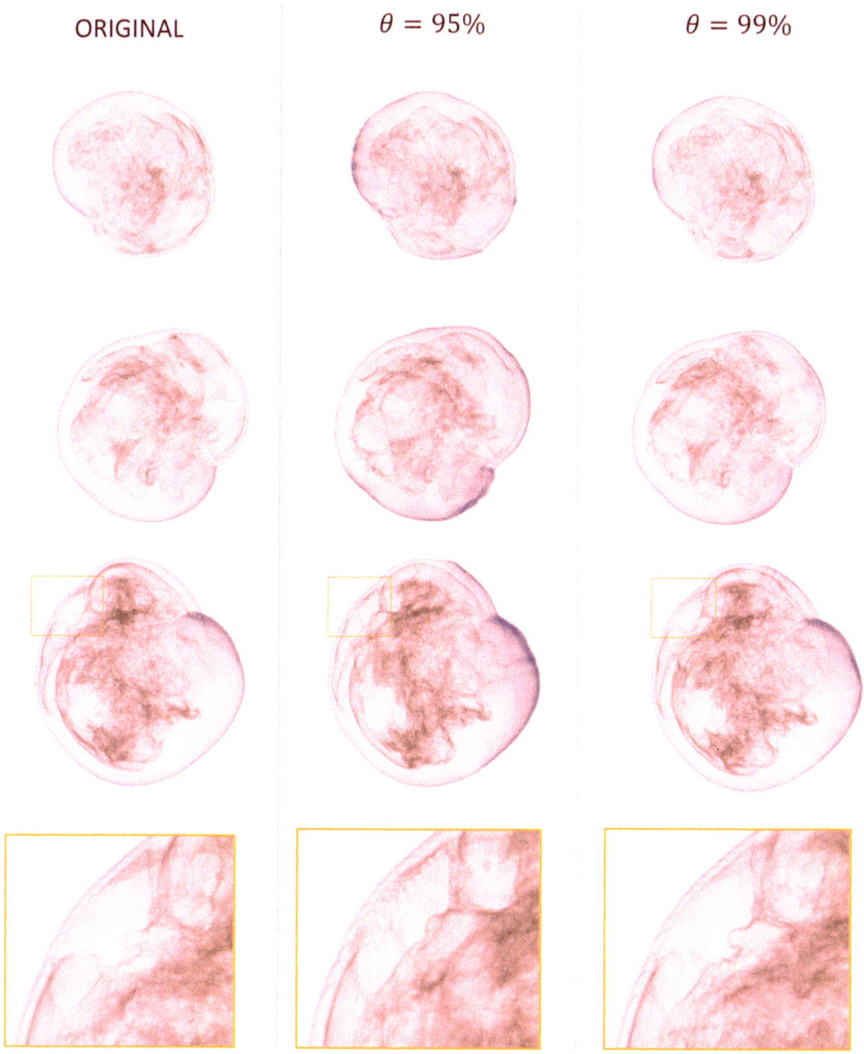

Fig. 12 Selected frames of the Supernova dataset. Each row shows an original frame from the dataset (left), compared to the frame reconstructed using 100 bands and $\Theta = 95\%$ (middle) and $\Theta = 99\%$ (right). The bottom row shows a zoomed in area of the visualizations in the third row

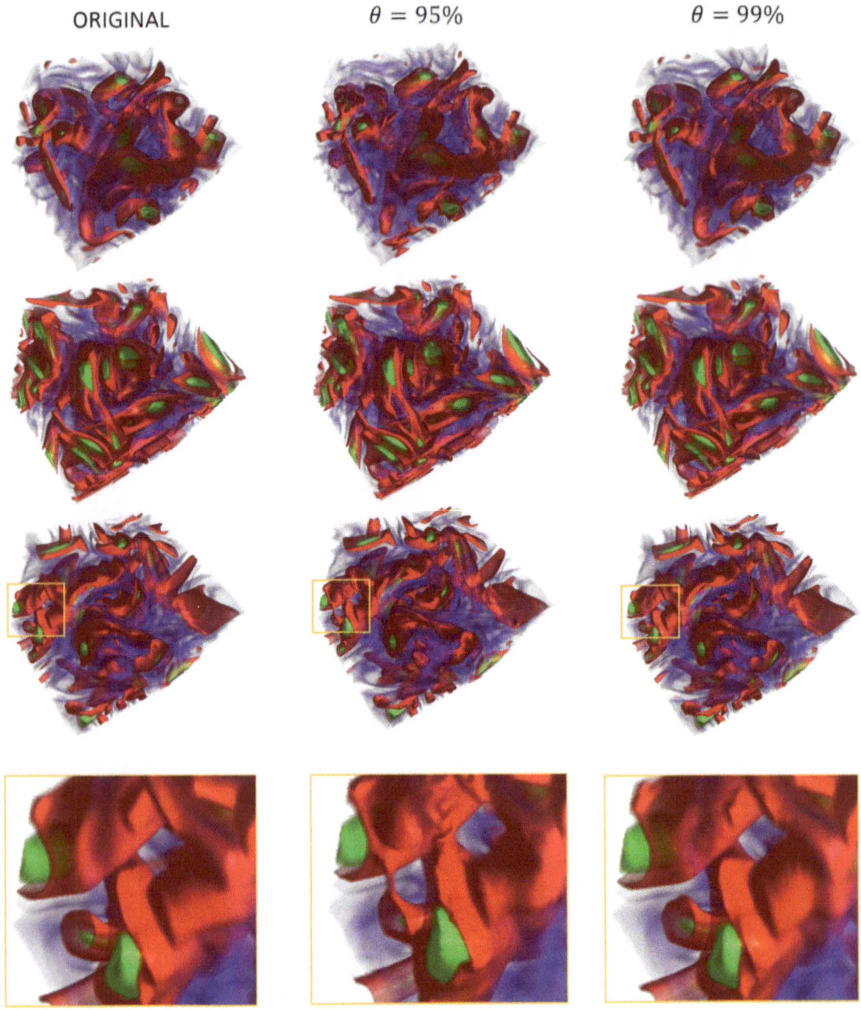

Fig. 13 Selected frames from the Turbulent Vortex dataset. Original frame (left) is compared to reconstructions using 100-band PCA, with $\Theta = 95\%$ (middle) and $\Theta = 99\%$ (right). The bottom row shows a zoomed-in area of the visualizations in the third row

Figure 10 shows the preserved information and compression ratio when reconstructing the supernova dataset for different number of bands and different Θ values. It is evident that increasing the number of bands improves the preserved information (with almost 4% from 20 bands to 100 bands) while the compression ratio is not significantly affected. This is because increasing the number of bands leads to better detection of background voxels. Figure 11 shows the relationship between compression ratio and preserved information using 100 bands for the *Supernova* and *Vortex* datasets respectively. We observed that better compression was achieved for

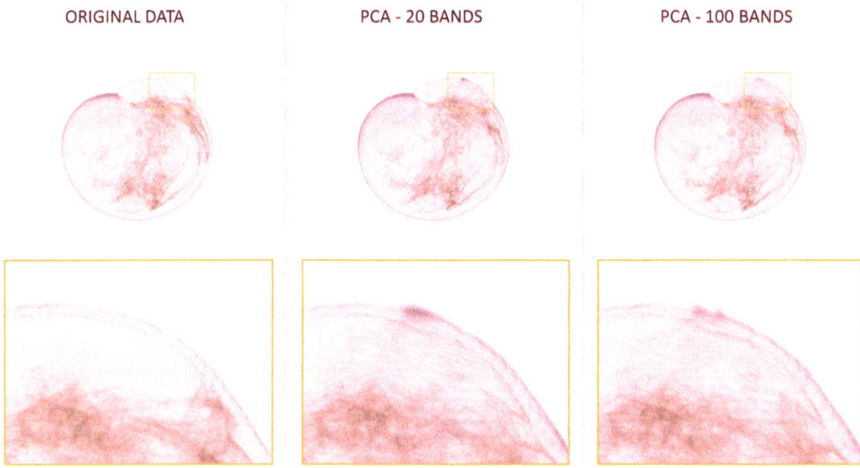

Fig. 14 Effect of bands on compression quality, Each row shows an original frame from the Supernova dataset (left) compared a reconstruction using 20 bands (middle) and 100 bands (right) with $\Theta = 99\%$ in each case. The bottom row shows a zoomed-in area

the *Supernova* but the preserved information was slightly better for the *Vortex*. It is likely that this is due to the fact that the background regions in the vortex data are much smaller than in the supernova.

Whilst, in this use case, we do not use PCA to reduce rendering complexity, band-based PCA appears to be an effective means for data compression in time-varying volume datasets.

6 Conclusions and Future Work

In this paper, we investigated the use of PCA for volume visualization. Our primary contribution is an adaptive decomposition technique that is able to reconstruct, in real-time, any volumetric model through a finite number of training images and generalize the eigenspace to produce high quality novel view images. We first extended the cell-based PCA approach initially proposed by [21] to volume data which is challenging due to its transparent nature and consequently its sensitivity to parallax errors. We further propose band-based PCA, a more generalized alternative, which we found to be less prone to cell-boundary artefacts at the cost of more random dithered noise.

Although both approaches necessitate a pre-computation stage, their run-time performance and capabilities are essentially independent of the complexity of the rendering process or of the volume data resolution. Thus, they could even be used in a minimal-spec standalone system to allow interactive rendering of high-resolution volumetric data, or data that has been visualized using complex rendering techniques not normally possible in real-time.

One clear limitation when using PCA for image-based rendering is that a change in transfer function (material colors and opacities) currently requires a change in the whole eigenspace. Where a suitable specific transfer function can be assumed, as in some practical scenarios such as in Medicine, the process of computing the eigenspace is done in a preprocessing step for the dataset. A potential generalization to this, which we would like to investigate in the future, might be to combine eigenspaces of different materials (e.g. flesh, bone, etc.) using composition techniques such as image level intermixing [25]. Previous authors have demonstrated that transfer function changes can be supported in a view-dependent image-based technique by multi-layering [27], and it would be interesting to see if similar strategies could be used to extend our approach.

As a second contribution, we extended the band-based PCA technique to directly reconstruct voxel data of full time-step frames in time-varying volume datasets. Results indicate that this can be used as an effective means of time-varying volume data compression.

Although we analyzed several aspects of PCA-reconstructed volume data, including rendering accuracy and information preservation, these measures were in the context of static volumes or individual frames within a time-varying dataset. Based on observing the datasets during an animated simulation sequence, we noted that it was particularly difficult to notice inaccuracies and artefacts whilst they were undergoing motion. Moreover a significant aspect that needs to be considered for time-varying data is the accuracy of the dynamic behavior (i.e. motion) of the dataset itself. However, there is limited previous literature on evaluating the perception of motion, and a complete analysis of this form was outside the scope of this paper. In future work, we plan to conduct perceptual user studies as well more direct feedback from expert users in specific user domains that employ volume data, in order to evaluate our techniques and, further, to gauge best-fit strategies and optimal parameter configurations to limit perceptible artefacts in the resulting visualizations.

Overall, PCA appears to be an interesting and viable alternative technique for image-based volume visualization and time-varying volume data compression. The reduction in computational complexity and compression of information provide potential advantages for applications such as in client-server visualization systems. Although our main motivation was to provide an alternative to a thin-client solution for volume visualization, the PCA-based approach could generally have advantages in volume compression and where run-time rendering complexity needs to be reduced at the cost of pre-processing computations.

Acknowledgements This research has been conducted with the financial support of Science Foundation Ireland (SFI) under Grant Number 13/IA/1895. The *VisMale Head* dataset is courtesy of the Visible Human Project at the U.S. National Library of Medicine. The *Tooth* was obtained from the Volume Library of Stefan Roettger. The *Chest* was obtained from the DICOM sample datasets provided by OsiriX. The *Supernova* data set is made available by Dr. John Blondin at the North Carolina State University through US Department of Energy's SciDAC Institute for Ultrascale Visuaization. The *Turbulent Vortex* dataset obtained from Time Varying Volume Data Reporsitory at UC Davis.

References

1. Bethel, W.: Visualization dot com. IEEE Comput. Graph. Appl. **20**(3), 17–20 (2000)
2. Broersen, A., van Liere, R., Heeren, R.M.: Comparing three pca-based methods for the visaulization of imaging spectroscopy data. In: Proceedings of the Fifth IASTED International Conference on Visualization, Imaging and Image Processing, pp. 540–545 (2005)
3. Callahan, S.P., Bavoil, L., Pascucci, V., Silva, C.T.: Progressive volume rendering of large unstructured grids. IEEE Trans. Vis. Comput. Graph. **12**(5), 1307–1314 (2006)
4. Chen, B., Kaufman, A., Tang, Q.: Image-based rendering of surfaces from volume data. In: Mueller, K., Kaufman, A.E. (eds) Volume Graphics 2001: Proceedings of the Joint IEEE TCVG and Eurographics Workshop in Stony Brook, pp. 279–295, New York, USA, 21–22 June 2001, Springer Vienna, Vienna (2001)
5. Choi, J.-J., Shin, Y.G.: Efficient image-based rendering of volume data. In: Computer Graphics and Applications, 1998. Pacific Graphics '98. Sixth Pacific Conference on, pp. 70–78, 226 (1998)
6. Engel, K., Ertl, T.: Texture-based volume visualization for multiple users on the world wide web. In: Virtual Environments, pp. 115–124. Springer, Berlin (1999)
7. Fout, N., Ma, K.L.: Transform coding for hardware-accelerated volume rendering. IEEE Trans. Vis. Comput. Graph. **13**(6), 1600–1607 (2007)
8. Frank, S., Kaufman, A. 2005: Distributed volume rendering on a visualization cluster. In: Ninth International Conference on Computer Aided Design and Computer Graphics (CAD-CG'05)
9. Gong, S., McKenna, S., Collins, J.J.: An investigation into face pose distributions. In: Proceedings of the Second International Conference on Automatic Face and Gesture Recognition, pp. 265–270. IEEE, Hoboken (1996)
10. Gourier, N., Hall, D., Crowley, J.L.: Estimating face orientation from robust detection of salient facial features. In: ICPR International Workshop on Visual Observation of Deictic Gestures, Citeseer (2004)
11. Grabner, M., Bischof, H., Zach, C., Ferko, A.: Multiple eigenspaces for hardware accelerated image based rendering. In: Proceedings of ÖAGM, pp. 111–118. http://www.vrvis.at/publications/PB-VRVis-2003-020
12. Hadwiger, M., Kniss, J.M., Rezk-salama, C., Weiskopf, D., Engel, K.: Real-time, vol. Graphics. A. K, Peters Ltd., Natick, MA, USA (2006)
13. Kirby, M., Sirovich, L.: Application of the karhunen-loeve procedure for the characterization of human faces. IEEE Trans. Pattern Anal. Mach. Intell. **12**(1), 103–108 (1990)
14. Knittel, G., Parys, R.: PCA-based seeding for improved vector quantization. In: Proceedings of the First International Conference on Computer Imaging Theory and Applications (VISIGRAPP 2009), pp. 96–99 (2009)
15. Kohlmann, P., Boskamp, T., Köhn, A., Rieder, C., Schenk, A., Link, F., Siems, U., Barann, M., Kuhnigk, J.-M., Demedts, D., Hahn, H.K.: Remote visualization techniques for medical imaging research and image-guided procedures. In: Linsen, L., Hamann, B., Hege, H.-C. (eds.) Visualization in Medicine and Life Sciences III: Towards Making an Impact, pp. 133–154. Springer International Publishing, Cham (2016)
16. Kroes, T., Post, F.H., Botha, C.P.: Exposure Render: an interactive photo-realistic volume rendering framework. PLoS ONE **8** (2013)
17. Leonardis, A., Bischof, H.: Multiple eigenspaces by mdl. In: Proceedings 15th International Conference on Pattern Recognition. ICPR-2000, vol. 1, pp. 233–237 (2000)
18. Liu, S., Wang, B., Thiagarajan, J.J., Bremer, P.T., Pascucci, V.: Multivariate volume visualization through dynamic projections. In: IEEE 4th Symposium on Large Data Analysis and Visualization (LDAV), pp. 35–42 (2014)
19. Meyer, M., Pfister, H., Hansen, C., Johnson, C., Meyer, M., Pfister, H., Hansen, C., Johnson, C.: Image-based volume rendering with opacity light fields, Technical report, University of Utah (2005)
20. Moser, M., Weiskopf, D.: Interactive volume rendering on mobile devices. In: Vision, Modeling, and Visualization VMV, vol. 8, pp. 217–226 (2008)

21. Nishino, K., Sato, Y., Ikeuchi, K.: Eigen-texture method: Appearance compression based on 3D model. In: IEEE Computer Society Conference on Computer Vision and Pattern Recognition, vol. 1. IEEE, Hoboken (1999)
22. Poliakov, A.V., , Albright, E., Corina, D., Ojemann, G., Martin, R., Brinkley, J.: Server-based approach to web visualization of integrated 3D medical image data. In: Proceedings of the AMIA Symposium, pp. 533–537 (2001)
23. Qi, X., Tyler, J.M.: A progressive transmission capable diagnostically lossless compression scheme for 3D medical image sets. Inf. Sci. **175**(3), 217–243 (2005)
24. Santhanam, A., Min, Y., Dou, T., Kupelian, P., Low, D.A.: A client-server framework for 3D remote visualization of radiotherapy treatment space. Front. Oncol. **3** (2013)
25. Schubert, N., Scholl, I.: Comparing GPU-based multi-volume ray casting techniques. Comput. Sci. Res. Dev. **26**(1), 39–50 (2011)
26. Takemoto, S., Nakao, M., Sato, T., Sugiura, T., Minato, K., Matsuda, T.: Interactive volume visualization of microscopic images using feature space reduction. BME **51**, U–6–U–6. http://ci.nii.ac.jp/naid/130004948124/en/ (2013)
27. Tikhonova, A., Correa, C.D., Ma, K.-L.: Explorable images for visualizing volume data. In: IEEE Pacific Visualization Symposium, pp. 177–184 (2010)
28. Turk, M., Pentland, A.: Eigenfaces for recognition. J. Cogn. Neurosci. **3**(1), 71–86 (1991)
29. Wang, Z., Bovik, A.C., Sheikh, H.R., Simoncelli, E.P.: Image quality assessment: from error visibility to structural similarity. IEEE Trans. Image Process. **13**(4), 600–612 (2004)
30. Yang, M.-H.: Kernel eigenfaces vs. kernel fisherfaces: Face recognition using kernel methods. In: fgr, vol. 2, p. 215 (2002)

Outlier-Resistant Data Processing with L1-Norm Principal Component Analysis

Panos P. Markopoulos, Sandipan Kundu, Shubham Chamadia, Nicholas Tsagkarakis and Dimitris A. Pados

Abstract Principal Component Analysis (PCA) has been a cornerstone of data analysis for more than a century, with important applications across most fields of science and engineering. However, despite its many strengths, PCA is known to have a major drawback: it is very sensitive to the presence of outliers among the processed data. To counteract the impact of outliers in data analysis, researchers have been long working on robust modifications of PCA. One of the most successful (and promising) PCA alternatives is *L1-PCA*. L1-PCA relies on the L1-norm of the processed data and, thus, tames any outliers that may exist in the dataset. Experimental studies in various applications have shown that L1-PCA (i) attains similar performance to PCA when the processed data are outlier-free and (ii) maintains sturdy resistance against outliers when the processed data are corrupted. Thus, L1-PCA is expected to play a significant role in the big-data era, when large datasets are often outlier corrupted. In this chapter, we present the theoretical foundations of L1-PCA, optimal and state-of-

P. P. Markopoulos (✉)
Department of Electrical and Microelectronic Engineering,
Rochester Institute of Technology, Rochester, NY 14623, USA
e-mail: panos@rit.edu

S. Kundu
Qualcomm Technologies, Inc., San Jose, CA 95110, USA
e-mail: sandipan@qti.qualcomm.com

S. Chamadia
Department of Anesthesia, Critical Care, and Pain Medicine,
Harvard Medical School, Massachusetts General Hospital,
Boston, MA 02114, USA
e-mail: schamadia@mgh.harvard.edu

N. Tsagkarakis
College of Engineering and Computing, University of South Carolina,
Columbia, SC 29208, USA
e-mail: tsagkara@mailbox.sc.edu

D. A. Pados
Computer and Electrical Engineering & Computer Science,
Florida Atlantic University, Boca Raton, FL 33431, USA
e-mail: dpados@fau.edu

G. R. Naik (ed.), *Advances in Principal Component Analysis*,
https://doi.org/10.1007/978-981-10-6704-4_6

the-art approximate algorithms for its implementation, and some numerical studies that demonstrate its favorable performance.

1 Introduction and Problem Formulation

Fundamentally, Principal Component Analysis (PCA) seeks orthogonal directions that span a subspace whereon data presence is maximized [2, 9, 13, 33]. These directions are defined by the, so called, Principal Components (PCs) of the data. In standard PCA, data presence is quantified by the aggregated squared L2-norm (or, Frobenius norm) of the projected data onto the sought-after subspace. Therefore, standard PCA is also known as L2-PCA. PCA has enjoyed great popularity over the past decades, due to several reasons, including its familiar low-cost implementation by means of Singular-Value Decomposition (SVD), its scalability (the kth principal component can be found in the nullspace of the first $k-1$ principal components), and the close approximation it attains to the true maximum-variance subspace when applied on clean/nominal data points.

In the big-data era, datasets are often contaminated by highly deviating samples, faulty measurements, and bursty-noise, often referred to as *outliers* [1] –a term used to describe that such points usually lie by far outside the nominal (sought-after) data subspace. Outliers appear in practice due to a variety of causes, including errors in data storage, or transcription, and intermittent sensor malfunctions. Of course, sporadic incoherences of the sensed environment and malevolent outlier insertion may also be causes of contamination/corruption of the processed dataset.

At the same time, standard PCA is observed to be highly sensitive to the presence of outliers [5]. Expectedly, by placing squared emphasis on the magnitude of all points benefits unfavorably points that lie in the dataset periphery; i.e., outliers.

To counteract the impact of outliers on PCA-based data processing, researchers have focused on alternative PCA formulations that seek to maximize data presence in the PC-spanned subspace by either (i) maximizing the aggregate L1-norm (sum of *absolute* values) of the projected data [6–8, 16–18, 20, 23, 24, 27, 29, 29–31, 37] or (ii) minimizing the aggregate absolute data representation error (i.e., L1-norm of error) [3, 4, 10, 14, 15, 36]. Due to their reliance on the L1-norm (in contrast to standard PCA's reliance on L2-norm), these methods are collectively referred to as "L1-PCA" methods. On the one hand, the general solution to error-minimization L1-PCA remains to date unknown and some approximate algorithms have been proposed in the literature. On the other hand, maximum-projection L1-PCA was recently shown to be equivalent to combinatorial optimization and, thus, two exact algorithms for its optimal solution were presented [23, 24]. In addition to the optimal solutions –and, in part, thanks to the insight they provided– several efficient suboptimal algorithms for maximum-projection L1-PCA have been also proposed in the literature [16, 17, 27, 30]. Maximum projection L1-PCA has found many important applications in the past five years, including image reconstruction [26], object recognition [12], reduced-rank filtering [21], Direction-of-Arrival (DoA) estimation [25, 28, 35], radar-based indoor human motion classification [22], video surveillance [19], and others. For the above reasons (namely, solvability and widespread research interest), in this chapter we focus specifically on maximum-projection L1-PCA, which we henceforth refer to simply as L1-PCA.

Consider data matrix $\mathbf{X} = [\mathbf{x}_1, \mathbf{x}_2, \ldots, \mathbf{x}_N] \in \mathbb{R}^{D \times N}$ of rank $d \leq \min\{D, N\}$. L1-PCA seeks a low-rank orthonormal data subspace basis $\mathbf{Q}_{L1} \in \mathbb{R}^{D \times K}$ of dimensionality $K < d$ that solves

$$\mathbf{Q}_{L1} = \underset{\mathbf{Q} \in \mathbb{R}^{D \times K},\ \mathbf{Q}^\top \mathbf{Q} = \mathbf{I}_K}{\arg\max} \left\| \mathbf{X}^\top \mathbf{Q} \right\|_1 . \tag{1}$$

In (1), $\|\cdot\|_1$ denotes the L1-norm of its matrix argument, equal to the summation of the absolute values of all its entries. It was recently shown that L1-PCA in the form of (1) is formally an NP-hard problem in *jointly asymptotic* N and d [24, 30]. In fact, as shown below, (1) was equivalently rewritten as an optimization problem over NK $\{\pm 1\}$-binary variables and, thus, it was solved exactly by exhaustive search with cost $\mathcal{O}(2^{NK})$. Moreover, [24] showed that for the special case of fixed data dimensionality d, (1) is not NP-hard and, in fact, it can be solved with polynomial cost $\mathcal{O}\left(N^{dK-K+1}\right)$; the corresponding polynomial-time algorithm of [24] is to date the fastest optimal L1-PCA calculator.

In the big-data era, very high data-dimension D and/or data-support size N may render the optimal algorithms of [24] practically inapplicable. To counteract this prohibitive computational-cost increase, authors in [16, 27] introduced recently L1-BF, a bit-flipping based, near-optimal algorithm for the calculation of $K \geq 1$ L1-PCs that has computational cost comparable to that of SVD –i.e., standard PCA [11]. L1-BF in [27] was accompanied by rigorous proof of convergence, detailed asymptotic complexity derivation, and theoretically established performance guarantees. Extensive numerical studies revealed that L1-BF algorithm outperforms all its suboptimal counterparts of comparable cost with respect to L1-PCA metric and at the same time retains high outlier-resistance similar to that of optimal L1-PCA.

In the following Sect. 2, we present the main results on the optimal solution of L1-PCA. Then, in Sect. 3, we present some of the most widely used algorithms for approximate L1-PCA, including the state-of-the-art L1-BF. Finally, Sect. 4 holds some numerical studies that illustrate the outlier-resistance of L1-PCA. Few concluding remarks are drawn in Sect. 5.

2 Exact Solution of L1-PCA

In this section, we provide the guidelines for solving L1-PCA optimally, as it was originally presented in [20, 24]. Consider matrix $\mathbf{C} \in \mathbb{R}^{m \times n}$ with $m > n$ that admits compact SVD $\mathbf{C} = \mathbf{U}\Sigma_{n \times n}\mathbf{V}^\top$. We define the unitary matrix $U(\mathbf{C}) = \mathbf{U}\mathbf{V}^\top$ and notice that

$$U(\mathbf{C}) = \underset{\substack{\mathbf{Q} \in \mathbb{R}^{m \times n} \\ \mathbf{Q}^\top \mathbf{Q} = \mathbf{I}_n}}{\text{argmin}} \|\mathbf{C} - \mathbf{Q}\|_F , \tag{2}$$

where $\|\cdot\|_F$ denotes the Frobenius norm of a matrix, equal to the square root of the summation of the squares of all its entries. That is, in accordance to the well-studies Procrustes Problem [34], $U(\mathbf{C})$ is the closest matrix to $\mathbf{C}$ that has orthonormal columns.

Let us next denote by $\|\cdot\|_*$ the nuclear norm of its matrix argument, which equals the summation of its singular values. In [20, 24] it was shown that, if

$$\mathbf{B}_{\text{opt}} = \underset{\mathbf{B}\in\{\pm 1\}^{N\times K}}{\text{argmax}} \ \|\mathbf{XB}\|_* , \tag{3}$$

then the optimal solution to (1) is given by

$$\mathbf{Q}_{L1} = U(\mathbf{XB}_{\text{opt}}). \tag{4}$$

In addition, [20, 24] showed that $\left\|\mathbf{Q}_{L1}^\top \mathbf{X}\right\|_1 = \left\|\mathbf{XB}_{\text{opt}}\right\|_*$ and $\mathbf{B}_{\text{opt}} = \text{sgn}(\mathbf{X}^\top \mathbf{Q}_{L1})$, where $\text{sgn}(\cdot)$ returns a matrix that contains the $\{\pm 1\}$-signs of the entries of its matrix argument. For $K = 1$, (3) becomes equivalent to a binary quadratic maximization problem of the form

$$\mathbf{b}_{\text{opt}} = \underset{\mathbf{b}\in\{\pm 1\}^{N}}{\text{argmax}} \ \|\mathbf{Xb}\|_2 . \tag{5}$$

This is because every $\mathbf{c} \in \mathbb{R}^m$ admits SVD $\mathbf{c} = (\mathbf{c}\|\mathbf{c}\|_2^{-1}) \cdot \|\mathbf{c}\|_2 \cdot 1$ and therefore, $\|\mathbf{c}\|_* = \|\mathbf{c}\|_2$. Accordingly, the L1-PC of $\mathbf{X}$ is given by

$$\mathbf{q}_{L1} = U(\mathbf{Xb}_{\text{opt}}) = \mathbf{Xb}_{\text{opt}} \left\|\mathbf{Xb}_{\text{opt}}\right\|_2^{-1} . \tag{6}$$

In addition, $\left\|\mathbf{X}^\top \mathbf{q}_{L1}\right\|_1 = \left\|\mathbf{Xb}_{\text{opt}}\right\|_* = \left\|\mathbf{Xb}_{\text{opt}}\right\|_2$ and $\mathbf{b}_{\text{opt}} = \text{sgn}(\mathbf{X}^\top \mathbf{q}_{L1})$.

In view of (3) and (4), both optimal algorithms seek first to obtain the optimal binary matrix $\mathbf{B}_{\text{opt}}$ and then they return the L1-PCs through (4). Therefore, L1-PCA is transformed into a combinatorial problem. For solving (3), the conceptually simplest approach is to perform an exhaustive search over its size-2^{NK} feasibility set, $\{\pm 1\}^{N\times K}$, and for every element in the set, say $\mathbf{B} \in \{\pm 1\}^{N\times K}$, evaluate $\|\mathbf{XB}\|_*$. Finally, this exhaustive-search method returns the element of $\{\pm 1\}^{N\times K}$ that attained the highest value to the metric. This method is certainly guaranteed to provide the exact solution of L1-PCA. However, clearly, it demands 2^{NK} nuclear-norm evaluations. Each nuclear-norm evaluation has N-by-K SVD cost. Therefore, keeping the dominant terms, this first algorithm has computational cost in $\mathcal{O}(2^{NK})$, exponential in N. In practice, taking advantage of the nuclear-norm invariability to negations and permutations of the columns of the matrix argument, the exhaustive-search algorithm searches in a size-$\binom{2^{N-1}+K-1}{K}$ subset of $\{\pm 1\}^{N\times K}$ wherein a solution to (3) is guaranteed to exist. Nevertheless, the asymptotic complexity of this search remains $\mathcal{O}\left(\binom{2^{N-1}+K-1}{K}\right) \equiv \mathcal{O}(2^{NK})$.

The second optimal algorithm in [20, 24] is, in principle, not exhaustive. Instead, it builds in a sophisticated way a subset $\mathcal{B}$ of $\{\pm 1\}^{N\times K}$ wherein a solution to (3) is proven to exist. Importantly, if d is considered to be a constant with respect to N (a very meaningful assumption in most applications where D is the number of measured features, or sensors), the cost to construct $\mathscr{B}$ and search exhaustively within it is $\mathscr{O}(N^{dK-K+1})$, polynomial in the number of data points, N.

2.1 Special Case: Non-negative Data

In the special case of non-negative data, such as images, the optimal calculation of the L1-PC of $\mathbf{X}$ is proven to be simple [26].

Let $\mathbf{X}$ consist of non-negative entries, so that $[\mathbf{X}]_{i,j} \geq 0 \; \forall i, j$. Then, $\mathbf{X}^\top \mathbf{X}$ also consists of non-negative entries. Therefore, the solution to (5) is, trivially, the all-ones vector $\mathbf{1}_N$. Accordingly, the L1-PC of $\mathbf{X}$ is given by $\mathbf{q}_{L1} = \mathbf{X}\mathbf{1}_N \|\mathbf{X}\mathbf{1}_N\|_2^{-1}$. That is, the L1-PC of $\mathbf{X}$ is simply the normalized vector on the direction of the mean of the data points. Arguably, this result, presented for the first time in the context of image reconstruction in [26], implies that the *direction* of the mean of non-negative data exhibits an L1-PCA-certified robustness against outliers.

3 Approximate Algorithms

Prior to the exact algorithm of [24], there were three approximate popular solvers for L1-PCA in the literature. In the sequel, we first present these three solvers and then we present in more detail the L1-BF approximate solver, which was introduced very recently and has been shown to attain state-of-the-art performance.

3.1 Sequential Fixed-Point Iterations [17]

The first approximate algorithm for L1-PCA was introduced by Kwak in 2008 [17, 18]. This algorithm first focuses on the $K = 1$ case. Initialized at some binary vector $\mathbf{b}^{(0)} \in \{\pm 1\}^N$, the algorithm conducts converging *fixed-point* iteration

$$\mathbf{b}^{(t)} = \operatorname{sgn}\left(\mathbf{X}^\top \mathbf{X}\mathbf{b}^{(t-1)}\right), \; t = 1, 2, \ldots. \tag{7}$$

Denoting by $\mathbf{b}_{\text{fp},1}$ the convergence point of iteration in (7), the "first" L1-PC is approximated by

$$\mathbf{q}_{\text{fp},1} = \mathbf{X}\mathbf{b}_{\text{fp}} \|\mathbf{X}\mathbf{b}_{\text{bf}}\|_2^{-1}. \tag{8}$$

At this point, it is worth noticing that, in contrast to what holds true in standard PCA, scalability does not hold in L1-PCA. That is, solving a ($K = 1$) problem in the nullspace of $\mathbf{q}_{\text{fp},1}$ does not offer the "second" L1-PCA of $\mathbf{X}$. This can be easily deduced by the form of the nuclear-norm problem in (3), which demands that all columns of $\mathbf{B}_{\text{opt}}$ are jointly calculated; thus, in turn, all columns of $\mathbf{Q}_{L1} = U(\mathbf{X}\mathbf{B}_{\text{opt}})$, the solution to size-$K$ L1-PCA must be jointly calculated as well. However despite this lack of scalability in L1-PCA, for ease in computation, [17] proposes that the "second" L1-PC is approximated by the solution to

$$\max_{\mathbf{q}\in\mathbb{R}^{D\times 1},\ \|\mathbf{q}\|_2=1} . \left\| \mathbf{X}^\top (\mathbf{I}_D - \mathbf{q}_{\text{fp},1}\mathbf{q}_{\text{fp},1}^\top)\mathbf{q} \right\|_1 , \tag{9}$$

pursued once again by means of fixed-point iterations. That is, [17] proposed that the second L1-PC is approximated by

$$\mathbf{q}_{\text{fp},2} = \mathbf{X}\mathbf{b}_{\text{fp},2} \|\mathbf{X}\mathbf{b}_{\text{fp},2}\|_2^{-1} \tag{10}$$

where $\mathbf{b}_{\text{fp},2}$ is the converging point of the iterations

$$\mathbf{b}^{(t)} = \text{sgn}\left(\mathbf{X}^\top (\mathbf{I}_D - \mathbf{q}_{\text{fp},1}\mathbf{q}_{\text{fp},1}^\top)\mathbf{X}\mathbf{b}^{(t-1)} \right), \ t = 1, 2, \ldots . \tag{11}$$

Accordingly, the Kth L1-PC is given by $\mathbf{q}_{\text{fp},K} = \mathbf{X}\mathbf{b}_{\text{fp},K} \|\mathbf{X}\mathbf{b}_{\text{fp},K}\|_2^{-1}$, where $\mathbf{b}_{\text{fp},K}$ is the converging point of the fixed-point iterations

$$\mathbf{b}^{(t)} = \text{sgn}\left(\mathbf{X}^\top (\mathbf{I}_D - \sum_{i=1}^{K-1} \mathbf{q}_{\text{fp},i}\mathbf{q}_{\text{fp},i}^\top)\mathbf{X}\mathbf{b}^{(t-1)} \right), \ t = 1, 2, \ldots . \tag{12}$$

Certainly, all K iterations must be run sequentially, as the formulation of the kth iteration demands the convergence of all previous $k-1$ iterations. At the end of the K-iteration, the algorithm of [17] approximated the solution to (1) by $\mathbf{Q}_{\text{fp}} = [\mathbf{q}_{\text{fp},1}, \mathbf{q}_{\text{fp},2}, \ldots, \mathbf{q}_{\text{fp},K}]$, which by construction satisfies $\mathbf{Q}_{\text{fp}}^\top \mathbf{Q}_{\text{fp}} = \mathbf{I}_K$. The computational complexity of the algorithm of [17] can be found to be $\mathcal{O}(N^2 DK)$.

3.2 Joint Fixed-Point Iterations ("non-greedy" approach) [32]

An alternative approximate L1-PCA calculator was proposed in [32]. Authors in [32] refer to the proposed algorithm as "non-greedy", because in contrast to the algorithm of [17] it tries to approximate all K columns of $\mathbf{Q}L_1$ jointly. Specifically, the algorithm of [32] initializes at an arbitrary binary matrix $\mathbf{B}^{(0)}$ and conducts the iterations

$$\mathbf{B}^{(t)} = \text{sgn}\left(\mathbf{X}^\top U \left(\mathbf{X}\mathbf{B}^{(t-1)} \right) \right), \quad t = 1, 2, 3, \ldots . \tag{13}$$

Then, if $\mathbf{B}_{\text{ng}}$ is the converging point of the iterations, the algorithm returns $\mathbf{Q}_{\text{ng}} = U(\mathbf{X}\mathbf{B}_{\text{ng}})$ as an approximation to the optimal $\mathbf{Q}_{L1}$. Certainly, for the single L1-PC case ($K = 1$), the algorithm in [32] coincides with that of the scheme presented in [17]. The computational cost to execute the scheme in [32] is $\mathcal{O}(N^2DK + NK^3)$.

3.3 *Semi-definite Programming Relaxation [30]*

A third approximate approach for solving (1), relying on a popular semi-definite programming (SDP) approach, was presented in [30]. Specifically, focusing on the $K = 1$ case, the authors in [30] notice that the binary quadratic form maximization in (5) can be equivalently rewritten as

$$\mathbf{Z}_{\text{opt}} = \underset{\mathbf{Z}\in\mathcal{S}_+^N,\ [\mathbf{Z}]_{n,n}=1\ \forall n,\ \text{rank}(\mathbf{Z})=1}{\arg\max} \operatorname{Tr}\left(\mathbf{Z}\mathbf{X}^\top\mathbf{X}\right), \tag{14}$$

where $\mathcal{S}_+^N$ is the set of positive semi-definite matrices in $\mathbb{R}^{N\times N}$. Then, $[\mathbf{Z}_{\text{opt}}]_{:,n}$ is an optimal solution to the problem in (5), $\mathbf{b}_{\text{opt}}$, for any $n \in \{1, 2, \ldots, N\}$. Expectedly, though (14) is as hard to solve as (5). Therefore, authors in [30] opt to relax (14) by removing the rank-1 constraint and forming instead, the convex SDP problem

$$\mathbf{Z}_{\text{sdp}} = \underset{\mathbf{Z}\in\mathcal{S}_+^N,\ [\mathbf{Z}]_{n,n}=1\ \forall n}{\arg\max} \operatorname{Tr}\left(\mathbf{Z}\mathbf{X}^\top\mathbf{X}\right). \tag{15}$$

Then, the algorithm factorizes $\mathbf{Z}_{\text{sdp}} = \mathbf{W}\mathbf{W}^T$ and generates L instances of the binary vector $\mathbf{b} = \text{sgn}\left(\mathbf{W}^\top\mathbf{r}\right)$ defined upon arbitrary values of $\mathbf{r}$ drawn from Gaussian distribution $\sim \mathcal{N}(\mathbf{0}_N, \mathbf{I}_N)$. This procedure is known as "Gaussian randomization". Then, an approximate L1-PC is designed as $\mathbf{q}_{\text{sdp}} = \mathbf{X}\mathbf{b}_{\text{sdp}}\|\mathbf{X}\mathbf{b}_{\text{sdp}}\|_2^{-1}$ with $\mathbf{b}_{\text{sdp}}$ being the one out of L randomized binary-vector instances that maximizes $\|\mathbf{X}\mathbf{b}\|_2$. For $K > 1$, the remaining L1-PCs are generated similar to [17] using the method of sequential nullspace projections. The computational complexity to solve the SDP (15) within ε accuracy is $\mathcal{O}(N^{3.5}\log(1/\varepsilon))$ and thus the overall computational cost of this approximate L1-PCA method turns out to be $\mathcal{O}(KN^{3.5}\log(1/\varepsilon) + KL(N^2 + DN))$. s

3.4 *Bit-Flipping Iterations [16, 27, 29]*

An efficient, low-cost, near-exact L1-PC solver was presented recently in [27, 29]. Similarly to the algorithms above, the algorithm is initialized at a binary matrix $\mathbf{B}^{(0)} \in \{\pm 1\}$ and conducts, until convergence, *optimal single-bit flipping* iterations. At the end of the iterations, it uses the convergence point, $\mathbf{B}_{\text{bf}}$, to form the approximate L1-PCs

$$\mathbf{Q}_{\text{bf}} = U\left(\mathbf{X}\mathbf{B}_{\text{bf}}\right). \tag{16}$$

This algorithm is known as *L1-PCA Bit-Flipping*, or *L1-BF*. The single-bit flipping iterations are defined as follows.

At the tth iteration step, $t = 1, 2, \ldots$, the algorithm finds the binary matrix $\mathbf{B}^{(t)}$ that (i) attains the highest possible value in the metric of (3) and (ii) it differs from $\mathbf{B}^{(t-1)}$ in exactly one entry (i.e., by a single bit flipping) the index of which does not belong to the used-bits memory set $\mathcal{W} \subseteq \{1, 2, \ldots, NK\}$. At $t = 1$, $\mathcal{W}$ is set empty. We observe that, flipping the (n, k)th bit of $\mathbf{B}^{(t-1)}$ we obtain the binary matrix $\mathbf{B}^{(t-1)} - 2B^{(t)}_{n,k}\mathbf{e}_{n,N}\mathbf{e}^{\top}_{k,K}$ where $\mathbf{e}_{m,N}$ denotes the mth column of N-order identity matrix $\mathbf{I}_N$. Thus, mathematically, the algorithm seeks for the solution to

$$s(n, k) = \underset{\substack{(m,l)\in\{1,2,\ldots,N\}\times\{1,2,\ldots,K\}\\ (l-1)N+m\notin\mathcal{W}}}{\arg\max} \left\| \mathbf{X}\mathbf{B}^{(t-1)} - 2B^{(t-1)}_{m,l}\mathbf{x}_m\mathbf{e}^{\top}_{l,K} \right\| \tag{17}$$

and defines the intermediate/temporary binary matrix $\mathbf{B}_{\text{temp}} = \mathbf{B}^{(t-1)} - 2B^{(t)}_{n,k}\mathbf{e}_{n,N}$ $\mathbf{e}^{\top}_{k,K}$. Under the two conditions above, the returned matrix $\mathbf{B}_{\text{temp}}$ may yield higher, or lower value than $\mathbf{B}^{(t-1)}$ to the objective metric in (3). If $\|\mathbf{X}\mathbf{B}_{\text{temp}}\|_* > \|\mathbf{X}\mathbf{B}^{(t-1)}\|_*$, then the algorithm sets $\mathbf{B}^{(t)} = \mathbf{B}_{\text{temp}}$ (flips the (n, k)th bit), inserts the index of the flipped bit, $(k-1)N + n$, to $\mathcal{W}$, and proceeds to the next iteration. If, however, $\|\mathbf{X}\mathbf{B}_{\text{temp}}\|_* \leq \|\mathbf{X}\mathbf{B}^{(t-1)}\|_*$, then certainly flipping the (n, k)th bit does not increase the metric of interest. Therefore, the algorithm resets the memory set $\mathcal{W}$ to empty and solves (17) again, obtaining a new pair (n, k) and a new $\mathbf{B}_{\text{temp}} = \mathbf{B}^{(t-1)} - 2B^{(t)}_{n,k}\mathbf{e}_{n,N}\mathbf{e}^{\top}_{k,K}$. If, for this new $\mathbf{B}_{\text{temp}}$ it holds that $\|\mathbf{X}\mathbf{B}_{\text{temp}}\|_* > \|\mathbf{X}\mathbf{B}^{(t-1)}\|_*$, then the algorithm sets $\mathbf{B}^{(t)} = \mathbf{B}_{\text{temp}}$ (flips the (n, k)th bit), inserts the index of the flipped bit, $(k-1)N + n$, to $\mathcal{W}$, and proceeds to the next iteration. If, on the other hand, it holds that $\|\mathbf{X}\mathbf{B}_{\text{temp}}\|_* \leq \|\mathbf{X}\mathbf{B}^{(t-1)}\|_*$, then the iterations terminate, returning $\mathbf{B}_{\text{bf}} = \mathbf{B}^{(t-1)}$.

3.4.1 Intelligent Initialization

To attain superior convergence point in fewer iterations, the authors in [29] proposed an intelligently selected initialization point $\mathbf{B}^{(0)}$. Specifically, they set

$$\mathbf{B}^{(0)} = \text{sgn}(\mathbf{v})\mathbf{1}^{\top}_{K}, \tag{18}$$

where $\mathbf{v}$ is the highest-singular-value right-hand singular vector of $\mathbf{X}$ and $\mathbf{1}_K$ is the all-one vector of length K. The motivation behind this initialization was that for $d = 1$ and $K = 1$, $\mathbf{b}_{\text{opt}} = \text{sgn}(\mathbf{v})$.

3.4.2 Complexity

To initialize the bit-flipping iterations, the algorithm calculates $\mathbf{v}$, by means of SVD of $\mathbf{X}$, with $\mathcal{O}(ND\min\{N, D\})$. Then, for the proposed initialization, $\mathbf{XB}^{(0)}$ and $\|\mathbf{XB}^{(0)}\|_*$ are calculated in the beginning of the first iteration with low complexity $\mathcal{O}(NdK)$. To find a solution to (17), the worst case cost is $\mathcal{O}(NK(K^2+d))$. Then, setting the maximum number of iterations to NK, $\mathbf{B}_{\text{bf}}$ is found with total worst-case cost $\mathcal{O}(ND\min\{N, D\} + N^2K^2(K^2+d))$. Calculation of $\mathbf{Q}_{\text{bf}}$ from $\mathbf{B}_{\text{bf}}$ costs an additional $\mathcal{O}(ND\min\{N, D\} + NDK)$. Since $K \leq d \leq \min\{N, D\}$, the total complexity of L1-BF is $\mathcal{O}(ND\min\{N, D\} + N^2K^2(K^2+d))$.

4 Numerical Studies

In this section, we present some numerical studies that compare the performances of L2-PCA and L1-PCA when applied on clean/nominal and on outlier-corrupted training data. Also, we provide numerical studies that compare all the approximate calculators presented above, with respect to the L1-PCA metric of interest.

4.1 Line-Fitting

To illustrate the robustness of L1-PCA against outliers, and juxtapose it with the sensitivity of standard PCA, we conduct the following line-fitting study. We generate data matrix $\mathbf{X}_{2\times 40}$, drawing each of its columns independently from the multi-variate

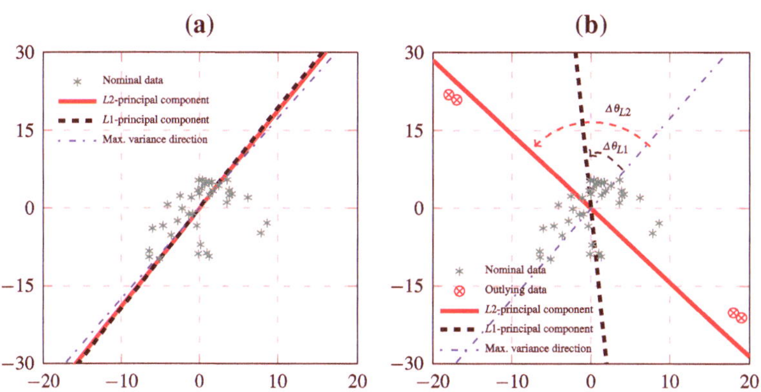

Fig. 1 **a** Principal component over the original clean data matrix $\mathbf{X}_{2\times 40}$ ($*$), **b** Principal component over data matrix $\mathbf{X}_{2\times 40}$ corrupted by 4 appended outliers ($\otimes$) (angular deviation $\Delta\theta_{L2}=95°$ and $\Delta\theta_{L1}=63°$)

Gaussian distribution $\mathcal{N}\left(\mathbf{0}_2, \mathbf{R} = \begin{bmatrix} 15 & 12 \\ 12 & 29 \end{bmatrix}\right)$. In Fig. 1a, we plot the data points in $\mathbf{X}$ on the plane ($*$). Together with these 40 points, we plot (i) the PC (i.e., L2-PC) of $\mathbf{X}$, (ii) the L1-PC (exact) of $\mathbf{X}$, and (iii) the maximum-variance line of the distribution, defined by the highest-eigenvalue eigenvector of the autocorrelation matrix $\mathbf{R}$. We observe that both the L1-PC and the L2-PC, calculated over the 40 given nominal points, approximate very well the maximum variance line of the distribution. Next, we consider that four (4) outlier points in $\mathbf{O} \in \mathbb{R}^{2\times 4}$ are appended to the dataset $\mathbf{X}$, forming the corrupted dataset $\mathbf{X}_{\text{corr}} = [\mathbf{X},\ \mathbf{O}] \in \mathbb{R}^{2\times 44}$. The outliers are also plotted on the plan in Fig. 1b ($\otimes$). We now calculate the L1-PC and L2-PC of the corrupted dataset $\mathbf{X}_{\text{corr}}$ and plot them in Fig. 1b. For reference, we plot again the ideal, sought-after maximum-variance line. We observe that, quite interestingly, the L1-PC stays much closer to the maximum variance line than the L2-PC, which is completely misled and points directly to the outliers.

4.2 Direction-of-Arrival (DoA) Estimation

We consider a uniform linear antenna array of $D = 7$ elements that collects $N = 30$ observations of a binary phase-shift-keying (BPSK) signal that impinge with direction-of-arrival (DoA) $\theta_1 = 60°$, with respect to the broadside, in the presence of additive white complex Gaussian noise. The nth received observation is of the form

$$\mathbf{x}_n = A b_n \mathbf{s}_{\theta_1} + \mathbf{v}_n, \quad n = 1,\ 2, \ldots,\ 30. \tag{19}$$

In (19), $\mathbf{s}_{\theta_1}$ is the array response vector, $\mathbf{v}_n \sim \mathcal{CN}(\mathbf{0}_7, \mathbf{I}_7)$ is the additive white Gaussian noise (AWGN), $A > 0$ accounts for the transmission power, and $b_n \in \{\pm 1\}$ is the Bernoulli equiprobable BPSK symbol. A is such that the signal-to-noise ratio (SNR) is $\text{SNR}_1 = 3$ dB. Next, we assume that 3 observations in $\mathbf{X} = [\mathbf{x}_1, \mathbf{x}_2, \ldots, \mathbf{x}_{30}]$ (without knowing which ones) are corrupted by 3 jammers transmitting from angles $-60°$, $-30°$ and $5°$, with SNR 9 dB each. This jammer corruption transforms the nominal dataset $\mathbf{X}$, into the corrupted dataset $\mathbf{X}_{\text{corr}} \in \mathbb{C}^{7\times 30}$ upon which the receiver must operate to estimate the DoA of the signal of interest. As a first step, in accordance to common practice, the receiver transforms $\mathbf{X}_{\text{corr}}$ to its real-valued version $\mathbf{X}'_{\text{corr}} = \left[\Re(\mathbf{X}^{\top}{}_{\text{corr}}),\ \Im(\mathbf{X}^{\top}{}_{\text{corr}})\right]^{\top} \in \mathbb{R}^{14\times 30}$, where $\Re(\cdot)$ and $\Im(\cdot)$ return the real and imaginary parts of their arguments, respectively. Then, it calculates the first PC (by L1-PCA and standard L2-PCA/SVD) of $\mathbf{X}'_{\text{corr}}$, $\mathbf{q}$, and uses it to form the familiar MUSIC DoA estimation spectrum [25]

$$P_{\mathbf{q}}(\theta) = \frac{1}{\mathbf{s}'^{T}_{\theta}\left(\mathbf{I} - \mathbf{q}\mathbf{q}^{T}\right)\mathbf{s}'_{\theta}}, \quad \theta \in \left(-\frac{\pi}{2}, \frac{\pi}{2}\right), \tag{20}$$

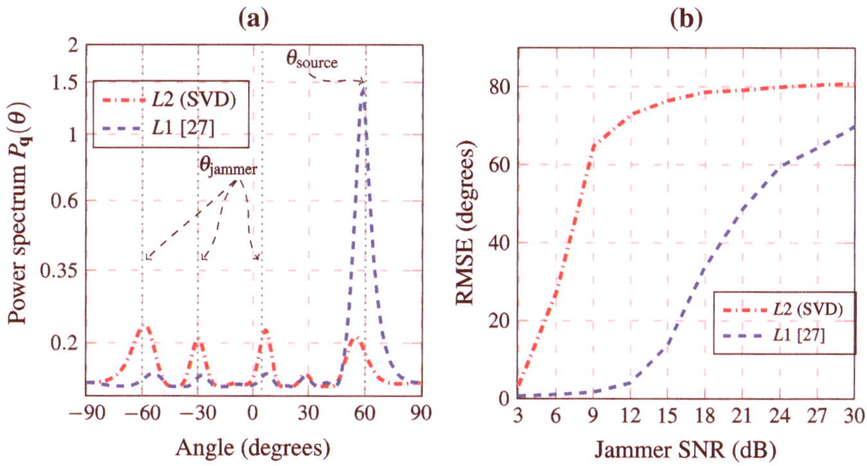

Fig. 2 MUSIC analysis: **a** Instantaneous spectrum with jammers located at $\theta_{\text{jammer}} = \{-60°, -30°, 5°\}$ and signal of interest at $\theta_{\text{source}} = 60°$. **b** Root-mean-square-error (RMSE) versus jammer SNR

where $\mathbf{s}'_\theta = \left[\Re(\mathbf{s}^\top{}_\theta),\ \ \Im(\mathbf{s}^\top{}_\theta)\right]^\top \in \mathbb{R}^{14\times 1}$. Finally, the receiver estimates the DoA of the signal of interest (equal to 60°) by the angle argument that yields the highest peak of the MUSIC spectrum. In Fig. 2, we plot a single realization of the MUSIC spectrum offered by the L1-PC and L2-PC. It is interesting to observe the corruption-resistance of L1-PCA (in contrast to L2-PCA), which is virtually unaffected by the jammers, despite the strong corruption of the processed dataset, it identifies the true active signal direction of arrival.

In Fig. 2b, we repeat the same experiment 2^{14} times and document average performance of the two methods. Specifically, denoting by $\hat{\theta}^{(m)}$ the estimated DoA at the m-th experiment, we plot the root-mean-squared-error

$$\text{RMSE} = \sqrt{\frac{1}{M}\sum_{m=1}^{M}\left(\theta_1 - \hat{\theta}^{(m)}\right)^2} \tag{21}$$

versus the SNR of the jamming sources. The instant jammers' locations are chosen independently across the experiments, uniformly in $\theta_j \in \left(-\frac{\pi}{2}, \frac{\pi}{2}\right)$. The corruption resistance of L1-PCA is once again clearly documented and the performance gap between L1-PCA and L2-PCA is striking.

4.3 L1-PCA Computation Accuracy

This experiment compares the performance of the four approximate L1-PCA calculators discussed above with respect to the L1-PCA metric, using as benchmark the optimum point attained by the exact algorithms of [24].

We generate 1000 arbitrary data matrices $\mathbf{X} \in \mathbb{R}^{3\times 8}$ with entries independently drawn from a standard Gaussian distribution $\mathcal{N}(0, 1)$. Then, we calculate the $K = 2$ L1-PCs of each matrix, by means of the five algorithms: (i) Sequential Fixed-Point Iterations [17], (ii) Joint Fixed-Point Iterations [32], (iii) Semi-Definite Programming Relaxation (SDP) [30], L1-PCA by Bit-Flipping (L1-BF) [29], and (v) the optimal one [24]. Then, for each approximate L1-PCA solution $\mathbf{Q}$, we measure the performance degradation ratio, with respect to the L1-PCA metric,

$$\Delta(\mathbf{Q}; \mathbf{X}) = \frac{\|\mathbf{X}^\top \mathbf{Q}_{L1}\|_1 - \|\mathbf{X}^\top \mathbf{Q}\|_1}{\|\mathbf{X}^\top \mathbf{Q}_{L1}\|_1}, \tag{22}$$

where $\mathbf{Q}_{L1}$ is the optimal solution. Over the 1000 tested data matrices, we calculate the empirical cumulative distribution function (CDF) of the performance degradation in (22), for each one of the four approximate algorithms, and plot it in Fig. 3a. We observe that L1-BF [29] returns the optimal solution with empirical probability 0.73 and, with probability 1, it suffers no more than 0.12 performance degradation.

Next, we run each algorithm on $NK = 16$ distinct initialization points, keeping the run that attains higher value to the L1-PCA metric (for SDP we consider $L = 16$ Gaussian randomization instances). Expectedly, the performance of all methods will be improved. In Fig. 3b, we plot again the empirical CDF of the performance degradation rations. Quite interestingly, with multiple initializations L1-BF returns the optimal solution with empirical probability 1 –i.e., L1-BF becomes, with probability 1, optimal L1-PCA calculator. Close to optimal, but inferior performance, is exhibited by the Joint Fixed-Point Iteration algorithm [32]. On the other hand, due to the sequential nullspace-projection approach they follow (that violates the no-scalability property of L1-PCA), Sequential Fixed-Point Iterations and SDP suffer from heavy performance degradation.

5 Conclusions

Research in wireless communications, signal processing, image processing, computer vision, and genomics, among other fields, has shown that L1-PCA (i) attains similar performance to PCA when the processed data are outlier-free and (ii) maintains sturdy resistance against outliers when the processed data are corrupted. Therefore, L1-PCA is expected to a play significant role in data analytics in the big-data era, when large datasets are often outlier corrupted. In this chapter, we presented the theoretical foundations of L1-PCA, optimal and state-of-the-art

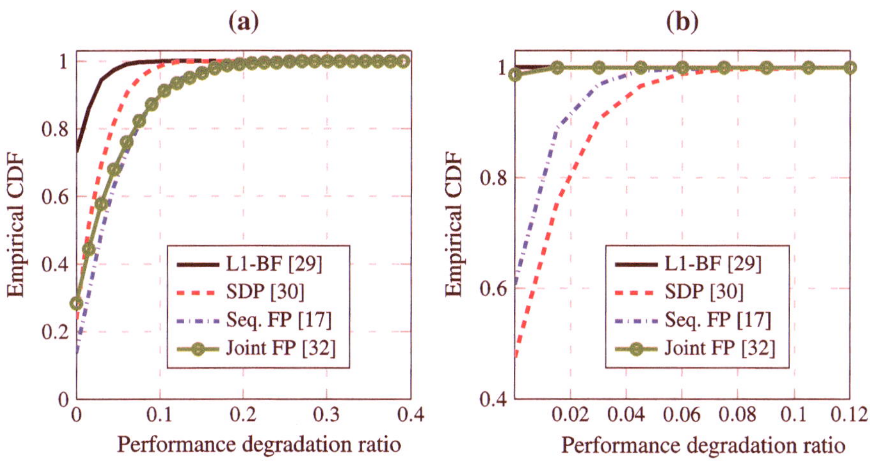

Fig. 3 Empirical CDF of $\Delta(\mathbf{Q}_{\text{ng}}; \mathbf{X})$, $\Delta(\mathbf{Q}_{\text{fp}}; \mathbf{X})$, $\Delta(\mathbf{Q}_{\text{sdp}}; \mathbf{X})$, and $\Delta(\mathbf{Q}_{\text{bf}}; \mathbf{X})$ ($D = 3$, $N = 8$, $K = 2$) for **a** $L = 1$ and **b** $L = NK = 16$ initializations

approximate algorithms for its implementation, and numerical studies that demonstrated its favorable performance, compared to standard PCA.

References

1. Barnett, V., Lewis, T.: Outliers in Statistical Data. Wiley, New York, NY (1994)
2. Bishop, C.M.: Pattern Recognition and Machine Learning. Springer, New York, NY (2006)
3. Brooks, J.P., Dulá, J.H.: The L1-norm best-fit hyperplane problem. Appl. Math. Lett. **26**, 51–55 (2013)
4. Brooks, J.P., Dulá, J.H., Boone, E.L.: A pure L1-norm principal component analysis. J. Comput. Stat. Data Anal. **61**, 83–98 (2013)
5. Candès, E.J., Li, X., Ma, Y., Wright, J.: Robust principal component analysis? J. ACM **58**(3), 37 (2011)
6. Chamadia, S., Pados, D.A.: Optimal sparse l1-norm principal-component analysis. In: Proceedings of IEEE International Conference on Acoustics, Speech and Signal Processing (ICASSP). IEEE, pp. 2686–2690 (2017a)
7. Chamadia, S., Pados, D.A.: Outlier processing via l1-principal subspaces. In: Proceedings of Florida Artificial Intelligence Research Society (FLAIRS), Marco Island, FL, pp 508–513 (2017b)
8. Ding, C., Zhou, D., He, X., Zha, H.: r_1-PCA: Rotational invariant L1-norm principal component analysis for robust subspace factorization. In: Proceedings of International Conference on Machine Learning (ICML 2006), Pittsburgh, PA, pp. 281–288 (2006)
9. Duda, R.O., Hart, P.E., Stork, D.G.: Pattern Classification, 2nd edn. Wiley, New York, NY (2001)
10. Eriksson, A., van den Hengel, A.: In: Efficient computation of robust low-rank matrix approximations in the presence of missing data using the L1 norm. In: Proceedings of IEEE Conference on Computer Vision Pattern Recognition (CVPR), San Francisco, CA, USA, pp. 771–778 (2010)

11. Golub, G.H.: Some modified matrix eigenvalue problems. SIAM Rev. **15**, 318–334 (1973)
12. Johnson, M., Savakis, A.: Fast L1-eigenfaces for robust face recognition. In: Proceedings of IEEE Western New York Image Signal Processing Workshop (WNYISPW), Rochester, NY, pp. 1–5 (2014)
13. Jolliffe, I.T.: Principal Component Analysis. Springer, New York, NY (1986)
14. Ke, Q., Kanade, T.: Robust subspace computation using L1 norm. Techincal report Internal Technical Report, Computer Science Department, Carnegie Mellon University, Pittsburgh, PA, USA, CMU-CS-03-172 (2003)
15. Ke, Q., Kanade, T.: Robust L1 norm factorization in the presence of outliers and missing data by alternative convex programming. In: Proceedings of IEEE Conference on Computer Vision and Pattern Recognition (CVPR), San Diego, CA, pp. 739–746 (2005)
16. Kundu, S., Markopoulos, PP., Pados, D.A.: Fast computation of the L1-principal component of real-valued data. In: Proceedings of IEEE International Conference on Acoustics, Speech and Signal Processing (ICASSP), Florence, Italy, pp. 8028–8032 (2014)
17. Kwak, N.: Principal component analysis based on L1-norm maximization. IEEE Trans. Patt. Anal. Mach. Intell. **30**, 1672–1680 (2008)
18. Kwak, N., Oh, J.: Feature extraction for one-class classification problems: enhancements to biased discriminant analysis. Patt. Recog. **42**, 17–26 (2009)
19. Liu, Y., Pados, D.A.: Compressed-sensed-domain L1-PCA video surveillance. IEEE Trans. Multimedia **18**(3), 351–363 (2016)
20. Markopoulos, P.: Optimal algorithms for L1-norm principal component analysis: new tools for signal processing and machine learning with few and/or faulty training data. Ph.D. thesis, State University of New York at Buffalo (2015)
21. Markopoulos, P.P.: Reduced-rank filtering on L1-norm subspaces. In: Proceedings of IEEE Sensor Array Multichannel Signal Processing Workshop (SAM), Rio de Janeiro, Brazil, pp.1–5 (2016)
22. Markopoulos, P.P., Ahmad, F.: Indoor human motion classification by L1-norm subspaces of micro-doppler signatures. In: Proceedings of IEEE Radar Conference (Radarcon), Seattle, WA, pp. 1807–1810 (2017)
23. Markopoulos, P.P., Karystinos, G.N., Pados, D.A.: Some options for L1-subspace signal processing. In: Proceedings of 10th International Sympoisum on Wireless Communication System (ISWCS), Ilmenau, Germany, pp. 622–626 (2013)
24. Markopoulos, P.P., Karystinos, G.N., Pados, D.A.: Optimal algorithms for L1-subspace signal processing. IEEE Trans. Signal Process. **62**, 5046–5058 (2014a)
25. Markopoulos, P.P., Tsagkarakis, N., Pados, D.A., Karystinos, G.N.: Direction finding with L1-norm subspaces. In: Proceedings of Commercial Sensing Conference on SPIE Defence Security Sensors (DSS), Baltimore MD, pp. 91-090J1–91-090J11 (2014b)
26. Markopoulos, P.P., Kundu, S., Pados, D.A.: L1-fusion: Robust linear-time image recovery from few severely corrupted copies. In: Proceedings of IEEE International Conference on Image Processing (ICIP), Quebec City, Canada, pp. 1225–1229 (2015)
27. Markopoulos, P.P., Kundu, S., Chamadia, S., Pados, D.A.: L1-norm principal-component analysis via bit flipping. In: Proceedings of IEEE International Conference on Machine Learning Applications (ICMLA), Anaheim, CA, pp. 326–332 (2016a)
28. Markopoulos, P.P., Tsagkarakis, N., Pados, D.A., Karystinos, G.N.: Direction-of-arrival estimation from l1-norm principal components. In: Proceedings of IEEE International Sympoisum on Phased Array Systems and Technology (PAST), Boston, MA, pp. 1–6 (2016b)
29. Markopoulos, P.P., Kundu, S., Chamadia, S., Pados, D.: Efficient l1-norm principal-component analysis via bit flipping. IEEE Transactions on Signal Processing (2016b)
30. McCoy, M., Tropp, J.A.: Two proposals for robust PCA using semidefinite programming. Electron. J. Stat. **5**, 1123–1160 (2011)
31. Meng, D., Zhao, Q., Xu, Z.: Improve robustness of sparse PCA by L1-norm maximization. Patt. Recog. **45**, 487–497 (2012)
32. Nie, F., Huang, H., Ding, C., Luo, D., Wang, H.: Robust principal component analysis with non-greedy L1-norm maximization. In: Proceedings of International Joint Conference on Artificial intelligence (IJCAI), Barcelona, Spain, pp. 1433–1438 (2011)

33. Pearson, K.: On lines and planes of closest fit to systems of points in space. Philos. Mag. **2**, 559–572 (1901)
34. Schönemann, P.H.: A generalized solution of the orthogonal procrustes problem. Psychometrika **31**(1), 1–10 (1966)
35. Tsagkarakis, N., Markopoulos, P.P., Pados, D.A.: Direction finding by complex L1-principal component analysis. In: Proceedings of IEEE 16th International Workshop on Signal Processing Advances in Wireless Communications (SPAWC), Stockholm, Sweden, pp. 475–479 (2015)
36. Tsagkarakis, N., Markopoulos, P.P., Pados, D.A.: On the l1-norm approximation of a matrix by another of lower rank. In: Proceedings of IEEE International Conference on Machine Learning Applications (IEEE ICMLA 2016), IEEE, pp. 768–773 (2016)
37. Wang, H.: Block principal component analysis with L1-norm for image analysis. Patt. Recog. Lett. **33**, 537–542 (2012)

Damage and Fault Detection of Structures Using Principal Component Analysis and Hypothesis Testing

Francesc Pozo and Yolanda Vidal

Abstract This chapter illustrates the application of principal component analysis (PCA) plus statistical hypothesis testing to online damage detection in structures, and to fault detection of an advanced wind turbine benchmark under actuators (pitch and torque) and sensors (pitch angle measurement) faults. A baseline pattern or PCA model is created with the healthy state of the structure using data from sensors. Subsequently, when the structure is inspected or supervised, new measurements are obtained and projected into the baseline PCA model. When both sets of data are compared, both univariate and multivariate statistical hypothesis testing is used to make a decision. In this work, both experimental results (with a small aluminum plate) and numerical simulations (with a well-known benchmark wind turbine) show that the proposed technique is a valuable tool to detect structural changes or faults.

1 Introduction

Principal component analysis (PCA) is a statistical technique that transforms a number of possibly correlated variables into a smaller number of uncorrelated variables called principal components. It is well-known that the basic idea behind the PCA is to reduce the dimension of the data, while retaining as much as possible the variation present in these data, see [1]. Applications of PCA can be found in a vast variety of fields from neuroscience to image processing. This chapter provides a thorough review to the application of PCA to detect structural changes (damages, structural health monitoring) or faults (in the sensors or in the actuators, condition monitor-

F. Pozo (✉) · Y. Vidal
CoDAlab, Department of Mathematics, Escola d'Enginyeria de Barcelona Est (EEBE), Universitat Politècnica de Catalunya (UPC), Eduard Maristany 10-14, 08019 Barcelona, Spain
e-mail: francesc.pozo@upc.edu

Y. Vidal
e-mail: yolanda.vidal@upc.edu

G. R. Naik (ed.), *Advances in Principal Component Analysis*,
https://doi.org/10.1007/978-981-10-6704-4_7

ing). First reviewing how data (from sensors) is usually represented, second showing how in this work is represented in a different manner, then reviewing the group-scaling processing of the data, and finally showing that PCA plus (univariate and multivariate) statistical hypothesis testing is a valuable tool to detect structural changes or faults.

In a standard application of the principal component analysis strategy in the field of structural health monitoring or condition monitoring, the projections onto the vectorial space spanned by the principal components (scores) allow a visual grouping or separation. In some other cases, two classical indices can be used for damage or fault detection, such as the Q index and the Hotelling's T^2 index, see [2]. However, when a visual grouping, clustering or separation cannot be performed with the scores a more powerful and reliable tool is needed to be able to detect a damage or a fault. The approaches proposed in this chapter for the damage or fault detection are based on a group scaling of the data and multiway principal component analysis (MPCA) combined with both univariate and multivariate statistical hypothesis testing [3–5].

On one hand, the basic premise of vibration based structural health monitoring feature selection is that damage will significantly alter the stiffness, mass or energy dissipation properties of a system, which, in turn, alter the measured dynamic response of that system. Subsequently, the structure to be diagnosed is excited by the same signal and the dynamic response is compared with the pattern, see [6]. In this chapter, these techniques will be applied to an experimental set-up with a smooth-raw aluminium plate.

On the other hand, in the fault detection case (condition monitoring), this chapter applies the techniques to an advanced wind turbine benchmark (numerical simulations). In this case, the only available excitation is the wind. Therefore, guided waves in wind turbines cannot be considered as a realistic scenario. In spite of that, the new paradigm described is based on the fact that, even with a different wind field, the fault detection strategy based on PCA and statistical hypothesis testing will be able to detect faults. A growing interest is being shown in offshore wind turbines, because they have enormous advantages compared to their onshore version including higher and steadier wind speed, and less restrictions due to remoteness to urban areas, see [7]. The main disadvantages of offshore wind energy farms are high construction costs, and operation and maintenance (O&M) costs because they must withstand rough weather conditions. The field of wind turbine O&M represents a growing research topic as they are the critical elements affecting profitability in the offshore wind turbine sector. We believe that PCA plus statistical hypothesis testing has a tremendous potential in this area. In fact, the work described in this chapter is only the beginning of a large venture. Future work will develop complete fault detection, isolation, and reconfigurable control strategies in response to faults based on efficient fault feature extraction by means of PCA.

This chapter is divided into five main sections. In Sect. 1 we introduce the scope of the chapter. Section 2 poses the experimental set-up and the reference wind turbine where the techniques will be applied and tested. The methodology is stated in Sect. 3. The obtained results are discussed and analyzed in Sect. 4. Finally, Sect. 5 draws the conclusions.

2 Experimental Set-Up and Reference Wind Turbine

The damage and fault detection strategies reviewed in this chapter will be applied to both an experimental set-up and a simulated wind turbine, as described in Sects. 2.1 and 2.2. On one hand, with respect to the experimental set-up, the analysis of changes in the vibrational properties of a small aluminum plate is used to explain, validate and test the damage detection strategies. As the aluminum plate will be always excited by the same signal, this experiment corresponds to guided waves in structures for structural health monitoring. On the other hand, we will address the problem of online fault detection of an advanced wind turbine benchmark under actuators (pitch and torque) and sensors (pitch angle measurements) faults of different type. In this case, the excitation signal is never the same, as it is given by the wind. Even in this case, with a different wind signal, the fault detection strategy will be able to detect the faults. More precisely, the key idea behind the detection strategy is the assumption that a change in the behavior of the overall system, even with a different excitation, has to be detected.

2.1 Experimental Set-Up

The small aluminium plate (25 cm $\times$ 25 cm $\times$ 0.2 cm) in Fig. 1 (top) is used to experimentally validate the proposed approach in this work. The plate is suspended by two elastic ropes in a metallic frame in order to isolate the environmental noise and remove boundary conditions (Fig. 2). Four piezoelectric transducer discs (PZT's) are attached on the surface, as can be seen in Fig. 1 (bottom). Each PZT is able to produce a mechanical vibration (Lamb waves in a thin plate) if some electrical excitation is applied (actuator mode). Besides, PZT's are able to detect time varying mechanical response data (sensor mode). In every phase of the experimental stage, just one PZT is used as actuator (exciting the plate) and the rest are used as sensors (and thus recording the dynamical response). A total number of 100 experiments were performed using the healthy structure: 50 for the baseline (BL) and 50 for testing (Un, which stands for *undamaged*, is an abbreviation used throughout the chapter). Additionally, nine damages (D1, D2, …, D9) were simulated adding different masses at different locations, see Fig. 1 (bottom). For each damage, 50 experiments were implemented, resulting in a total number of 450 experiments. The excitation is a sinusoidal signal of 112 KHz modulated by a Hamming window, as illustrated in Fig. 3 (top). An example of the signal collected by PZT2 is shown in Fig. 3 (bottom).

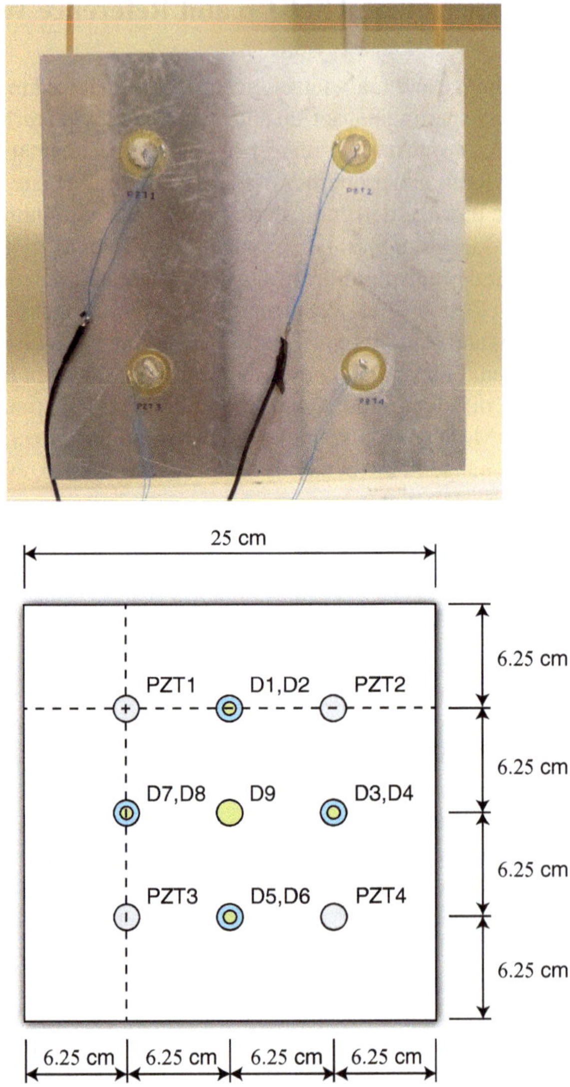

Fig. 1 Aluminium plate (top). Dimensions and piezoelectric transducers location (bottom)

2.2 *Reference Wind Turbine*

The National Renewable Energy Laboratory (NREL) offshore 5-MW baseline wind turbine [8] is used in the simulations of the fault detection strategy. This model is used as a reference by research teams throughout the world to standardize baseline offshore wind turbine specifications and to quantify the benefits of advanced land- and sea-based wind energy technologies. In this work, the wind turbine is operated

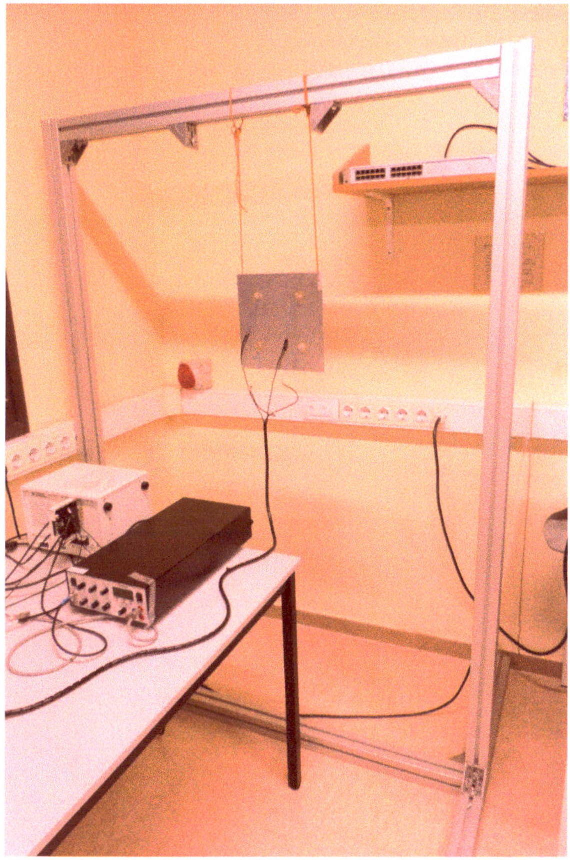

Fig. 2 The plate is suspended by two elastic ropes in a metallic frame.

in its onshore version and in the above-rated wind-speed range. The main properties of this turbine are listed in Table 1.

In this chapter, the proposed fault detection method is SCADA-data based, that is, it uses data already collected at the wind turbine controller. In particular, Table 2 presents assumed available data on a MW-scale commercial wind turbine that is used in this work by the fault detection method.

The reference wind turbine has a conventional variable-speed, variable blade-pitch-to-feather configuration. In such wind turbines, the conventional approach for controlling power-production operation relies on the design of two basic control systems: a generator-torque controller and a rotor-collective blade-pitch controller. In this work, the baseline torque and pitch controllers are utilized, but the generator-converter and the pitch actuators are modeled and implemented externally; i.e., apart from the embedded FAST code. This will facilitate to model different type of faults on the generator and the pitch actuator. The next subsections recall these models and also the wind model used to generate the wind data.

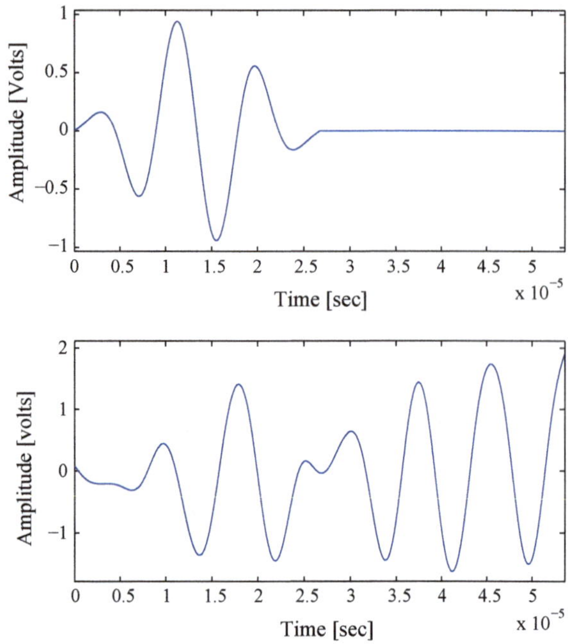

Fig. 3 Excitation signal (top) and, dynamic response recorded by PZT 2 (bottom)

Table 1 Gross properties of the wind turbine [8]

Reference wind turbine	
Rated power	5 MW
Number of blades	3
Rotor/Hub diameter	126, 3 m
Hub Height	90 m
Cut-In, Rated, Cut-Out Wind Speed	3, 11.4, 25 m/s
Rated generator speed	1173.7 rpm
Gearbox ratio	97

2.2.1 Wind Modeling

The TurbSim stochastic inflow turbulence tool (National Wind Technology Center, Boulder, Colorado, USA) [9] has been used. It provides the ability to drive design code (e.g., FAST) simulations of advanced turbine designs with simulated inflow turbulence environments that incorporate many of the important fluid dynamic features known to adversely affect turbine aeroelastic response and loading.

The generated wind data has the following characteristics: Kaimal turbulence model with intensity set to 10%, logarithmic profile wind type, mean speed is set to 18.2 m/s and simulated at hub height, and the roughness factor is set to 0.01 m.

In this work, every simulation is ran with a different wind data set.

Table 2 Assumed available measurements. These sensors are representative of the types of sensors that are available on a MW-scale commercial wind turbine

Number	Sensor type	Symbol	Units
1	Generated electrical power	$P_{e,m}$	kW
2	Rotor speed	$\omega_{r,m}$	rad/s
3	Generator speed	$\omega_{g,m}$	rad/s
4	Generator torque	$\tau_{c,m}$	Nm
5	First pitch angle	$\beta_{1,m}$	deg
6	Second pitch angle	$\beta_{2,m}$	deg
7	Third pitch angle	$\beta_{3,m}$	deg
8	Fore-aft acceleration at tower bottom	$a^b_{fa,m}$	m/s^2
9	Side-to-side acceleration at tower bottom	$a^b_{ss,m}$	m/s^2
10	Fore-aft acceleration at mid-tower	$a^m_{fa,m}$	m/s^2
11	Side-to-side acceleration at mid-tower	$a^m_{ss,m}$	m/s^2
12	Fore-aft acceleration at tower top	$a^t_{fa,m}$	m/s^2
13	Side-to-side acceleration at tower top	$a^t_{ss,m}$	m/s^2

2.2.2 Generator-Converter Actuator Model and Pitch Actuator Model

The generator-converter and the pitch actuators are modeled apart from the embedded FAST code, with the objective to ease the model of different type of faults on these parts of the wind turbine.

On one hand, the generator-converter can be modeled by a first-order differential system [10]:

$$\frac{\tau_r(s)}{\tau_c(s)} = \frac{\alpha_{gc}}{s + \alpha_{gc}}$$

where τ_r and τ_c are the real generator torque and its reference (given by the controller), respectively, and we set $\alpha_{gc} = 50$ [8]. The power produced by the generator, $P_e(t)$, can be modeled by [10]:

$$P_e(t) = \eta_g \omega_g(t) \tau_r(t)$$

where η_g is the efficiency of the generator and ω_g is the generator speed. In the numerical experiments, $\eta_g = 0.98$ is used [10].

On the other hand, the three pitch actuators are modeled as a second-order linear differential equation, pitch angle $\beta_i(t)$, and its reference $u(t)$ (given by the collective-pitch controller) [10]:

$$\frac{\beta_i(s)}{u(s)} = \frac{\omega_n^2}{s^2 + 2\xi\omega_n s + \omega_n^2}, \quad i = 1, 2, 3 \tag{1}$$

where ω_n and ξ are the natural frequency and the damping ratio, respectively. In the fault free case, these values are set to $\omega_n = 11.11$ rad/s, and $\xi = 0.6$.

2.2.3 Fault Description

In this chapter, the different faults proposed in the fault tolerant control benchmark [11] will be considered, as gathered in Table 3. These faults selected by the benchmark cover different parts of the wind turbine, different fault types and classes, and different levels of severity.

Usually, pitch systems use either an electric or a fluid power actuator. However, the fluid power subsystem has lower failure rates and better capability of handling extreme loads than the electrical systems. Therefore, fluid power pitch systems are preferred on multi-MW size and offshore turbines. However, general issues such as leakage, contamination, component malfunction and electrical faults make current systems work sub-optimal [12]. In this work, faults in the pitch actuator are considered in the hydraulic system, which result in changed dynamics due to either a high air content in oil (fault 1) or a drop in pressure in the hydraulic supply system due to pump wear (fault 2) or hydraulic leakage (fault 3) [13], as well as pitch position sensor faults (faults 5–7).

Pump wear (fault 2) is an irreversible slow process over the years that results in low pump pressure. As this wear is irreversible, the only possibility to fix it is to replace the pump, which will happen after pump wear reaches certain level. Meanwhile, the pump will still be operating and the system dynamics is slowly changing, while the turbine structure should be able to withstand the effects of this fault. Pump wear after approximately 20 years of operation might result in pressure reduction to 75% of the rated pressure, which is reflected by the faulty natural frequency $\omega_n = 7.27$ rad/s and a fault damping ratio of $\xi = 0.75$.

Table 3 Fault scenarios

Fault	Type	Description
1	Pitch actuator	Change in dynamics: high air content in oil ($\omega_n = 5.73$ rad/s, $\xi = 0.45$)
2	Pitch actuator	Change in dynamics: pump wear ($\omega_n = 7.27$ rad/s, $\xi = 0.75$)
3	Pitch actuator	Change in dynamics: hydraulic leakage ($\omega_n = 3.42$ rad/s, $\xi = 0.9$)
4	Generator speed sensor	Scaling (gain factor equal to 1.2)
5	Pitch angle sensor	Stuck (fixed value equal to 5 deg)
6	Pitch angle sensor	Stuck (fixed value equal to 10 deg)
7	Pitch angle sensor	Scaling (gain factor equal to 1.2)
8	Torque actuator	Offset (offset value equal to 2000 Nm)

Hydraulic leakage (fault 3) is another irreversible incipient fault but is introduced considerably faster than the pump wear. Leakage of pitch cylinders can be internal or external [12]. When this fault reaches a certain level, system repair is necessary, and if the leakage is too fast (normally due to external leakage), it will lead to a pressure drop and the preventive procedure is deployed to shut down the turbine before the blade is stuck in undesired position (if the hydraulic pressure is too low, the hydraulic system will not be able to move the blades that will cause the actuator to be stuck in its current position resulting in blade seize). The fast pressure drop is easy to detect (even visually as it is normally related to external leakage) and requires immediate reaction; however, the slow hydraulic leakage reduces the dynamics of the pitch system, and for a reduction of 50% of the nominal pressure the natural frequency under this fault condition is reduced to $\omega_n = 3.42$ rad/s and the corresponding damping ratio is $\xi = 0.9$. In this work, the slow (internal) hydraulic leakage is studied.

On the contrary to pump wear and hydraulic leakage, high air content in the oil (fault 1) is an incipient reversible process, which means that the air content in the oil may disappear without any necessary repair to the system. The nominal value of the air content in the oil is 7%, whereas the high air content in the oil corresponds to 15%. The effect of such a fault is expressed by the new natural frequency $\omega_n = 5.73$ rad/s and the damping ratio of $\xi = 0.45$ (corresponding to the high air content in the oil).

The generator speed measurement is done using encoders. The gain factor fault (fault 4) is introduced when the encoder reads more marks on the rotating part than actually present, which can happen as a result of dirt or other false markings on the rotating part.

Faults in the pitch position measurement (pitch position sensor fault) are also advised. This is one of the most important failure modes found on actual systems [12, 14]. The origin of these faults is either electrical or mechanical, and it can result in either a fixed value (faults 5 and 6) or a changed gain factor (fault 7) on the measurements. In particular, the fixed value fault should be easy to detect, and, therefore, it is important that a fault detection, isolation, and accommodation scheme be able to deal with this fault. If not handled correctly, these faults will influence the pitch reference position because the pitch controller is based on these pitch position measurements.

Finally, a converter torque offset fault is considered (fault 8). It is difficult to detect this fault internally (by the electronics of the converter controller). However, from a wind turbine level, it is possible to be detected, isolated, and accommodated because it changes the torque balance in the wind turbine power train.

3 Fault Detection Strategy

The overall fault detection strategy is based on principal component analysis and statistical hypothesis testing. A baseline pattern or PCA model is created with the healthy state of the structure (plate or wind turbine) to study. When the current state has to be diagnosed, the collected data is projected using the PCA model. The final diagnosis is performed using statistical hypothesis testing.

The main paradigm of vibration based structural health monitoring is based on the basic idea that a change in physical properties due to structural changes or damage will cause detectable changes in dynamical responses. This idea is illustrated in Fig. 4, where the healthy structure is excited by a signal to create a pattern. Subsequently, the structure to be diagnosed is excited by the same signal and the dynamic response is compared with the pattern. The scheme in Fig. 4 is also know as guided waves in structures for structural health monitoring [6].

However, in the case of wind turbines, the only available excitation is the wind. Therefore, guided waves in wind turbines for SHM as in Fig. 4 cannot be considered as a realistic scenario. In spite of that, the new paradigm described in Fig. 5 is based on the fact that, even with a different wind field, the fault detection strategy based on PCA and statistical hypothesis testing will be able to detect some damage, fault or misbehavior. More precisely, the key idea behind the detection strategy is the

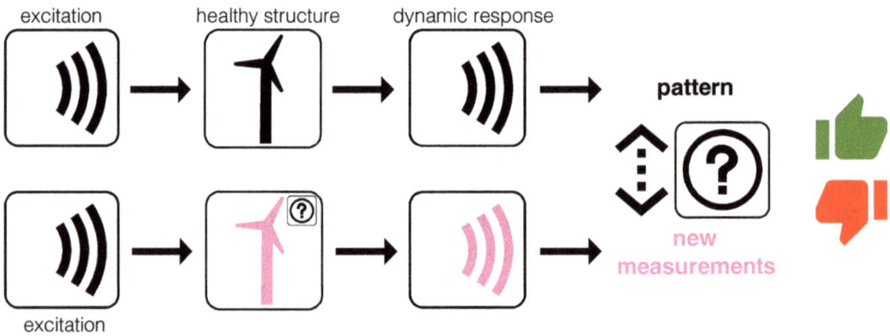

Fig. 4 Guided waves in structures for structural health monitoring. The healthy structure is excited by a signal and the dynamic response is measured to create a baseline pattern. Then, the structure to diagnose is excited by the same signal and the dynamic response is also measured and compared with the baseline pattern. A significant difference in the pattern would imply the existence of a fault

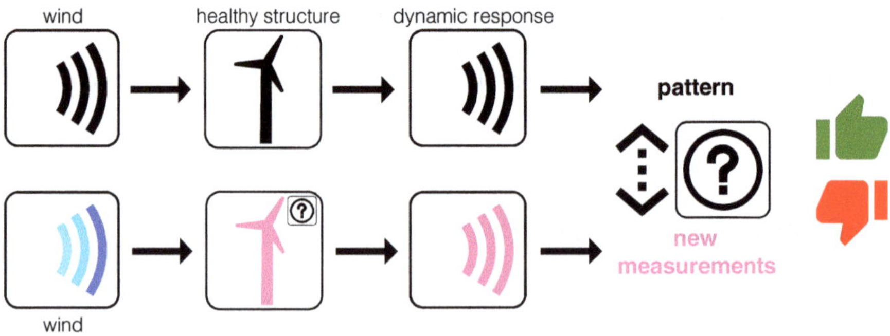

Fig. 5 Even with a different wind field, the fault detection strategy is able to detect some damage, fault or misbehavior

assumption that a change in the behavior of the overall system, even with a different excitation, has to be detected. The results presented in Sects. 4.3 and 4.4 confirm this hypothesis.

3.1 Data Driven Baseline Modeling Based on PCA

Classical approaches to the application of principal component analysis can be summarized in the following example. Let us assume that we have N sensors or variables that are measuring during $(L-1)\Delta$ seconds, where Δ is the sampling time and $L \in \mathbb{N}$. The discretized measures of each sensor can be arranged as a column vector $\mathbf{x}^i = (x_1^i, x_2^i, \ldots, x_L^i)^T$, $i = 1, \ldots, N$ so we can build up a $L \times N$ matrix as follows:

$$X = \left(\mathbf{x}^1 \middle| \mathbf{x}^2 \middle| \cdots \middle| \mathbf{x}^N\right) = \begin{pmatrix} x_1^1 & x_1^2 & \cdots & x_1^N \\ x_2^1 & x_2^2 & \cdots & x_2^N \\ \vdots & \vdots & \ddots & \vdots \\ x_i^1 & x_i^2 & \cdots & x_i^N \\ \vdots & \vdots & \ddots & \vdots \\ x_L^1 & x_L^2 & \cdots & x_L^N \end{pmatrix} \in \mathscr{M}_{L\times N}(\mathbb{R}) \tag{2}$$

It is worth noting that each column in matrix X in Eq. (2) represents the measures of a single sensor or variable.

However, when multiway principal component analysis is applied to data coming from N sensors at L discretization instants and n experimental trials, the information can be stored in an unfolded $n \times (N \times L)$ matrix as follows:

$$\mathbf{X} = \left(\begin{array}{cccc|ccc|c|ccc} x_{11}^1 & x_{12}^1 & \cdots & x_{1L}^1 & x_{11}^2 & \cdots & x_{1L}^2 & \cdots & x_{11}^N & \cdots & x_{1L}^N \\ \vdots & \vdots & \ddots & \vdots & \vdots & \ddots & \vdots & \ddots & \vdots & \ddots & \vdots \\ x_{i1}^1 & x_{i2}^1 & \cdots & x_{iL}^1 & x_{i1}^2 & \cdots & x_{iL}^2 & \cdots & x_{i1}^N & \cdots & x_{iL}^N \\ \vdots & \vdots & \ddots & \vdots & \vdots & \ddots & \vdots & \ddots & \vdots & \ddots & \vdots \\ x_{n1}^1 & x_{n2}^1 & \cdots & x_{nL}^1 & x_{n1}^2 & \cdots & x_{nL}^2 & \cdots & x_{n1}^N & \cdots & x_{nL}^N \end{array}\right) \tag{3}$$

In this case, a column in matrix $\mathbf{X}$ in Eq. (3) no longer represents the values of a variable at different time instants but the measurements of a variable at one particular time instant in the whole set of experimental trials. The work by Mujica et al. [2] presents one of the first applications of multiway principal component analysis (MPCA) for damage assessment in structures using two different measures or distances (Q and T indices). One of the advantages of the classical approach of principal component analysis is that the largest components (in absolute value) of the unit eigenvector related to the largest eigenvalue gives direct information on the most important sensors installed in the structure [15, 16]. This information is no longer available when multiway principal component analysis is applied to the collected data [16]. Another important difference between the classical approach and MPCA lies on normalization. On one hand, to apply the PCA in its classical version, each column vector is normalized to have zero mean and unit variance. On the other hand (MPCA), the normalization has to take into account that several columns contain the information of the same sensor. In this case, several strategies can be applied, such as autoscaling or group scaling. In this work we use the so-called *group scaling*, that it is detailed in Sect. 3.1.3.

3.1.1 Guides Waves in Structures for Structural Health Monitoring: Data Collection

Let us address the PCA modeling by measuring, from a healthy structure, N sensors at L discretization instants and n experimental trials. In this case, since we consider guided waves, the structure is excited by the same signal at each experimental trial. This way, the collected data can be arranged in matrix form as follows:

$$\mathbf{X}_{\text{GW}} = \left(\begin{array}{cccc|ccc|c|ccc} x_{11}^1 & x_{12}^1 & \cdots & x_{1L}^1 & x_{11}^2 & \cdots & x_{1L}^2 & \cdots & x_{11}^N & \cdots & x_{1L}^N \\ \vdots & \vdots & \ddots & \vdots & \vdots & \ddots & \vdots & \ddots & \vdots & \ddots & \vdots \\ x_{i1}^1 & x_{i2}^1 & \cdots & x_{iL}^1 & x_{i1}^2 & \cdots & x_{iL}^2 & \cdots & x_{i1}^N & \cdots & x_{iL}^N \\ \vdots & \vdots & \ddots & \vdots & \vdots & \ddots & \vdots & \ddots & \vdots & \ddots & \vdots \\ x_{n1}^1 & x_{n2}^1 & \cdots & x_{nL}^1 & x_{n1}^2 & \cdots & x_{nL}^2 & \cdots & x_{n1}^N & \cdots & x_{nL}^N \end{array}\right) \tag{4}$$

In this way, each row vector represents, for a particular experimental trial, the measurements from all the sensors at every specific time instant. Similarly, each column vector represents measurements from one sensor at one specific time instant in the whole set of experimental trials. The number of rows of matrix $\mathbf{X}_{\text{GW}}$ in Eq. (4), n, is defined by the number of experimental trials. The number of columns of matrix $\mathbf{X}_{\text{GW}}$, $N \cdot L$, is the number of sensors (N) times the number of discretization instants (L).

3.1.2 Condition Monitoring of Wind Turbines: Data Collection

In the case of wind turbines, the excitation comes from different wind fields. Therefore, instead of considering different *experimental trials* as in Sect. 3.1.1, we will first measure, from a healthy wind turbine, a sensor during $(nL - 1)\Delta$ seconds, where Δ is the sampling time and $n, L \in \mathbb{N}$. The discretized measures of the sensor are a real vector

$$\left(x_{11}\ x_{12} \cdots x_{1L}\ x_{21}\ x_{22} \cdots x_{2L} \cdots x_{n1}\ x_{n2} \cdots x_{nL}\right) \in \mathbb{R}^{nL} \tag{5}$$

where the real number x_{ij}, $i = 1, \ldots, n$, $j = 1, \ldots, L$ corresponds to the measure of the sensor at time $((i-1)L + (j-1))\,\Delta$ seconds. This collected data can be arranged in matrix form as follows:

$$\begin{pmatrix} x_{11} & x_{12} & \cdots & x_{1L} \\ \vdots & \vdots & \ddots & \vdots \\ x_{i1} & x_{i2} & \cdots & x_{iL} \\ \vdots & \vdots & \ddots & \vdots \\ x_{n1} & x_{n2} & \cdots & x_{nL} \end{pmatrix} \in \mathscr{M}_{n \times L}(\mathbb{R}) \tag{6}$$

where $\mathscr{M}_{n \times L}(\mathbb{R})$ is the vector space of $n \times L$ matrices over $\mathbb{R}$. When the measures are obtained from $N \in \mathbb{N}$ sensors also during $(nL-1)\Delta$ seconds, the collected data, for each sensor, can be arranged in a matrix as in Eq. (6). Finally, all the collected data coming from the N sensors is disposed in a matrix $\mathbf{X}_{\mathrm{WT}} \in \mathscr{M}_{n \times (N \cdot L)}$ as follows:

$$\begin{aligned} \mathbf{X}_{\mathrm{WT}} &= \left(\begin{array}{cccc|ccc|c|ccc} x_{11}^1 & x_{12}^1 & \cdots & x_{1L}^1 & x_{11}^2 & \cdots & x_{1L}^2 & \cdots & x_{11}^N & \cdots & x_{1L}^N \\ \vdots & \vdots & \ddots & \vdots & \vdots & \ddots & \vdots & \ddots & \vdots & \ddots & \vdots \\ x_{i1}^1 & x_{i2}^1 & \cdots & x_{iL}^1 & x_{i1}^2 & \cdots & x_{iL}^2 & \cdots & x_{i1}^N & \cdots & x_{iL}^N \\ \vdots & \vdots & \ddots & \vdots & \vdots & \ddots & \vdots & \ddots & \vdots & \ddots & \vdots \\ x_{n1}^1 & x_{n2}^1 & \cdots & x_{nL}^1 & x_{n1}^2 & \cdots & x_{nL}^2 & \cdots & x_{n1}^N & \cdots & x_{nL}^N \end{array}\right) \\ &= \left(\mathbf{X}_{\mathrm{WT}}^1 \middle| \mathbf{X}_{\mathrm{WT}}^2 \middle| \cdots \middle| \mathbf{X}_{\mathrm{WT}}^N \right) \end{aligned} \tag{7}$$

where the superindex $k = 1, \ldots, N$ of each element x_{ij}^k in the matrix represents the number of sensor.

It is worth noting that in both approaches —guided waves for structural health monitoring and condition monitoring of wind turbines— the structure of matrices $\mathbf{X}_{\mathrm{GW}}$ and $\mathbf{X}_{\mathrm{WT}}$ in Eqs. (4) and (7), respectively, are completely equivalent. Therefore, in the rest of the chapter, we will simply refer to these matrices as $\mathbf{X}$.

The objective of the principal component analysis, as a pattern recognition technique, is to find a linear transformation orthogonal matrix $\mathbf{P} \in \mathscr{M}_{(N \cdot L) \times (N \cdot L)}(\mathbb{R})$ that will be used to transform or project the original data matrix $\mathbf{X}$ according to the subsequent matrix product:

$$\mathbf{T} = \mathbf{X}\mathbf{P} \in \mathscr{M}_{n \times (N \cdot L)}(\mathbb{R}) \tag{8}$$

where $\mathbf{T}$ is a matrix having a diagonal covariance matrix.

3.1.3 Group Scaling

Since the data in matrix $\mathbf{X}$ come from several sensors and could have different scales and magnitudes, it is required to apply a preprocessing step to rescale the data using the mean of all measurements of the sensor at the same column and the standard deviation of all measurements of a sensor [17].

More precisely, for $k = 1, 2, \ldots, N$ we define

$$\mu_j^k = \frac{1}{n}\sum_{i=1}^{n} x_{ij}^k, \; j = 1, \ldots, L, \tag{9}$$

$$\mu^k = \frac{1}{nL}\sum_{i=1}^{n}\sum_{j=1}^{L} x_{ij}^k, \tag{10}$$

$$\sigma^k = \sqrt{\frac{1}{nL}\sum_{i=1}^{n}\sum_{j=1}^{L}(x_{ij}^k - \mu^k)^2} \tag{11}$$

where μ_j^k is the mean of the measures placed at the same column, that is, the mean of the n measures of sensor k in matrix $\mathbf{X}^k$ —that corresponds to the n measures of sensor k at the j-th discretization instant for the whole set of experimental trials (guided waves) or the measures of sensor k at time instants $((i-1)L + (j-1))\,\Delta$ seconds, $i = 1, \ldots, n$ (wind turbine)—; μ^k is the mean of all the elements in matrix $\mathbf{X}^k$, that is, the mean of all the measures of sensor k; and σ^k is the standard deviation of all the measures of sensor k. Therefore, the elements x_{ij}^k of matrix $\mathbf{X}$ are scaled to define a new matrix $\check{\mathbf{X}}$ as

$$\check{x}_{ij}^k := \frac{x_{ij}^k - \mu_j^k}{\sigma^k}, \; i = 1, \ldots, n, \; j = 1, \ldots, L, \; k = 1, \ldots, N. \tag{12}$$

When the data are normalized using Eq. (12), the scaling procedure is called variable scaling or group scaling [18].

For the sake of clarity, and throughout the rest of the chapter, the scaled matrix $\check{\mathbf{X}}$ is renamed as simply $\mathbf{X}$. The mean of each column vector in the scaled matrix $\mathbf{X}$ can be computed as

$$\frac{1}{n}\sum_{i=1}^{n}\check{x}_{ij}^k = \frac{1}{n}\sum_{i=1}^{n}\frac{x_{ij}^k - \mu_j^k}{\sigma^k} = \frac{1}{n\sigma^k}\sum_{i=1}^{n}\left(x_{ij}^k - \mu_j^k\right) \tag{13}$$

$$= \frac{1}{n\sigma^k}\left[\left(\sum_{i=1}^{n} x_{ij}^k\right) - n\mu_j^k\right] \tag{14}$$

$$= \frac{1}{n\sigma^k}\left(n\mu_j^k - n\mu_j^k\right) = 0 \tag{15}$$

Since the scaled matrix $\mathbf{X}$ is a mean-centered matrix, it is possible to calculate its covariance matrix as follows:

$$\mathbf{C}_{\mathbf{X}} = \frac{1}{n-1}\mathbf{X}^T\mathbf{X} \in \mathcal{M}_{(N\cdot L)\times(N\cdot L)}(\mathbb{R}) \tag{16}$$

The covariance matrix $\mathbf{C_X}$ is a $(N \cdot L) \times (N \cdot L)$ symmetric matrix that measures the degree of linear relationship within the data set between all possible pairs of columns. At this point it is worth noting that each column can be viewed as a virtual sensor and, therefore, each column vector $\mathbf{X}(:, j) \in \mathbb{R}^n$, $j = 1, \ldots, N \cdot L$, represents a set of measurements from one virtual sensor.

The subspaces in PCA are defined by the eigenvectors and eigenvalues of the covariance matrix as follows:

$$\mathbf{C_X P} = \mathbf{P}\Lambda \tag{17}$$

where the columns of $\mathbf{P} \in \mathscr{M}_{(N \cdot L) \times (N \cdot L)}(\mathbb{R})$ are the eigenvectors of $\mathbf{C_X}$. The diagonal terms of matrix $\Lambda \in \mathscr{M}_{(N \cdot L) \times (N \cdot L)}(\mathbb{R})$ are the eigenvalues λ_i, $i = 1, \ldots, N \cdot L$, of $\mathbf{C_X}$ whereas the off-diagonal terms are zero, that is,

$$\Lambda_{ii} = \lambda_i, \ i = 1, \ldots, N \cdot L \tag{18}$$

$$\Lambda_{ij} = 0, \ i, j = 1, \ldots, N \cdot L, \ i \neq j \tag{19}$$

The eigenvectors p_j, $j = 1, \ldots, N \cdot L$, representing the columns of the transformation matrix $\mathbf{P}$ are classified according to the eigenvalues in descending order and they are called the principal components or the loading vectors of the data set. The eigenvector with the highest eigenvalue, called the first principal component, represents the most important pattern in the data with the largest quantity of information.

Matrix $\mathbf{P}$ is usually called the principal components of the data set or loading matrix and matrix $\mathbf{T}$ is the transformed or projected matrix to the principal component space, also called score matrix. Using all the $N \cdot L$ principal components, that is, in the full dimensional case, the orthogonality of $\mathbf{P}$ implies $\mathbf{PP}^{\mathrm{T}} = \mathbf{I}$, where $\mathbf{I}$ is the $(N \cdot L) \times (N \cdot L)$ identity matrix. Therefore, the projection can be inverted to recover the original data as

$$\mathbf{X} = \mathbf{TP}^T \tag{20}$$

However, the objective of PCA is, as said before, to reduce the dimensionality of the data set $\mathbf{X}$ by selecting only a limited number $\ell < N \cdot L$ of principal components, that is, only the eigenvectors related to the ℓ highest eigenvalues. Thus, given the reduced matrix

$$\hat{\mathbf{P}} = (p_1 | p_2 | \cdots | p_\ell) \in \mathscr{M}_{N \cdot L \times \ell}(\mathbb{R}) \tag{21}$$

matrix $\hat{\mathbf{T}}$ is defined as

$$\hat{\mathbf{T}} = \mathbf{X}\hat{\mathbf{P}} \in \mathscr{M}_{n \times \ell}(\mathbb{R}) \tag{22}$$

Note that opposite to $\mathbf{T}$, $\hat{\mathbf{T}}$ is no longer invertible. Consequently, it is not possible to fully recover $\mathbf{X}$ although $\hat{\mathbf{T}}$ can be projected back onto the original $N \cdot L$−dimensional space to get a data matrix $\hat{\mathbf{X}}$ as follows:

$$\hat{\mathbf{X}} = \hat{\mathbf{T}}\hat{\mathbf{P}}^T \in \mathscr{M}_{n\times(N\cdot L)}(\mathbb{R}) \tag{23}$$

The difference between the original data matrix $\mathbf{X}$ and $\hat{\mathbf{X}}$ is defined as the residual error matrix $\mathbf{E}$ or $\tilde{\mathbf{X}}$ as follows:

$$\mathbf{E} = \mathbf{X} - \hat{\mathbf{X}} \tag{24}$$

or, equivalenty,

$$\mathbf{X} = \hat{\mathbf{X}} + \mathbf{E} = \hat{\mathbf{T}}\hat{\mathbf{P}}^T + \mathbf{E} \tag{25}$$

The residual error matrix $\mathbf{E}$ describes the variability not represented by the data matrix $\hat{\mathbf{X}}$, and can also be expressed as

$$\mathbf{E} = \mathbf{X}(\mathbf{I} - \hat{\mathbf{P}}\hat{\mathbf{P}}^T) \tag{26}$$

Even though the real measures obtained from the sensors as a function of time represent physical magnitudes, when these measures are projected and the scores are obtained, these scores no longer represent any physical magnitude [3]. The key aspect in this approach is that the scores from different experiments can be compared with the reference pattern to try to detect a different behavior.

3.2 Fault Detection Based on Univariate Hypothesis Testing

The current structure to diagnose—in Sects. 3.2 and 3.3 we will refer to a *structure* as a generic noun for both the aluminium plate, the wind turbine or more complex mechanical systems—is subjected to the same excitation (guided waves) or to a wind field (wind turbines) as described in Sects. 3.1.1 and 3.1.2. When the measures are obtained from $N \in \mathbb{N}$ sensors at L discretization instants and ν experimental trials (guides waves) or during $(\nu L - 1)\Delta$ seconds (wind turbines), a new data matrix $\mathbf{Y}$ is constructed as in Eqs. (4) and (7), respectively:

$$\mathbf{Y} = \left(\begin{array}{cccc|ccc|c|ccc} y_{11}^1 & y_{12}^1 & \cdots & y_{1L}^1 & y_{11}^2 & \cdots & y_{1L}^2 & \cdots & y_{11}^N & \cdots & y_{1L}^N \\ \vdots & \vdots & \ddots & \vdots & \vdots & \ddots & \vdots & \ddots & \vdots & \ddots & \vdots \\ y_{i1}^1 & y_{i2}^1 & \cdots & y_{iL}^1 & y_{i1}^2 & \cdots & y_{iL}^2 & \cdots & y_{i1}^N & \cdots & y_{iL}^N \\ \vdots & \vdots & \ddots & \vdots & \vdots & \ddots & \vdots & \ddots & \vdots & \ddots & \vdots \\ y_{\nu 1}^1 & y_{\nu 2}^1 & \cdots & y_{\nu L}^1 & y_{\nu 1}^2 & \cdots & y_{\nu L}^2 & \cdots & y_{\nu 1}^N & \cdots & y_{\nu L}^N \end{array}\right) \in \mathscr{M}_{\nu\times(N\cdot L)}(\mathbb{R}) \tag{27}$$

It is worth remarking that the natural number ν (the number of rows of matrix $\mathbf{Y}$) is not necessarily equal to n (the number of rows of $\mathbf{X}$), but the number of columns of $\mathbf{Y}$ must agree with that of $\mathbf{X}$; that is, in both cases the number N of sensors and the number of samples per row must be equal.

Before the collected data arranged in matrix $\mathbf{Y}$ is projected into the new space spanned by the eigenvectors in matrix $\mathbf{P}$ in Eq. (17), the matrix has to be scaled to define a new matrix $\check{\mathbf{Y}}$ as in Eq. (12):

$$\check{y}_{ij}^k := \frac{y_{ij}^k - \mu_j^k}{\sigma^k}, \; i = 1, \ldots, \nu, \; j = 1, \ldots, L, \; k = 1, \ldots, N, \tag{28}$$

where μ_j^k and σ^k are defined in Eqs. (9) and (11), respectively.

The projection of each row vector

$$r^i = \check{\mathbf{Y}}(i, :) \in \mathbb{R}^{N \cdot L}, \; i = 1, \ldots, \nu \tag{29}$$

of matrix $\check{\mathbf{Y}}$ into the space spanned by the eigenvectors in $\hat{\mathbf{P}}$ is performed through the following vector to matrix multiplication:

$$t^i = r^i \cdot \hat{\mathbf{P}} \in \mathbb{R}^{\ell}. \tag{30}$$

For each row vector r^i, $i = 1, \ldots, \nu$, the first component of vector t^i is called the *first score* or *score* 1; similarly, the second component of vector t^i is called the *second score* or *score* 2, and so on.

In a standard application of the principal component analysis strategy in the field of structural health monitoring, the scores allow a visual grouping or separation [2]. In some other cases, as in [19], two classical indices can be used for damage detection, such as the Q index (also known as SPE, *square prediction error*) and the Hotelling's T^2 index. The Q index of the ith row y_i^T of matrix $\check{\mathbf{Y}}$ is defined as follows:

$$Q_i = y_i^T (\mathbf{I} - \hat{\mathbf{P}}\hat{\mathbf{P}}^T) y_i. \tag{31}$$

The T^2 index of the ith row y_i^T of matrix $\check{\mathbf{Y}}$ is defined as follows:

$$T_i^2 = y_i^T (\hat{\mathbf{P}} \Lambda^{-1} \hat{\mathbf{P}}^T) y_i \tag{32}$$

In this case, however, it can be observed in Fig. 6—where the projection onto the two first principal components of samples coming from the healthy and faulty wind turbines are plotted—that a visual grouping, clustering or separation cannot be performed. A similar conclusion is deducted from Fig. 7. In this case, the plot of the natural logarithm of indices Q and T^2—defined in Eqs. (31) and (32)—of samples coming from the healthy and faulty wind turbines does not allow any visual grouping. A visual separation is neither possible from Fig. 8, where the first score for baseline experiments of the healthy aluminium plate are plotted together with testing experiments with several damages. Some strategies can be found in the literature

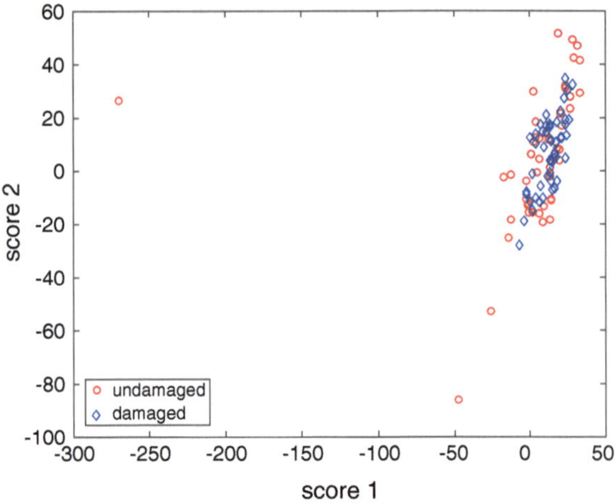

Fig. 6 Projection onto the two first principal components of samples coming from the healthy wind turbine (red, circle) and from the faulty wind turbine (blue, diamond)

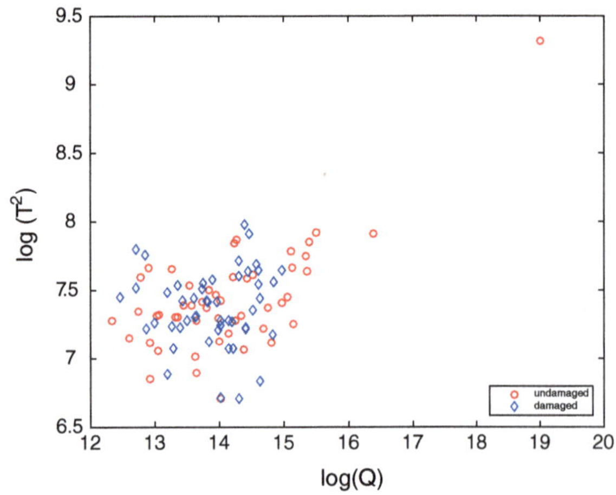

Fig. 7 Natural logarithm of indices Q and T^2 of samples coming from the healthy wind turbine (red, circle) and from the faulty wind turbine (blue, diamond)

with the objective to overcome these difficulties. For instance, principal component analysis together with self-organizing maps SOM [20], a robust version of principal component analysis (RPCA) in the presence of outliers [21] or even nonlinear PCA (NPCA) or hierarchical PCA (HPCA) [22]. Some of these approaches have a high computational cost that can lead to delays in the damage or fault diagnosis [16]. Therefore, the methodologies reviewed in this work can be seen as a powerful and reliable tool with less computational cost with the aim of online damage and fault detection of structures using principal component analysis.

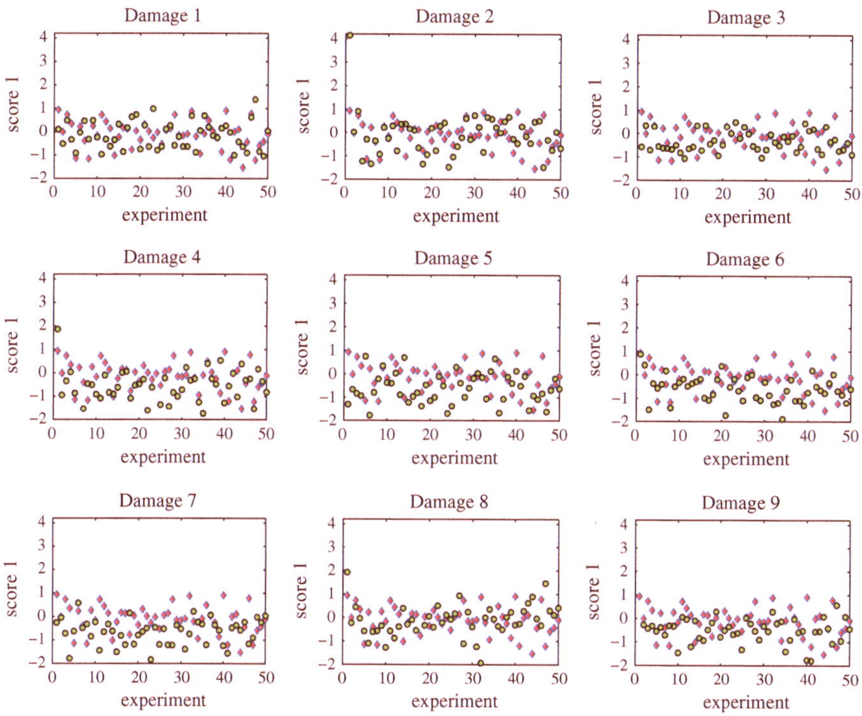

Fig. 8 First score for baseline experiments (diamonds) and testing experiments (circles)

3.2.1 The Random Nature of the Scores

Since the dynamic response of a structure (guided waves) and the turbulent wind (wind turbine) can be considered as a random process, the dynamic response of the structure (aluminium plate and wind turbine) can be considered as a stochastic process and the measurements in r^i are also stochastic. Therefore, each component of t^i in Eq. (30) acquires this stochastic nature and it will be regarded as a random variable to construct the stochastic approach in this chapter.

3.2.2 Test for the Equality of Means

The objective of the present work is to examine whether the current structure to diagnosed is healthy or subjected to a damage (aluminium plate) or to a fault as those described in Table 3 (wind turbine). To achieve this end, we have a PCA model (matrix $\hat{\mathbf{P}}$ in Eq. (21)) built as in Sect. 3.1.3 with data coming from a structure or a wind turbine in a full healthy state. For each principal component $j = 1, \ldots, \ell$, the baseline sample is defined as the set of n real numbers computed as the j−th component of the vector to matrix multiplication $\mathbf{X}(i, :) \cdot \hat{\mathbf{P}}$. Note that n is the number of rows of matrix $\mathbf{X}$ in Eq. (7). That is, we define the baseline sample as the set of numbers $\{\tau_j^i\}_{i=1,\ldots,n}$ given by

$$\tau_j^i := (\mathbf{X}(i,:) \cdot \hat{\mathbf{P}})(j) = \mathbf{X}(i,:) \cdot \hat{\mathbf{P}} \cdot \mathbf{e}_j, \; i = 1, \ldots, n, \tag{33}$$

where $\mathbf{e}_j$ is the j−th vector of the canonical basis.

Similarly, and for each principal component $j = 1, \ldots, \ell$, the sample of the current structure to diagnose is defined as the set of ν real numbers computed as the j−th component of the vector t^i in Eq. (30). Note that ν is the number of rows of matrix $\mathbf{Y}$ in Eq. (27). That is, we define the sample to diagnose as the set of numbers $\{t_j^i\}_{i=1,\ldots,\nu}$ given by

$$t_j^i := t^i \cdot \mathbf{e}_j, \; i = 1, \ldots, \nu. \tag{34}$$

As said before, the goal of this chapter is to obtain a damage and fault detection method such that when the distribution of the current sample is related to the distribution of the baseline sample a healthy state is predicted and otherwise a damage or fault is detected. To that end, a test for the equality of means will be performed. Let us consider that, for a given principal component, (a) the baseline sample is a random sample of a random variable having a normal distribution with unknown mean μ_X and unknown standard deviation σ_X; and (b) the random sample of the current structure is also normally distributed with unknown mean μ_Y and unknown standard deviation σ_Y. Let us finally consider that the variances of these two samples are not necessarily equal. As said previously, the problem that we will consider is to determine whether these means are equal, that is, $\mu_X = \mu_Y$, or equivalently, $\mu_X - \mu_Y = 0$. This statement leads immediately to a test of the hypotheses

$$H_0 : \mu_X - \mu_Y = 0 \text{ versus} \tag{35}$$

$$H_1 : \mu_X - \mu_Y \neq 0 \tag{36}$$

that is, the null hypothesis is "the sample of the structure to be diagnosed is distributed as the baseline sample" and the alternative hypothesis is "the sample of the structure to be diagnosed is not distributed as the baseline sample". In other words, if the result of the test is that the null hypothesis is not rejected, the current structure is categorized as healthy. Otherwise, if the null hypothesis is rejected in favor of the alternative, this would indicate the presence of some damage or faults in the structure.

The test is based on the Welch-Satterthwaite method [23], which is outlined below. When random samples of size n and ν, respectively, are taken from two normal distributions $N(\mu_X, \sigma_X)$ and $N(\mu_Y, \sigma_Y)$ and the population variances are unknown, the random variable

$$\mathscr{W} = \frac{(\bar{X} - \bar{Y}) + (\mu_X - \mu_Y)}{\sqrt{\left(\frac{S_X^2}{n} + \frac{S_Y^2}{\nu}\right)}} \tag{37}$$

can be approximated with a t-distribution with ρ degrees of freedom, that is

$$\mathscr{W} \hookrightarrow t_\rho, \tag{38}$$

where

$$\rho = \left\lfloor \frac{\left(\frac{s_X^2}{n} + \frac{s_Y^2}{\nu}\right)^2}{\frac{(s_X^2/n)^2}{n-1} + \frac{(s_Y^2/\nu)^2}{\nu-1}} \right\rfloor \tag{39}$$

and where $\bar{X}$, $\bar{Y}$ is the sample mean as a random variable; S^2 is the sample variance as a random variable; s^2 is the variance of a sample; and $\lfloor\cdot\rfloor$ is the floor function.

The value of the standardized test statistic using this method is defined as

$$t_{\text{obs}} = \frac{\bar{x} - \bar{y}}{\sqrt{\left(\frac{s_X^2}{n} + \frac{s_Y^2}{\nu}\right)}} \tag{40}$$

where $\bar{x}$, $\bar{y}$ is the mean of a particular sample. The quantity t_{obs} is the damage or fault indicator. We can then construct the following test:

$$|t_{\text{obs}}| \leq t^\star \implies \text{Fail to reject } H_0 \tag{41}$$

$$|t_{\text{obs}}| > t^\star \implies \text{Reject } H_0, \tag{42}$$

where $t^\star$ is such that

$$\mathbb{P}\left(t_\rho < t^\star\right) = 1 - \frac{\alpha}{2} \tag{43}$$

where $\mathbb{P}$ is a probability measure and α is the chosen risk (significance) level for the test. More precisely, the null hypothesis is rejected if $|t_{\text{obs}}| > t^\star$ (this would indicate the existence of a damage or fault in the structure). Otherwise, if $|t_{\text{obs}}| \leq t^\star$ there is no statistical evidence to suggest that both samples are normally distributed but with different means, thus indicating that no damage or fault in the structure has been found. This idea is represented in Fig. 9.

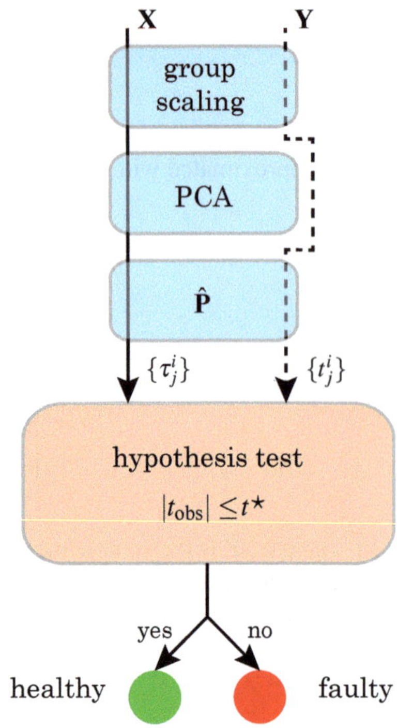

Fig. 9 Fault detection will be based on testing for significant changes in the distributions of the baseline sample and the sample coming from the wind turbine to diagnose

3.3 Fault Detection Based on Multivariate Hypothesis Testing

In this section, the projections onto the first components —the so-called *scores*— are used for the construction of the multivariate random samples to be compared and consequently to obtain the structural damage or fault indicator, as it is illustrated in Figs. 10 (guided waves) and 11 (wind turbine).

3.3.1 Multivariate Random Variables and Multivariate Random Samples

As in Sect. 3.2, the current structure to diagnose is subjected to the same excitation (guided waves) or to a wind field (wind turbines). The time responses recorded by the sensors are arranged in a matrix $\mathbf{Y} \in \mathscr{M}_{\nu \times (N \cdot L)}(\mathbb{R})$ as in Eq. (27). The rows of matrix $\mathbf{Y}$ are called $r^i \in \mathbb{R}^{N \cdot L}$, $i = 1, \ldots, \nu$, as in Eq. (29), where N is the number of sensors, L is the number of discretization instants and ν is the number of experimental trials (guides waves) or the number of rows of matrix $\mathbf{Y}$ in Eq. (27). Selecting the jth principal component, v_j, $j = 1, \ldots, \ell$, the projection of the recorded data onto this principal component is the dot product

Fig. 10 The structure to be diagnosed is subjected to a predefined number of experiments and a data matrix $\mathbf{X}_{\mathrm{GW}}$ is constructed. This matrix is projected onto the baseline PCA model $\mathbf{P}$ to obtain the projections onto the first components $\mathbf{T}$

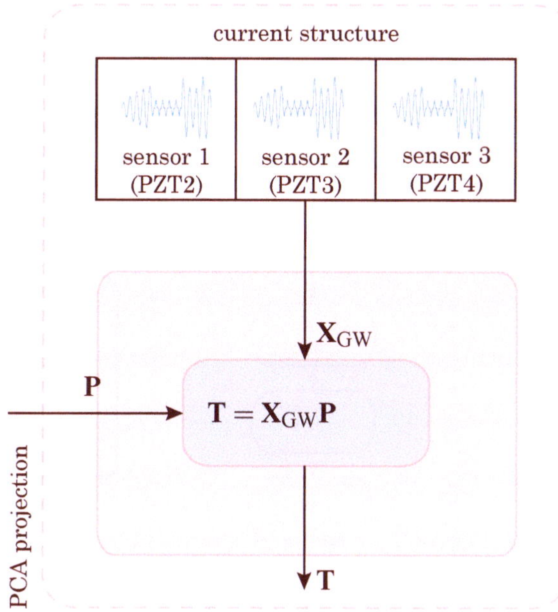

$$t_j^i = r^i \cdot v_j \in \mathbb{R}, \; i = 1, \ldots, \nu, \; j = 1, \ldots, \ell \tag{44}$$

as in Eq. (34).

Since the dynamic behaviour of a structure depends on some indeterminacy, its dynamic response can be considered as a stochastic process and the measurements in r^i are also stochastic. On the one hand, t_j^i acquires this stochastic nature and it will be regarded as a random variable to construct the stochastic approach in this section. On the other hand, an s-dimensional random vector can be defined by considering the projections onto several principal components as follows

$$\mathbf{t}^i_{j_1,\ldots,j_s} = \left[t^i_{j_1} \; t^i_{j_2} \; \cdots \; t^i_{j_s} \right]^T \in \mathbb{R}^s, \tag{45}$$
$$i = 1, \ldots, \nu, \; s \in \mathbb{N}, \; j_1, \ldots, j_s \in \{1, \ldots, \ell\}, \; j_\alpha \neq j_\beta \text{ if } \alpha \neq \beta.$$

The set of s-dimensional vectors $\left\{ \mathbf{t}^i_{j_1,\ldots,j_s} \right\}_{i=1,\ldots,\nu}$ can be seen as a realization of a multivariate random sample of the variable $\mathbf{t}_{j_1,\ldots,j_s}$. When the realization is performed on the healthy structure, the baseline sample is denoted as the set of s-dimensional vectors

$$\left\{ \tau^i_{j_1,\ldots,j_s} \right\}_{i=1,\ldots,n},$$

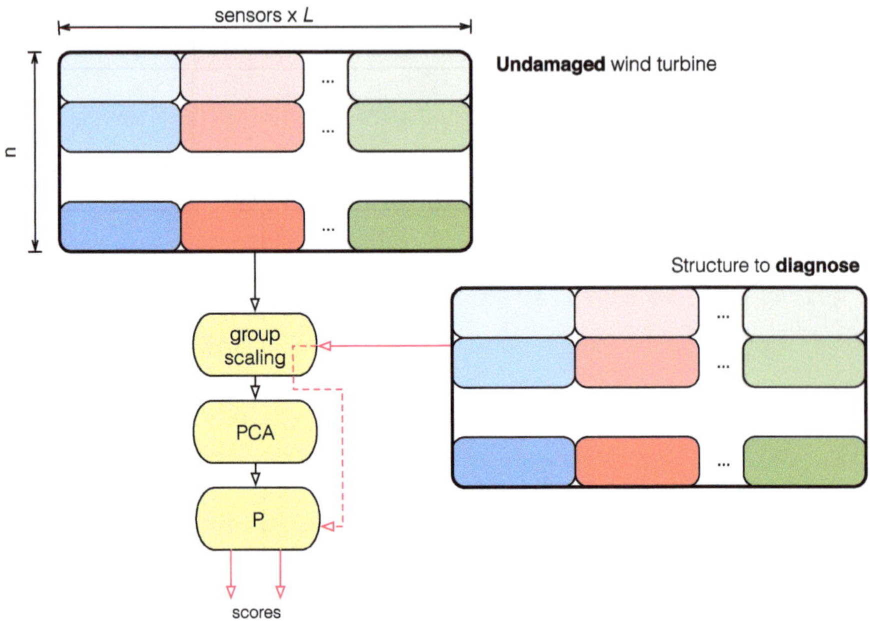

Fig. 11 The current wind turbine to diagnose is subjected to a wind field. Then the collected data is projected into the new space spanned by the eigenvectors in matrix **P**

where n is the number of rows of matrix $\mathbf{X}$ in Eqs. (4) (guides waves) and (7) (wind turbine). As an example, in the case of the aluminium plate experimental set-up, in Fig. 12 two three-dimensional samples are represented; one is the three-dimensional baseline sample (left) and the other is referred to damages 1 to 3 (right). This illustrating example refers to actuator phase 1 and the first, second and third principal components. More precisely, Fig. 12 (right) depicts the values of the multivariate random variable $\mathbf{t}_{1,2,3}$. The diagnosis sample is formed by 20 experiments and the baseline sample is made by 100 experiments.

3.3.2 Detection Phase and Testing for Multivariate Normality

In this work, the framework of multivariate statistical inference is used with the objective of the classification of structures in healthy or damaged. With this goal, a test for multivariate normality is first performed. A test for the plausibility of a value for a normal population mean vector is then performed.

Many statistical tests and graphical approaches are available to check the multivariate normality assumption [24]. But there is no a single most powerful test and it is recommended to perform several tests to assess the multivariate normality. Let us consider the three most widely used multivariate normality tests. That is: (i) Mardia's;

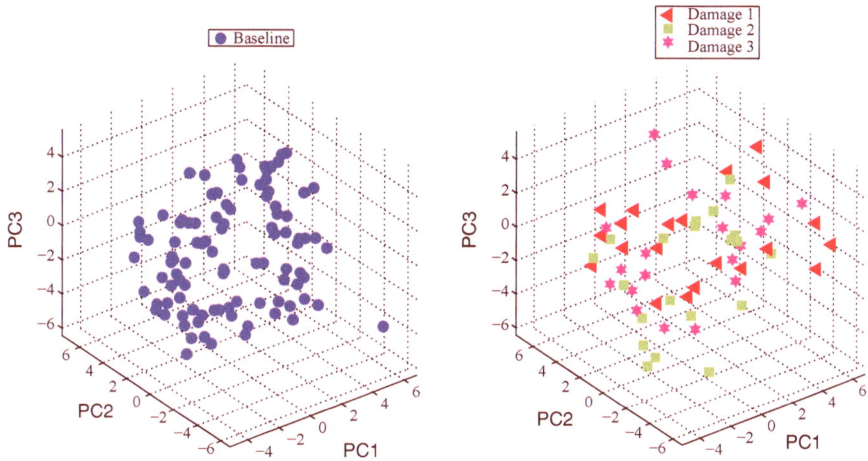

Fig. 12 Baseline sample (left) and sample from the structure to be diagnosed (right)

(ii) Henze-Zirkler's; and (iii) Royston's multivariate normality tests. We include a brief explanation of these methods for the sake of completeness.

Mardia's test

Mardia's test is based on multivariate extensions of skewness ($\hat{\gamma}_{1,s}$) and kurtosis ($\hat{\gamma}_{2,s}$) measures [24, 25]:

$$\hat{\gamma}_{1,s} = \frac{1}{\nu^2} \sum_{i=1}^{\nu} \sum_{j=1}^{\nu} m_{ij}^3,$$

$$\hat{\gamma}_{2,s} = \frac{1}{\nu} \sum_{i=1}^{\nu} m_{ij}^2,$$

where $m_{ij} = (x_i - \bar{x})^T S^{-1} \left(x_j - \bar{x}\right)$, $i, j = 1, \ldots, \nu$ is the squared Mahalanobis distance, S is the variance-covariance matrix, s is the number of variables and ν is the sample size. The test statistic for skewness, $(\nu/6)\, \hat{\gamma}_{1,s}$, is approximately χ^2 distributed with $s\,(s+1)\,(s+2)\,/6$ degrees of freedom. Similarly, the test statistic for kurtosis, $\hat{\gamma}_{2,s}$, is approximately normally distributed with mean $s\,(s+2)$ and variance $8s\,(s+2)\,/\nu$. For multivariate normality, both p-values of skewness and kurtosis statistics should be greater than 0.05.

For small samples, the power and the type I error could be violated. Therefore, Mardia introduced a correction term into the skewness test statistic [26], usually when $\nu < 20$, in order to control type I errors. The corrected skewness statistic for small samples is $(\nu k/6)\, \hat{\gamma}_{1,s}$, where

$$k = (s+1)\,(\nu+1)\,(\nu+3)\,/\,(\nu\,(\nu+1)\,(s+1) - 6)\,.$$

This statistic is also χ^2 distributed with $s\,(s+1)\,(s+2)\,/6$ degrees of freedom.

Henze-Zirkler's test

The Henze-Zirkler's test is based on a non-negative functional distance that measures the distance between two distribution functions [25, 27]. If the data is multivariate normal distributed, the test statistic *HZ* in Eq. (46) is approximately lognormally distributed. It proceeds to calculate the mean, variance and smoothness parameter. Then, mean and variance are lognormalized and the *p*-value is estimated. The test statistic of Henze-Zirkler's multivariate normality test is

$$HZ = \frac{1}{\nu}\sum_{i=1}^{\nu}\sum_{j=1}^{\nu} e^{-\frac{\beta^2}{2}D_{ij}} - 2\left(1+\beta^2\right)^{-\frac{s}{2}}\sum_{i=1}^{\nu} e^{-\frac{\beta^2}{2\left(1+\beta^2\right)}D_i} + \nu\left(1+\beta^2\right)^{-\frac{s}{2}}, \quad (46)$$

where s is the number of variables,

$$\beta = \frac{1}{\sqrt{2}}\left(\frac{\nu(2s+1)}{4}\right)^{\frac{1}{s+4}},$$
$$D_{ij} = \left(x_i - x_j\right)^T S^{-1}\left(x_i - x_j\right),\ i,j = 1,\ldots,\nu,$$
$$D_i = \left(x_i - \bar{x}\right)^T S^{-1}\left(x_i - \bar{x}\right) = m_{ii},\ i = 1,\ldots,\nu.$$

D_i gives the squared Mahalanobis distance of the ith observation to the centroid and D_{ij} gives the Mahalanobis distance between the ith and the jth observations. If data are multivariate normal distributed, the test statistic (*HZ*) is approximately lognormally distributed with mean μ and variance σ^2 as given below:

$$\mu = 1 - \frac{a^{-\frac{s}{2}}\left(1 + s\beta^{\frac{2}{a}} + s(s+2)\beta^4\right)}{2a^2},$$
$$\sigma^2 = 2\left(1+4\beta^2\right)^{-\frac{s}{2}} + \frac{2a^{-s}\left(1+2s\beta^4\right)}{a^2} + \frac{3s(s+2)\beta^8}{4a^4}$$
$$- 4w_\beta^{-\frac{s}{2}}\left(1 + \frac{3s\beta^4}{2w_\beta} + \frac{s(s+2)\beta^8}{2w_\beta^2}\right),$$

where $a = 1 + 2\beta^2$ and $w_\beta = \left(1+\beta^2\right)\left(a + 3\beta^2\right)$. Hence, the lognormalized mean and variance of the *HZ* statistic can be defined as follows:

$$\mu_{\log} = \ln\left(\sqrt{\frac{\mu^4}{\sigma^2+\mu^2}}\right),$$
$$\sigma^2_{\log} = \ln\left(\frac{\sigma^2+\mu^2}{\sigma^2}\right).$$

By using the lognormal distribution parameters, $\mu_{\log}$ and $\sigma^2_{\log}$, we can test the significance of multivariate normality. The Wald test statistic for multivariate normality is given in the following equation:

$$z = \frac{\ln(HZ) - \mu_{\log}}{\sqrt{\sigma^2_{\log}}}. \tag{47}$$

Royston's test

Royston's test uses the Shapiro-Wilk/Shapiro-Francia statistic to test multivariate normality [25]. If kurtosis of the data is greater than 3, then it uses the Shapiro-Francia test for leptokurtic distributions. Otherwise, it uses the Shapiro-Wilk test for platykurtic distributions. The Shapiro-Wilk test statistic is:

$$W = \frac{\left(\sum_{i=1}^{\nu} \left(a_i \cdot x_{(i)}\right)\right)^2}{\sum_{i=1}^{\nu} (x_i - \mu)^2},$$

where $x_{(i)}$ is the ith order statistic, that is, the ith-smallest number in the sample, μ is the mean, $a = \frac{\mathbf{m}^T V^{-1}}{\sqrt{\mathbf{m}^T V^{-1} V^{-1} \mathbf{m}}}$, V is the covariance matrix of the order statistics of a sample of s standard normal random variables with expectation vector $\mathbf{m}$. Let W_j be the Shapiro-Wilk/Shapiro-Francia test statistic for the jth variable, $j = 1, \ldots, s$, and Z_j be the values obtained from the normality transformation proposed by [28]:

$$\begin{aligned} &\text{if} \quad 4 \leq \nu \leq 11 \quad \text{then} \quad x = \nu \quad \text{and} \quad w_j = -\ln\left(\gamma - \ln\left(1 - W_j\right)\right) \\ &\text{if} \quad 12 \leq \nu \leq 2000 \quad \text{then} \quad x = \ln(\nu) \quad \text{and} \quad w_j = \ln\left(1 - W_j\right). \end{aligned}$$

Then transformed values of each random variable can be obtained from the following equation:

$$Z_j = \frac{w_j - \mu}{\sigma}, \tag{48}$$

where γ, μ and σ are derived from the polynomial approximations given in equations [28]:

$$\begin{aligned} \text{if} \quad 4 \leq \nu \leq 11 \quad \gamma &= -2.273 + 0.459x, \\ \mu &= 0.544 - 0.39978x + 0.025054x^2 - 0.0006714x^3, \\ \ln(\sigma) &= 1.3822 - 0.77857x + 0.062767x^2 - 0.0020322x^3, \\ \text{if} \quad 12 \leq \nu \leq 2000 \quad \mu &= -1.5861 - 0.31082x - 0.083751x^2 + 0.0038915x^3, \\ \ln(\sigma) &= -0.4803 - 0.082676x + 0.0030302x^2. \end{aligned}$$

The Royston's test statistic for multivariate normality is then defined as follows:

$$H = \frac{\varepsilon \sum_{j=1}^{s} \psi_j}{s} \sim \chi^2_{\varepsilon},$$

where ε is the equivalent degrees of freedom (edf) and $\Phi(\cdot)$ is the cumulative distribution function for standard normal distribution such that,

$$\varepsilon = s/\left(1 + (s-1)\bar{c}\right),$$
$$\psi_j = \left(\Phi^{-1}\left(\Phi\left(-Z_j\right)/2\right)\right)^2, \quad j = 1, 2, ..., s.$$

Another extra term $\bar{c}$ has to be calculated in order to continue with the statistical significance of Royston's test statistic. Let R be the correlation matrix and r_{ij} be the correlation between ith and jth variables. Then, the extra term can be found by using equation:

$$\bar{c} = \sum_{i=1}^{s} \sum_{j \neq i} \frac{c_{ij}}{s(s-1)}, \tag{49}$$

where

$$c_{ij} = g\left(r_{ij}, \nu\right) \tag{50}$$

with the boundaries of $g(\cdot)$ as $g(0, \nu) = 0$ and $g(1, \nu) = 1$. The function $g(\cdot)$ is defined as follows:

$$g(r, \nu) = r^{\lambda}\left(1 - \frac{\mu}{\xi(\nu)}(1-r)^{\mu}\right). \tag{51}$$

The unknown parameters μ, λ and ξ were estimated from a simulation study conducted by [28]. He found $\mu = 0.715$ and $\lambda = 5$ for sample size $10 \leq \nu \leq 2000$ and ξ is a cubic function which can be obtained as follows:

$$\xi(\nu) = 0.21364 + 0.015124 \ln^2(\nu) - 0.0018034 \ln^3(\nu). \tag{52}$$

Quantile-quantile plot

Apart from the multivariate normality tests, some *visual* representations can also be used to test for multivariate normality. The quantile–quantile (Q–Q) plot is a widely used graphical approach to evaluate the agreement between two probability distributions [24, 25]. Each axis refers to the quantiles of probability distributions to be compared, where one of the axes indicates theoretical quantiles (hypothesized quantiles) and the other indicates the observed quantiles. If the observed data fit hypothesized distribution, the points in the Q–Q plot will approximately lie on the bisectrix $y = x$. The sample quantiles for the Q–Q plot are obtained as follows. First we rank the observations $y_1, y_2, \ldots, y_\nu$ and denote the ordered values by $y_{(1)}, y_{(2)}, \ldots, y_{(\nu)}$; thus $y_{(1)} \leq y_{(2)} \leq \cdots \leq y_{(\nu)}$. Then the point $y_{(i)}$ is the i/ν sample quantile. The fraction i/ν is often changed to $(i - 0.5)/\nu$ as a continuity correction. With this convention, $y_{(i)}$ is designated as the $(i - 0.5)/\nu$ sample quantile. The population quantiles for the Q–Q plot are similarly defined corresponding to $(i - 0.5)/\nu$. If we denote these by $q_1, q_2, \ldots, q_\nu$, then q_i is the value below which a proportion $(i - 0.5)/\nu$ of the observations in the population lie; that is, $(i - 0.5)/\nu$ is the probability of getting an observation less than or equal to q_i. Formally, q_i can be found for the standard normal random variable Y with distribution $N(0, 1)$ by solving

$$\Phi(q_i) = P(Y < q_i) = \frac{i - 0.5}{\nu} \tag{53}$$

which would require numerical integration or tables of the cumulative standard normal distribution, $\Phi(x)$. Another benefit of using $(i - 0.5)/\nu$ instead of i/ν is that $\nu/\nu = 1$ would make $q_\nu = +\infty$. The population need not have the same mean and variance as the sample, since changes in mean and variance merely change the slope and intercept of the plotted lie in the Q–Q plot. Therefore, we use the standard normal distribution, and the q_i values can easily be found from a table of cumulative standard normal probabilities. We then plot the pairs $(q_i, y_{(i)})$ and examine the resulting Q–Q plot for linearity.

Contour plot

In addition to Q–Q plots, creating perspective and contour plots can be also useful [24, 25]. The perspective plot is an extension of the univariate probability distribution curve into a three-dimensional probability distribution surface related with bivariate distributions. It also gives information about where data are gathered and how two variables are correlated with each other. It consists of three dimensions where two dimensions refer to the values of the two variables and the third dimension, which is likely in univariate cases, is the value of the multivariate normal probability density function. Another alternative graph, which is called the *contour plot*, involves the projection of the perspective plot into a two-dimensional space and this can be used for checking multivariate normality assumption. Figure 13 shows the contour plot for bivariate normal distribution with mean $\begin{pmatrix} 0 & 0 \end{pmatrix}^T \in \mathbb{R}^2$ and covariance matrix $\begin{pmatrix} 0.250 & 0.375 \\ 0.375 & 1.000 \end{pmatrix} \in \mathcal{M}_{2\times 2}(\mathbb{R})$. For bivariate normally distributed data, we expect to obtain a three-dimensional bell-shaped graph from the perspective plot. Similarly, in the contour plot, we can observe a similar pattern.

3.3.3 Testing a Multivariate Mean Vector

The objective of this section is to determine whether the distribution of the multivariate random samples that are obtained from the structure to be diagnosed (undamaged or not, faulty or not) is connected to the distribution of the baseline. To this end, a

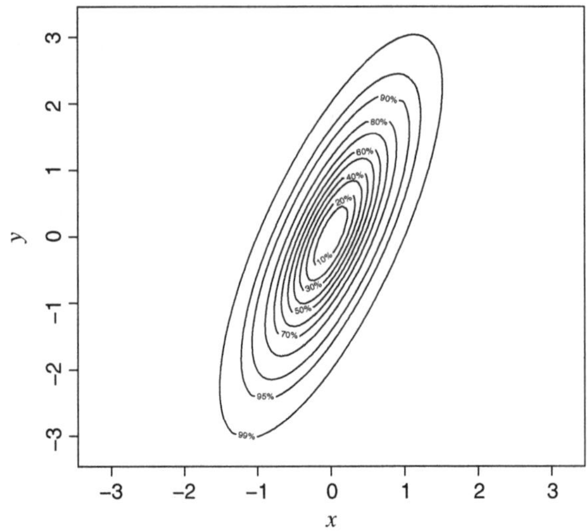

Fig. 13 Contour plot for a bivariate normal distribution. The ellipses denote places of equal probability for the distribution and provide confidence regions with different probabilities

test for the plausibility of a value for a normal population mean vector will be performed. Let $s \in \mathbb{N}$ be the number of principal components that will be considered jointly. We will also consider that: (a) the baseline projection is a multivariate random sample of a multivariate random variable following a multivariate normal distribution with known population mean vector $\boldsymbol{\mu}_{\mathrm{h}} \in \mathbb{R}^s$ and known variance-covariance matrix $\boldsymbol{\Sigma} \in \mathscr{M}_{s\times s}(\mathbb{R})$; and (b) the multivariate random sample of the structure to be diagnosed also follows a multivariate normal distribution with unknown multivariate mean vector $\boldsymbol{\mu}_{\mathrm{c}} \in \mathbb{R}^s$ and known variance-covariance matrix $\boldsymbol{\Sigma} \in \mathscr{M}_{s\times s}(\mathbb{R})$.

As said previously, the problem that we will consider is to determine whether a given s-dimensional vector $\boldsymbol{\mu}_{\mathrm{c}}$ is a plausible value for the mean of a multivariate normal distribution $N_s(\boldsymbol{\mu}_{\mathrm{h}}, \boldsymbol{\Sigma})$. This statement leads immediately to a test of the hypothesis

$$\begin{aligned} H_0 &: \boldsymbol{\mu}_{\mathrm{c}} = \boldsymbol{\mu}_{\mathrm{h}} \text{ versus} \\ H_1 &: \boldsymbol{\mu}_{\mathrm{c}} \neq \boldsymbol{\mu}_{\mathrm{h}}, \end{aligned}$$

that is, the null hypothesis is 'the multivariate random sample of the structure to be diagnosed is distributed as the baseline projection' and the alternative hypothesis is 'the multivariate random sample of the structure to be diagnosed is not distributed as the baseline projection'. In other words, if the result of the test is that the null hypothesis is not rejected, the current structure is categorized as healthy. Otherwise, if the null hypothesis is rejected in favor of the alternative, this would indicate the presence of some structural changes or faults in the structure.

The test is based on the statistic T^2—also called Hotelling's T^2—and it is summarized below. When a multivariate random sample of size $\nu \in \mathbb{N}$ is taken from a multivariate normal distribution $N_s(\boldsymbol{\mu}_{\mathrm{h}}, \boldsymbol{\Sigma})$, the random variable

$$T^2 = \nu \left(\bar{\mathbf{X}} - \boldsymbol{\mu}_{\rm h} \right)^T \mathbf{S}^{-1} \left(\bar{\mathbf{X}} - \boldsymbol{\mu}_{\rm h} \right)$$

is distributed as

$$T^2 \hookrightarrow \frac{(\nu - 1)s}{\nu - s} F_{s,\nu-s},$$

where $F_{s,\nu-s}$ denotes a random variable with an F-distribution with s and $\nu - s$ degrees of freedom, $\bar{\mathbf{X}}$ is the sample vector mean as a multivariate random variable; and $\frac{1}{n}\mathbf{S} \in \mathscr{M}_{s \times s}(\mathbb{R})$ is the estimated covariance matrix of $\bar{\mathbf{X}}$.

At the α level of significance, we reject H_0 in favor of H_1 if the observed

$$t^2_{\rm obs} = \nu \left(\bar{\mathbf{x}} - \boldsymbol{\mu}_{\rm h} \right)^T \mathbf{S}^{-1} \left(\bar{\mathbf{x}} - \boldsymbol{\mu}_{\rm h} \right)$$

is greater than $\frac{(\nu-1)s}{\nu-s} F_{s,\nu-s}(\alpha)$, where $F_{s,\nu-s}(\alpha)$ is the upper (100α)th percentile of the $F_{s,\nu-s}$ distribution. In other words, the quantity $t^2_{\rm obs}$ is the damage or fault indicator and the test is summarized as follows:

$$t^2_{\rm obs} \leq \frac{(\nu - 1)s}{\nu - s} F_{s,\nu-s}(\alpha) \implies \text{Fail to reject } H_0, \tag{54}$$

$$t^2_{\rm obs} > \frac{(\nu - 1)s}{\nu - s} F_{s,\nu-s}(\alpha) \implies \text{Reject } H_0, \tag{55}$$

where $F_{s,\nu-s}(\alpha)$ is such that

$$\mathbb{P}\left(F_{s,\nu-s} > F_{s,\nu-s}(\alpha) \right) = \alpha,$$

where $\mathbb{P}$ is a probability measure. More precisely, we fail to reject the null hypothesis if $t^2_{\rm obs} \leq \frac{(\nu-1)s}{\nu-s} F_{s,\nu-s}(\alpha)$, thus indicating that no structural changes or faults in the structure have been found. Otherwise, the null hypothesis is rejected in favor of the alternative hypothesis if $t^2_{\rm obs} > \frac{(\nu-1)s}{\nu-s} F_{s,\nu-s}(\alpha)$, thus indicating the existence of some structural changes or faults in the structure.

4 Results

In this section, the damage and fault detection strategies described in Sects. 3.2 and 3.3 are applied to both an aluminium plate and a simulated wind turbine. The experimental results of the damage detection strategy applied to the aluminium plate using the univariate and multivariate hypothesis testing are presented in Sects. 4.1 and 4.2, respectively. Similarly, the simulation results of the fault detection strategy applied to the wind turbine using the univariate and multivariate hypothesis testing are presented in Sects. 4.3 and 4.4, respectively.

4.1 Aluminum Plate and Univariate HT

Some experiments were performed in order to test the methods presented in Sect. 3.2. In these experiments, four piezoelectric transducer discs (PZTs) were attached to the surface of a thin aluminum plate, with dimensions 25 cm × 25 cm × 0.2 cm. Those PZTs formed a square with 144 mm per side. The plate was suspended by two elastic ropes, being isolated from environmental influences. Figures 1 (left) and 2 shows the plate hanging on the elastic ropes.

The experiments are performed in 4 independent phases: (i) piezoelectric transducer 1 (PZT1) is configured as actuator and the rest of PZTs as sensors; (ii) PZT2 as actuator; (iii) PZT3 as actuator; and (iv) PZT4 as actuator. In order to analyze the influence of each projection to the PCA model (score), the results of the first three scores have been considered. In this way, a total of 12 scenarios were examined. For each scenario, a total of 50 samples of 10 experiments each one (5 for the undamaged structure and 5 for the damaged structure with respect to each of the 9 different types of damages) plus the baseline are used to test for the equality of means, with a level of significance $\alpha = 0.30$ (the choice of this level of significance will be later on). Each set of 50 testing samples are categorized as follows: (i) number of samples from the healthy structure (undamaged sample) which were classified by the hypothesis test as 'healthy' (fail to reject H_0); (ii) undamaged sample classified by the test as 'damaged' (reject H_0); (iii) samples from the damaged structure (damaged sample) classified as 'healthy'; and (iv) damaged sample classified as 'damaged'. The results for the 12 different scenarios presented in Table 5 are organized according to the scheme in Table 4. It can be stressed from each scenario in Table 5 that the sum of the columns is constant: 5 samples in the first column (undamaged structure) and 45 samples in the second column (damaged structure).

In this table, it is worth noting that two kinds of misclassification are presented which are denoted as follows:

1. Type I error (or *false positive*), when the structure is healthy but the null hypothesis is rejected and therefore classified as damaged. The probability of committing a type I error is α, the level of significance.
2. Type II error (or *false negative*), when the structure is damaged but the null hypothesis is not rejected and therefore classified as healthy. The probability of committing a type II error is called β.

Table 4 Scheme for the presentation of the results in Table 5

	Undamaged sample (H_0)	Damaged sample (H_1)
Fail to reject H_0	Correct decision	Type II error (missing fault)
Reject H_0	Type I error (false alarm)	Correct decision

Table 5 Categorization of the samples with respect to presence or absence of damage and the result of the test, for each of the four phases and the three scores

	PZT1 act.		PZT2 act.		PZT3 act.		PZT4 act.	
	H_0	H_1	H_0	H_1	H_0	H_1	H_0	H_1
Score 1								
Fail to reject H_0	4	15	3	0	2	4	3	23
Reject H_0	1	30	2	45	3	41	2	22
Score 2								
Fail to reject H_0	4	0	1	4	4	5	5	7
Reject H_0	1	45	4	41	1	40	0	38
Score 3								
Fail to reject H_0	4	2	4	1	4	6	3	6
Reject H_0	1	43	1	44	1	39	2	39

4.1.1 Sensitivity and Specificity

Two statistical measures can be employed here to study the performance of the test: *the sensitivity* and *the specificity*. The sensitivity, also called as the power of the test, is defined, in the context of this work, as the proportion of samples from the damaged structure which are correctly identified as such. Thus, the sensitivity can be computed as $1 - \beta$. The specificity of the test is defined, also in this context, as the proportion of samples from the undamaged structure that are correctly identified and can be expressed as $1 - \alpha$.

The sensitivity and the specificity of the test with respect the 50 samples in each scenario have been included in Table 7. For each scenario in this table, the results are organized as shown in Table 6.

It is worth noting that type I errors are frequently considered to be more serious than type II errors. However, in this application a type II error is related to a *missing fault* whereas a type I error is related to a *false alarm*. In consequence, type II errors should be minimized. Therefore a small level of significance of 1, 5% or even 10% would lead to a reduced number of *false alarms* but to a higher rate of *missing faults*. That is the reason of the choice of a level of significance of 30% in the hypothesis test.

The results show that the sensitivity of the test $1 - \beta$ is close to 100%, as desired, with an average value of 86.58%. The sensitivity with respect to the projection onto

Table 6 Relationship between type I and type II errors

	Undamaged sample (H_0)	Damaged sample (H_1)
Fail to reject H_0	Specificity ($1 - \alpha$)	False negative rate (β)
Reject H_0	False positive rate (α)	Sensitivity ($1 - \beta$)

Table 7 Sensitivity and specificity of the test for each scenario

	PZT1 act.		PZT2 act.		PZT3 act.		PZT4 act.	
	H_0	H_1	H_0	H_1	H_0	H_1	H_0	H_1
Score 1								
Fail to reject H_0	0.80	0.33	0.60	0.00	0.40	0.09	0.60	0.51
Reject H_0	0.20	0.67	0.40	1.00	0.60	0.91	0.40	0.49
Score 2								
Fail to reject H_0	0.80	0.00	0.20	0.09	0.80	0.11	1.00	0.16
Reject H_0	0.20	1.00	0.80	0.91	0.20	0.89	0.00	0.84
Score 3								
Fail to reject H_0	0.80	0.04	0.80	0.02	0.80	0.13	0.60	0.13
Reject H_0	0.20	0.96	0.20	0.98	0.20	0.87	0.40	0.87

the second and third component (second and third score) is increased, in mean, to a 91.50%. The average value of the specificity is 68.33%, which is very close to the expected value of $1 - \alpha = 70\%$.

4.1.2 Reliability of the Results

The results in Table 9 are computed using the scheme in Table 8. This table is based on the Bayes' theorem [29], where $P(H_1|\text{accept } H_0)$ is the proportion of samples from the damaged structure that have been incorrectly classified as healthy (*true rate of false negatives*) and $P(H_0|\text{accept } H_1)$ is the proportion of samples from the undamaged structure that have been incorrectly classified as damaged (*true rate of false positives*).

Since these two true rates are not a function of the accuracy of the test alone, but also a function of the actual rate or frequency of occurrence within the test population, some of the results are not as good as desired. The results in Table 9 can be improved without affecting the results in Table 7 by considering an equal number of samples from the healthy structure and from the damaged structure.

Table 8 Relationship between proportion of false negative and false positives

	Undamaged sample (H_0)	Damaged sample (H_1)
Fail to reject H_0	$P(H_0|\text{accept}H_0)$	True rate of false negatives $P(H_1|\text{accept } H_0)$
Reject H_0	True rate of false positives $P(H_0|\text{accept } H_1)$	$P(H_1|\text{accept}H_1)$

Table 9 True rate of false positives and false negatives

	PZT1 act.		PZT2 act.		PZT3 act.		PZT4 act.	
	H_0	H_1	H_0	H_1	H_0	H_1	H_0	H_1
Score 1								
Fail to reject H_0	0.21	0.79	1.00	0.00	0.33	0.67	0.12	0.88
Reject H_0	0.03	0.97	0.04	0.96	0.07	0.93	0.08	0.92
Score 2								
Fail to reject H_0	1.00	0.00	0.20	0.80	0.44	0.56	0.42	0.58
Reject H_0	0.02	0.98	0.09	0.91	0.02	0.98	0.00	1.00
Score 3								
Fail to reject H_0	0.67	0.33	0.80	0.20	0.40	0.60	0.33	0.67
Reject H_0	0.02	0.98	0.02	0.98	0.03	0.97	0.05	0.95

4.1.3 The Receiver Operating Curves (ROC)

An additional study has been developed based on the ROC curves to determine the overall accuracy of the proposed method. These curves represent the trade-off between the *false positive rate* and the *sensitivity* in Table 6 for different values of the level of significance that is used in the statistical hypothesis testing. Note that the false positive rate is defined as the complementary of the specificity, and therefore these curves can also be used to visualize the close relationship between specificity and sensitivity. It can also be remarked that the sensitivity is also called true positive rate or probability of detection [30]. More precisely, for each scenario and for a given level of significance the pair of numbers

$$(\text{false positive rate, sensitivity}) \in [0, 1] \times [0, 1] \subset \mathbb{R}^2 \tag{56}$$

is plotted. We have considered 49 levels of significance within the range [0.2, 0.98] and with a difference of 0.02. Therefore, for each scenario 49 connected points are depicted, as can be seen in Fig. 14.

The placement of these points can be interpreted as follows. Since we are interested in minimizing the number of false positives while we maximize the number of true positives, these points must be placed in the upper-left corner as much as possible. However, this is impossible because there is also a relationship between the level of significance and the false positive rate. Therefore, a method can be considered acceptable if those points lie within the upper-left half-plane.

As said before, the ROC curves for all possible scenarios are depicted in Fig. 14. On one hand, in phase 1 (PZT1 as actuator) and phase 4 (PZT4 as actuator), the first score (diamonds) presents the worst performance because some points are very close to the diagonal or even below it. However, in the same phases, second and third scores present better results. It may be surprising that the results related to the first

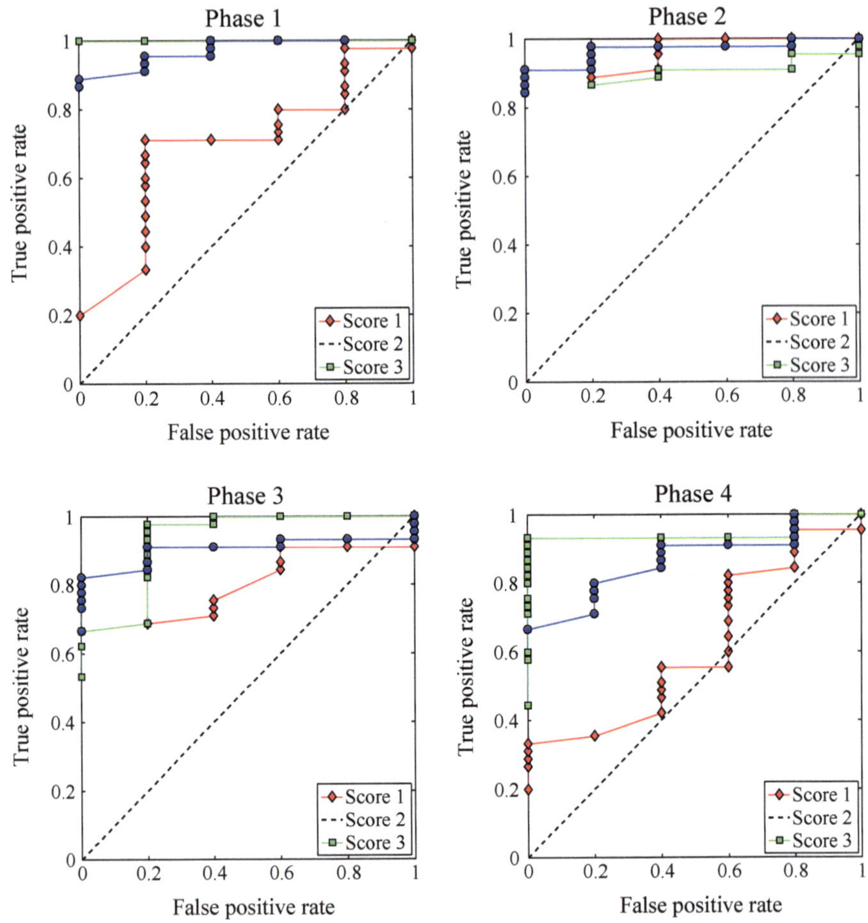

Fig. 14 The ROC curves for the three scores for each phase

score are not as good as those related to the rest of scores, but in Sect. 4.1.4 this will be justified. On the other hand, all scores in phases 2 and 3 present a very good performance to detect damages.

The curves are similar to stepped functions because we have considered 5 samples from the undamaged structure and therefore the possible values for the false positive rate (the values in the x-axis) are 0, 0.2, 0.4, 0.6, 0.8 and 1. Finally, we can say that the ROC curves provide a statistical assessment of the efficacy of a method and can be used to visualize and compare the performance of multiple scenarios.

4.1.4 Analysis and Discussion

Although the first score has the highest proportion of variance, it is not possible to visually separate between the baseline and the test. Each of the subfigures in Fig. 8 shows the comparison between the first score of the baseline experiments and the test experiment for each damage. A similar comparison can be found in Fig. 15 where all the observation points (first score of each experiment) are depicted in a single chart. The rest of the scores neither allow a visual grouping.

One of the scenarios with the worst results is the one that considers the PZT1 as actuator and the first score, because the false negative rate is 33%, the false positive rate is 20% and the true rate of false negatives is 79% (see Tables 7 and 9). These results, which are extracted from Table 5, are illustrated for each state of the structure separately in Fig. 15. Just one of the five samples of the healthy structure has been wrongly rejected (false alarm) whereas all the samples of the structure with damage D1 have been wrongly not rejected (missing fault). Only one of the five samples of the structure with damage D2 has been correctly rejected (correct decision). In this case, however, the bad result can be due to the lack of normality (Fig. 16). This lack of normality leads to results that cannot be reliable. In fact, these samples should not have been used for a hypothesis test. The samples of the structure with damage D7 are not normally distributed, although in this case the results are right. This problem can be solved by repeating the test excluding experiments with those damages (D2 and D7) or eliminating the outliers.

Contrary to what may seem reasonable, the projection on the first component (which represents the larger variance of the original data) is not always the best option to detect and distinguish damages. This fact can be explained because the PCA model is built using the data from the healthy structure and, therefore, the first component captures the maximal variance of these data. However, when new data are projected in this model, there is no longer guarantee of the existence of maximal variance in these new data.

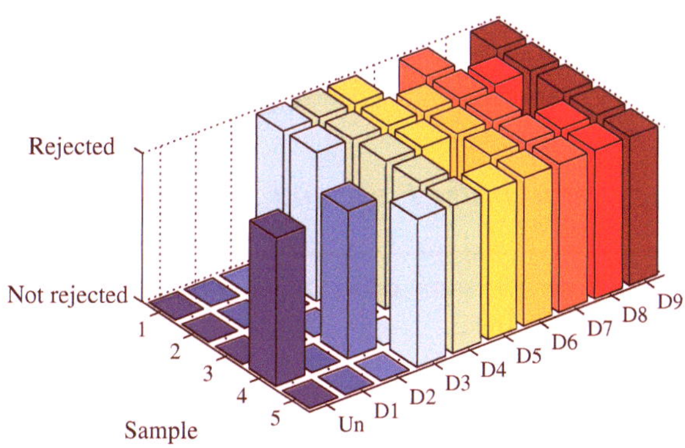

Fig. 15 Results of the hypothesis test considering the first score and PZT1 as actuator

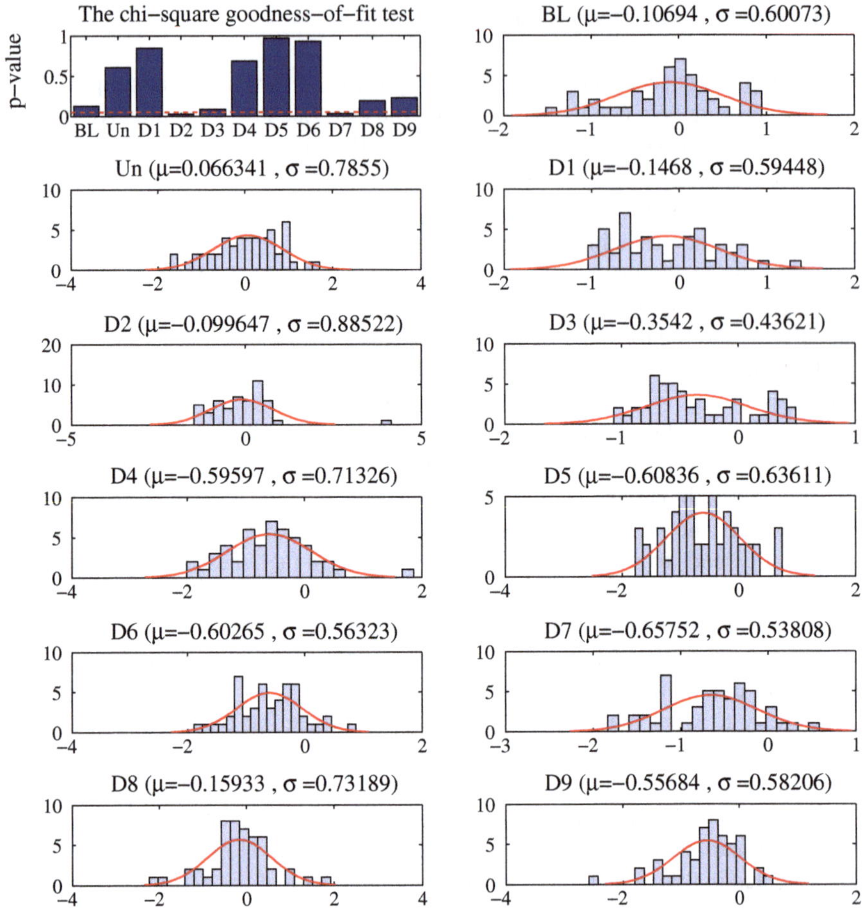

Fig. 16 Results of the chi-square goodness-of-fit test applied to the samples described Sect. 4.1. 'BL' stands for baseline projection, 'Un' for the sample obtained from the undamaged structure and 'Di' for the damage number i, where $i = 1, 2, \ldots, 9$. It can be shown by observing the upper-left barplot diagram that all the samples are normally distributed except those corresponding to damages D2 and D7

4.2 *Aluminum Plate and Multivariate HT*

As in Sect. 4.1, some experiments were performed in order to test the method presented in Sect. 3.3.

In this case, 500 experiments were performed over the healthy structure, and another 500 experiments were performed over the damaged structure with 5 damage types (100 experiments per damage type). Figure 17 shows the position of damages 1 to 5 (D1 to D5). As excitation, a 50 kHz sinusoidal signal modulated by a hamming

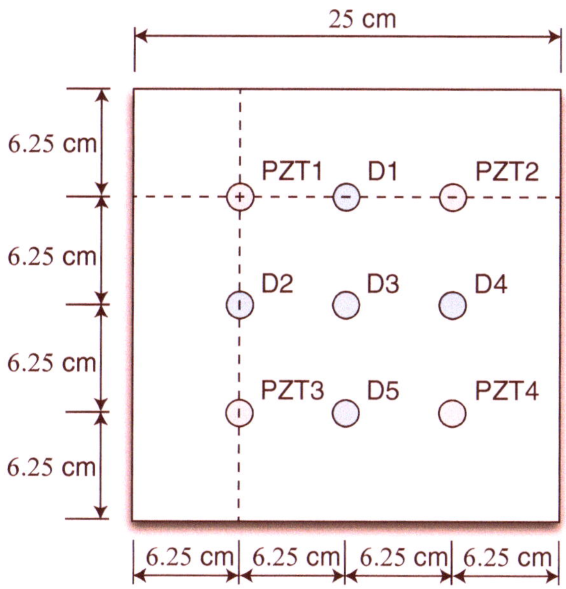

Fig. 17 Dimensions and piezoelectric transducers location

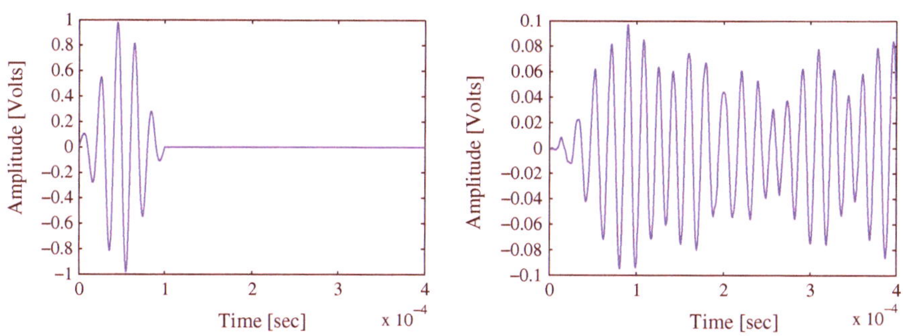

Fig. 18 Excitation signal (left) and dynamic response recorded by PZT 1 (right)

window were used. Figure 18 shows the excitation signal and an example of the signal collected by PZT 1.

4.2.1 Multivariate Normality

As said in Sect. 4.1, the experiments are performed in 4 independent phases: (i) piezoelectric transducer 1 (PZT1) is configured as actuator and the rest of PZTs as sensors; (ii) PZT2 as actuator; (iii) PZT3 as actuator; and (iv) PZT4 as actuator. In order to analyze the influence of each set of projections to the PCA model (score),

Table 10 Results of the multivariate normality tests when considering the first three principal components (PC1–PC3) in the four actuator phases. "–" means that all the tests rejected multivariate normality, "+" means that at least one test indicated multivariate normality while the subindex shows the tests that indicated normality: 1 (Mardia's test), 2 (Henze-Zirkler's test) or 3 (Royston's test)

	PZT1 act.	PZT2 act.	PZT3 act.	PZT4 act.
Undamaged (baseline)	–	$+_{2}$	$+_{2}$	–
Undamaged (first set to test)	–	$+_{1,2,3}$	$+_{2}$	–
Undamaged (second set to test)	$+_{1}$	$+_{1,2}$	–	–
Undamaged (third set to test)	–	$+_{2}$	–	$+_{2,3}$
Undamaged (fourth set to test)	–	–	–	$+_{1,2,3}$
Undamaged (fifth set to test)	–	–	$+_{1}$	$+_{1,3}$
D1	$+_{1,2,3}$	$+_{2}$	$+_{1,2}$	$+_{3}$
D2	$+_{1,2,3}$	$+_{1,2,3}$	$+_{1}$	$+_{1,3}$
D3	$+_{1,2,3}$	$+_{2}$	$+_{1,2}$	–
D4	$+_{2}$	$+_{2,3}$	–	$+_{3}$
D5	$+_{1,2,3}$	–	$+_{1}$	–

the results of scores 1 to 3 (jointly), scores 1 to 5 (jointly) and scores 1 to 10 (jointly) have been considered. In this way, a total of 12 scenarios were examined.

The multivariate normality tests described in Sect. 3.3.2 were performed for all the data. We summarize in Table 10 the results of the multivariate normality test when considering the first three principal components (PC1–PC3) for all the actuator phases.

Some examples of Q–Q plots for the data we consider in this paper are shown on Fig. 19. It can be observed that the points are distributed closely following the bisectrix, thus indicating the multivariate normality of the data as stated in Table 10.

Moreover, some other examples of contour plots for the data we consider in this Section are given in Figs. 20 and 21. These plots are similar to the contour plot of the bivariate normal distribution in Fig. 13.

Finally, the univariate normality for each principal component and for each actuator phase is also tested. The results are presented in Table 11. As it can be observed, the univariate data is normally distributed in most of the cases. However, this does not imply multivariate normality.

4.2.2 Type I and Type II Errors

For each scenario, a total of 50 samples of 20 experiments each one (25 for the undamaged structure and 5 for the damaged structure with respect to each of the 5 different types of damages) plus the baseline are used to test for the plausibility of a value for a normal population mean vector, with a level of significance $\alpha = 0.60$. Each set of 50 testing samples are categorized as follows: (i) number of samples from the healthy structure (undamaged sample) which were classified by the hypothesis

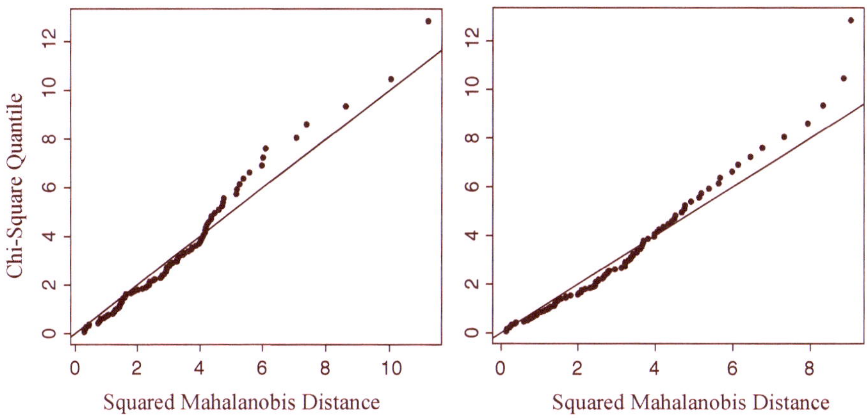

Fig. 19 Q–Q plots corresponding to: (i) fourth set of undamaged data to test, using the first three principal components (PC1–PC3) in the actuator phase 4 (left) and (ii) damage 2 data, using the first three principal components (PC1–PC3) in the actuator phase 1 (right). The points of these Q–Q plots are close to the line $y = x$ thus indicating the multivariate normality of the data

Fig. 20 Contour plot for undamaged case (fourth set to test), PZT4 act., PC1–PC2. The contour lines are similar to ellipses of normal bivariate distribution from Fig. 13 that means that the distribution in this case is normal

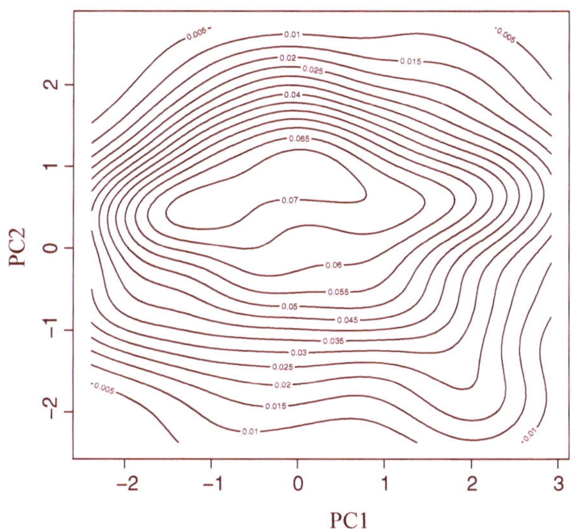

test as 'healthy' (fail to reject H_0); (ii) undamaged sample classified by the test as 'damaged' (reject H_0); (iii) samples from the damaged structure (damaged sample) classified as 'healthy'; and (iv) damaged sample classified as 'damaged'. The results for the 12 different scenarios presented in Table 12 are organized according to the scheme in Table 4. It can be stressed from each scenario in Table 12 that the sum of the columns is constant: 25 samples in the first column (undamaged structure) and 25 more samples in the second column (damaged structure).

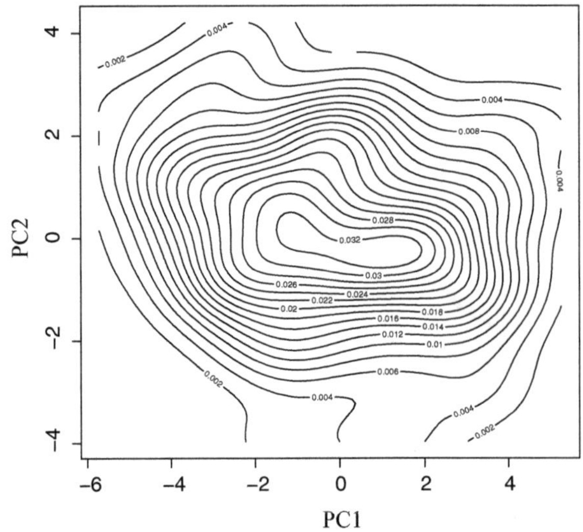

Fig. 21 Contour plot for case D3, PZT1 act., PC1–PC2. The contour lines are similar to ellipses of normal bivariate distribution from Fig. 13 that means that the distribution in this case is normal

Table 11 Results of univariate normality tests when considering the first five principal components separately in the four actuator phases. "−" means lack of normality while "+" means normality

	PZT1 act.	PZT2 act.	PZT3 act.	PZT4 act.
Undamaged (baseline)	−+−++	−++++	−++++	−++−+
Undamaged (first set to test)	−−−+−	−+++−	−++−+	+−+++
Undamaged (second set to test)	−++++	−++++	−++++	−++−+
Undamaged (third set to test)	−−+++	−++++	−++++	−++++
Undamaged (fourth set to test)	−+−++	−++++	−+++−	−++−+
Undamaged (fifth set to test)	−+−++	−++++	−++++	+++++
D1	−++++	−++−+	−++−−	+++++
D2	−++++	−++++	−+++−	+++++
D3	+++++	−++++	−++++	+++++
D4	−++++	+++−+	−++++	−++++
D5	++++−	−++++	−+−+−	−++++

As in Sect. 4.1, in Table 12 two kinds of misclassification are presented: (i) type I errors (or *false positive*), when the structure is healthy but the null hypothesis is rejected and therefore classified as damaged; and (ii) type II errors (or *false negative*), when the structure is damaged but the null hypothesis is not rejected and therefore classified as healthy.

It can be observed from Table 12 that Type I errors (false alarms) appear only when we consider scores 1 to 3 (jointly) and scores 1 to 5 (jointly), while in the last case (scores 1 to 10), all the decisions are correct.

Table 12 Categorization of the samples with respect to presence or absence of damage and the result of the test, for each of the four phases and considering the first score, the second score, scores 1 to 3 (jointly), scores 1 to 5 (jointly) and scores 1 to 10 (jointly)

	PZT1 act.		PZT2 act.		PZT3 act.		PZT4 act.	
	H_0	H_1	H_0	H_1	H_0	H_1	H_0	H_1
Score 1								
Fail to reject H_0	22	13	21	7	18	13	22	12
Reject H_0	3	12	4	18	7	12	3	13
Score 2								
Fail to reject H_0	21	2	24	18	18	5	22	14
Reject H_0	4	23	1	7	7	20	3	11
Scores 1–3								
Fail to reject H_0	24	0	24	13	25	9	24	4
Reject H_0	1	25	1	12	0	16	1	21
Scores 1–5								
Fail to reject H_0	21	0	23	0	21	0	20	0
Reject H_0	4	25	2	25	4	25	5	25
Scores 1–10								
Fail to reject H_0	25	0	25	0	25	0	25	0
Reject H_0	0	25	0	25	0	25	0	25

4.2.3 Sensitivity and Specificity

The sensitivity and the specificity of the test with respect to the 50 samples in each scenario have been included in Table 13. For each scenario in this table, the results are organized as shown in Table 6.

It is worth noting that type I errors are frequently considered to be more serious than type II errors. However, in this application a type II error is related to a *missing fault* whereas a type I error is related to a *false alarm*. In consequence, type II errors should be minimized. Therefore a small level of significance of $1, 5\%$ or even 10% would lead to a reduced number of *false alarms* but to a higher rate of *missing faults*. That is the reason of the choice of a level of significance of 60% in the hypothesis test.

The results show that the sensitivity of the test $1 - \beta$ is close to 100%, as desired, with an average value of 78%. The sensitivity with respect to Score 1 to 5 and Score 1 to 10 is increased, in mean, to a 100%. The average value of the specificity is 90%.

4.2.4 Reliability of the Results

The results in Table 14 are computed using the scheme in Table 8. As in Sect. 4.1.2, Table 14 is based on the Bayes' theorem [29], where $P(H_1|\text{accept } H_0)$ is the propor-

Table 13 Sensitivity and specificity of the test for each scenario

	PZT1 act.		PZT2 act.		PZT3 act.		PZT4 act.	
	H_0	H_1	H_0	H_1	H_0	H_1	H_0	H_1
Score 1								
Fail to reject H_0	0.88	0.52	0.84	0.28	0.72	0.52	0.88	0.48
Reject H_0	0.12	0.48	0.16	0.72	0.28	0.48	0.12	0.52
Score 2								
Fail to reject H_0	0.84	0.08	0.96	0.72	0.72	0.20	0.88	0.56
Reject H_0	0.16	0.92	0.04	0.28	0.28	0.80	0.12	0.44
Scores 1–3								
Fail to reject H_0	0.96	0.00	0.96	0.52	1.00	0.36	0.96	0.16
Reject H_0	0.04	1.00	0.04	0.48	0.00	0.64	0.04	0.84
Scores 1–5								
Fail to reject H_0	0.84	0.00	0.92	0.00	0.84	0.00	0.80	0.00
Reject H_0	0.16	1.00	0.08	1.00	0.16	1.00	0.20	1.00
Scores 1–10								
Fail to reject H_0	1.00	0.00	1.00	0.00	1.00	0.00	1.00	0.00
Reject H_0	0.00	1.00	0.00	1.00	0.00	1.00	0.00	1.00

Table 14 True rate of false positives and false negatives

	PZT1 act.		PZT2 act.		PZT3 act.		PZT4 act.	
	H_0	H_1	H_0	H_1	H_0	H_1	H_0	H_1
Score 1								
Fail to reject H_0	0.63	0.37	0.75	0.25	0.58	0.42	0.65	0.35
Reject H_0	0.20	0.80	0.18	0.82	0.37	0.63	0.19	0.81
Score 2								
Fail to reject H_0	0.91	0.09	0.57	0.43	0.78	0.22	0.61	0.39
Reject H_0	0.15	0.85	0.13	0.88	0.26	0.74	0.21	0.79
Scores 1–3								
Fail to reject H_0	1.00	0.00	0.65	0.35	0.74	0.26	0.86	0.14
Reject H_0	0.04	0.96	0.08	0.92	0.00	1.00	0.05	0.95
Scores 1–5								
Fail to reject H_0	1.00	0.00	1.00	0.00	1.00	0.00	1.00	0.00
Reject H_0	0.14	0.86	0.07	0.93	0.14	0.86	0.17	0.83
Scores 1–10								
Fail to reject H_0	1.00	0.00	1.00	0.00	1.00	0.00	1.00	0.00
Reject H_0	0.00	1.00	0.00	1.00	0.00	1.00	0.00	1.00

tion of samples from the damaged structure that have been incorrectly classified as healthy (*true rate of false negatives*) and $P(H_0|\text{accept } H_1)$ is the proportion of samples from the undamaged structure that have been incorrectly classified as damaged (*true rate of false positives*).

4.2.5 The Receiver Operating Characteristic (ROC) Curves

An additional study has been developed based on the ROC curves to determine the overall accuracy of the proposed method. More precisely, for each scenario and for a given level of significance the pair of numbers

$$(\text{false positive rate, sensitivity}) \in [0, 1] \times [0, 1] \subset \mathbb{R}^2 \tag{57}$$

is plotted. We have considered 99 levels of significance within the range [0.01, 0.99] and with a difference of 0.01. Therefore, for each scenario 99 connected points are depicted, as can be seen in Figs. 22, 23 and 24 when we consider scores 1 to 3 (jointly), scores 1 to 5 (jointly) and scores 1 to 10 (respectively).

As said before, the ROC curves for the 12 possible scenarios are depicted in Figs. 22, 23 and 24. The best performance is achieved for the case of scores 1 to 3 in phase 1 (Fig. 22) because all of the points are placed in the upper-left corner. In phases 2–4, the points lie in the upper left half-plain but not in the corner, which represents a very good behavior of the proposed method. When we consider the case of scores 1 to 5 (jointly) in Fig. 23 and the case of scores 1 to 10 (jointly) in Fig. 24 it

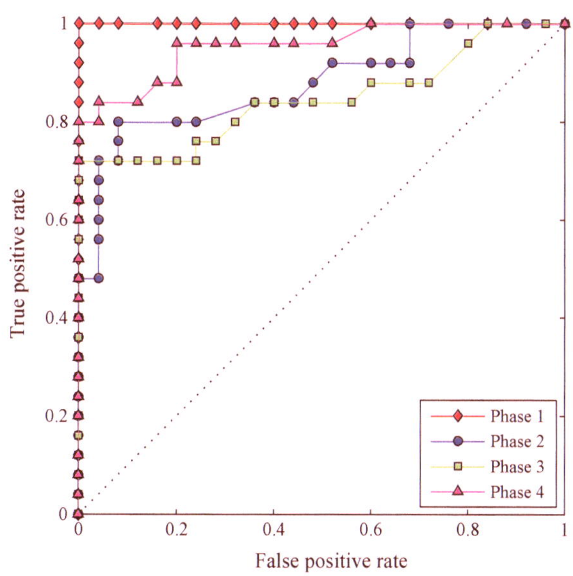

Fig. 22 The receiver operating characteristic (ROC) curves for the scores 1 to 3 (jointly) in the four actuator phases

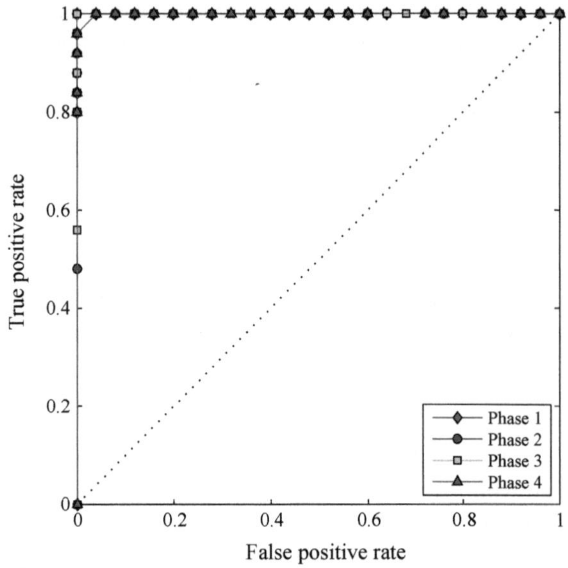

Fig. 23 The receiver operating characteristic (ROC) curves for the scores 1 to 5 (jointly) in the four actuator phases

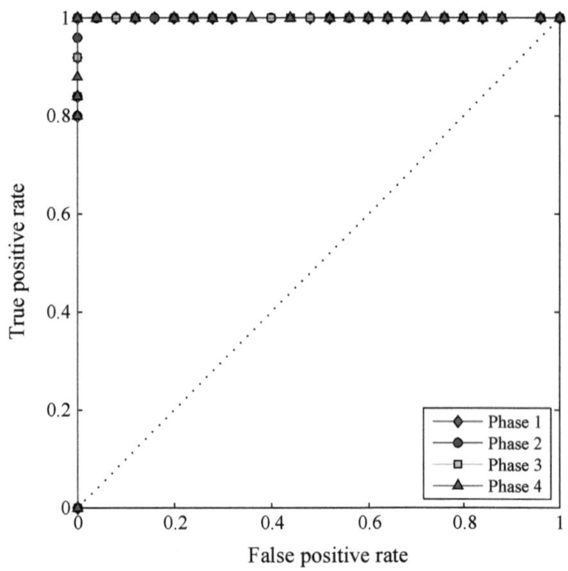

Fig. 24 The receiver operating characteristic (ROC) curves for the scores 1 to 10 (jointly) in the four actuator phases

can be observed that the area under the ROC curves is close to 1 in all of the actuator phases thus representing an excellent test.

4.2.6 Analysis and Discussion

Multivariate tests allow to get better results in damage detection than univariate tests. This is perfectly illustrated in Fig. 25 where a correct or wrong detections is represented as a function of the level of significance α used in the test. We can clearly characterize four different regions:

- $0 < \alpha \leq 0.13$. In this region, *all* the five univariate tests and the multivariate statistical inference **pass** (correct decision).
- $0.13 < \alpha \leq 0.62$. In this region, *some* of the five univariate tests **fail** (wrong decision) while the multivariate statistical inference **pass** (correct decision).
- $0.62 < \alpha \leq 0.71$. In this region, *all* the five univariate tests **fail** (wrong decision) while the multivariate statistical inference **pass** (correct decision).
- $0.71 < \alpha < 1$. In this region, *all* the five univariate tests and the multivariate statistical inference **fail** (wrong decision).

It is worth noting that in the region $0.62 < \alpha \leq 0.71$, that is, when the level of significance lies within the range $(0.62, 0.71]$ the multivariate statistical inference using scores 1 to 5 (jointly) is able to offer a correct decision even though all of the univariate tests make a wrong decision.

The scenarios with the best results are those that considers scores 1 to 10, because the false negative rate is 0% and the false positive rate is 0% for all the actuator phases. The results for scores 1 to 5 (jointly) are quite good, because the false negative rate is 0% for all actuators and the false positive rate is 7–17%.

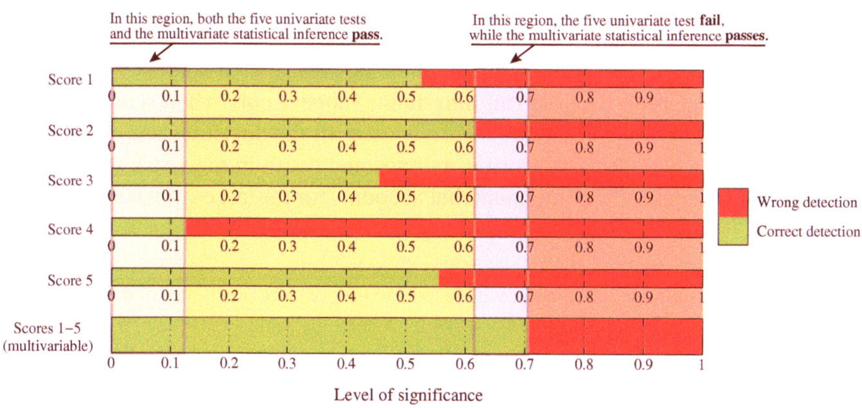

Fig. 25 Multivariate tests allow to get better results in damage detection that univariate tests. A correct or wrong detection is represented as a function of the level of significance where four regions can be identified

4.3 Wind Turbine and Univariate HT

To validate the fault detection strategy presented in Sect. 3.2 using a simulated wind turbine, we first consider a total of 24 samples of $\nu = 50$ elements each, according to the following distribution:

- 16 samples of a healthy wind turbine; and
- 8 samples of a faulty wind turbine with respect to each of the eight different fault scenarios described in Table 3.

In the numerical simulations in this section, each sample of $\nu = 50$ elements is composed by the measures obtained from the $N = 13$ sensors detailed in Table 2 during $(\nu \cdot L - 1)\Delta = 312.4875$ seconds, where $L = 500$ and the sampling time $\Delta = 0.0125$ s. The measures of each sample are then arranged in a $\nu \times (N \cdot L)$ matrix as in Eq. (27).

4.3.1 Type I and Type II Errors

For the first three principal components (score 1 to score 3), these 24 samples plus the baseline sample of $n = 50$ elements are used to test for the equality of means, with a level of significance $\alpha = 0.36$ (the choice of this level of significance will be justified in Sect. 4.3.2). Each sample of $\nu = 50$ elements is categorized as follows: (i) number of samples from the healthy wind turbine (healthy sample) which were classified by the hypothesis test as 'healthy' (fail to reject H_0); (ii) faulty sample classified by the test as "faulty" (reject H_0); (iii) samples from the faulty structure (faulty sample) classified as "healthy"; and (iv) faulty sample classified as "faulty". The results for the first four principal components presented in Table 15 are organized according to the scheme in Table 4. It can be stressed from each principal component in Table 15 that the sum of the columns is constant: 16 samples in the first column (healthy wind turbine) and 8 more samples in the second column (faulty wind turbine).

It can be observed from Table 15 that, in the numerical simulations, Type I errors (false alarms) and Type II errors (missing faults) appear only when scores 2, 3 or 4 are considered, while when the first score is used all the decisions are correct. The better performance of the first score is an expected result in the sense that the first principal component is the component that accounts for the largest possible variance.

Table 15 Categorization of the samples with respect to the presence or absence of damage and the result of the test for each of the four scores when the size of the samples to diagnose is $\nu = 50$

	Score 1		Score 2		Score 3		Score 4	
	H_0	H_1	H_0	H_1	H_0	H_1	H_0	H_1
Fail to reject H_0	16	0	12	1	11	5	9	1
Reject H_0	0	8	4	7	5	3	7	7

4.3.2 Sensitivity and Specificity

The sensitivity and the specificity of the test with respect to the 24 samples and for each of the first four principal components have been included in Table 16. For each principal component in this table, the results are organized as shown in Table 6.

The results in Table 16 show that the sensitivity of the test $1-\gamma$ is close to 100%, as desired, with an average value of 78.00%. The sensitivity with respect to the first, second and fourth principal component is increased, in mean, to a 91.33%. The average value of the specificity is 75.00%, which is very close to the expected value of $1-\alpha = 64\%$.

4.3.3 Reliability of the Results

The results in Table 17 are computed using the scheme in Table 8. This table is based on the Bayes' theorem [29], where $P(H_1|\text{accept } H_0)$ is the proportion of samples from the faulty wind turbine that have been incorrectly classified as healthy (*true rate of false negatives*) and $P(H_0|\text{accept } H_1)$ is the proportion of samples from the healthy wind turbine that have been incorrectly classified as faulty (*true rate of false positives*).

4.3.4 The Receiver Operating Curves (ROC)

The ROC curves are also used in this section to determine the overall accuracy of the proposed method for the fault detection in wind turbines. We have considered 49 levels of significance within the range [0.02, 0.98] and with a difference of 0.02.

Table 16 Sensitivity and specificity of the test for each of the four scores when the size of the samples to diagnose is $\nu = 50$

	Score 1		Score 2		Score 3		Score 4	
	H_0	H_1	H_0	H_1	H_0	H_1	H_0	H_1
Fail to reject H_0	1.00	0.00	0.75	0.13	0.69	0.62	0.56	0.13
Reject H_0	0.00	1.00	0.25	0.87	0.31	0.38	0.44	0.87

Table 17 True rate of false positives and false negatives for each of the four scores when the size of the samples to diagnose is $\nu = 50$

	Score 1		Score 2		Score 3		Score 4	
	H_0	H_1	H_0	H_1	H_0	H_1	H_0	H_1
Fail to reject H_0	1.00	0.00	0.92	0.08	0.69	0.31	0.90	0.10
Reject H_0	0.00	1.00	0.36	0.64	0.62	0.38	0.50	0.50

Therefore, for each of the first four principal components, 49 connected points are depicted, as can be seen in Fig. 26.

The results presented in Fig. 26, particularly with respect to score 1, are quite remarkable. The overall behavior of scores 2 and 4 are also acceptable, while the results of score 3 cannot be considered, in this case, as satisfactory.

In Figs. 27 and 28 a further study is performed. While in Fig. 26 we present the ROCs when the size of the samples to diagnose is $\nu = 50$, in Fig. 27 the reliability of the method is analyzed in terms of 48 samples of $\nu = 25$ elements each and in Fig. 28 the reliability of the method is analyzed in terms of 120 samples of $\nu = 10$ elements each. The effect of reducing the number of elements in each sample is the reduction in the total time needed for a diagnostic. More precisely, if we keep $L = 500$, when the size of the samples is reduced from $\nu = 50$ to $\nu = 25$ and $\nu = 10$, the total time needed for a diagnostic is reduced from about 312 s to 156 and 62 s, respectively. Another effect of the reduction in the number of elements in each sample is a slight deterioration of the overall accuracy of the detection method. However, the results of scores 1 and 2 in Figs. 27 and 28 are perfectly acceptable.

A very interesting alternative to keep a very good performance of the method without almost no degradation in its accuracy is by reducing L –the number of time instants per row per sensor— instead of reducing the number of elements per sample ν. This way, if we keep $\nu = 50$, when L is reduced from 500 to 50, the total time needed for a diagnostic is reduced from about 312 s to 31 s. Finally, with the goal to reduce the computational effort of the fault detection method, a sensor selection algorithm can be applied [16].

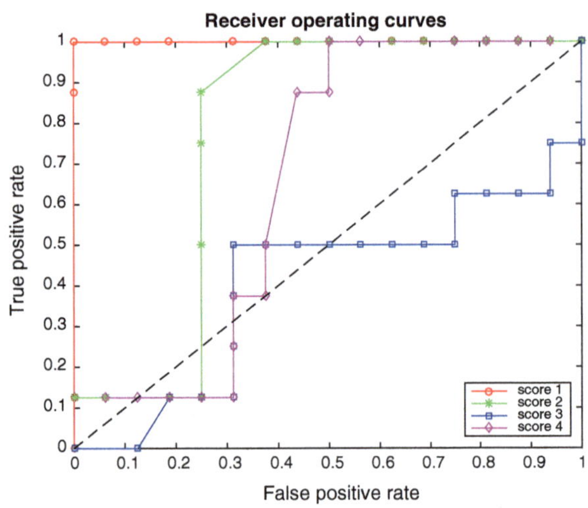

Fig. 26 The Receiver Operating Curves (ROCs) for the four scores when the size of the samples to diagnose is $\nu = 50$

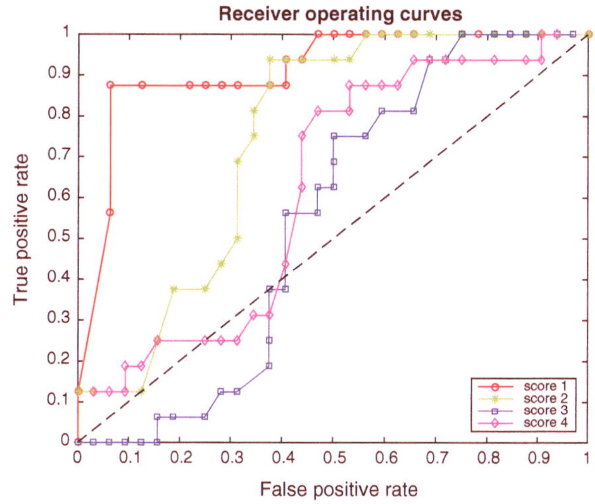

Fig. 27 The ROCs for the four scores when the size of the samples to diagnose is $\nu = 25$

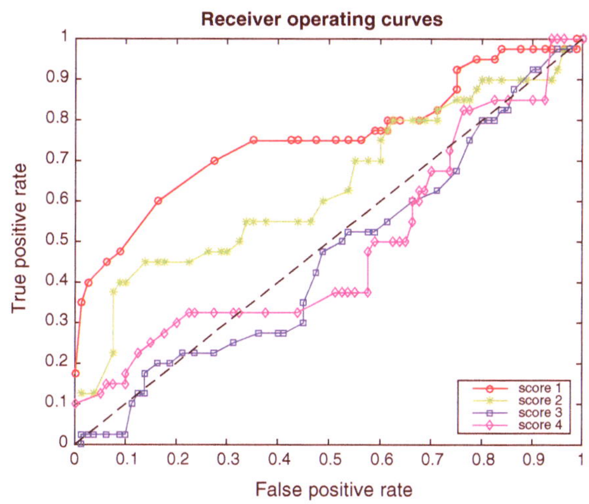

Fig. 28 The ROCs for the four scores when the size of the samples to diagnose is $\nu = 10$

4.4 Wind Turbine and Multivariate HT

To validate the fault detection strategy presented in Sect. 3.3 using a simulated wind turbine, we consider —as in Sect. 4.3— a total of 24 samples of $\nu = 50$ elements each, according to the following distribution:

- 16 samples of a healthy wind turbine; and
- 8 samples of a faulty wind turbine with respect to each of the eight different fault scenarios described in Table 3.

In the numerical simulations in this section, each sample of $\nu = 50$ elements is composed by the measures obtained from the $N = 13$ sensors detailed in Table 2 during $(\nu \cdot L - 1)\Delta = 312.4875$ seconds, where $L = 500$ and the sampling time $\Delta = 0.0125$ seconds. The measures of each sample are then arranged in a $\nu \times (N \cdot L)$ matrix as in Eq. (27).

For the sake of comparison, univariate and multivariate hypothesis testing are performed, as described in Sects. 3.2 and 3.3, respectively. On one hand, and with respect to the univariate HT, and for the first three principal components (score 1 to score 3), these 24 samples plus the baseline sample of $n = 50$ elements are used to test for the equality of means, with a level of significance $\alpha = 0.10$. On the other hand, the same 24 samples plus the baseline sample are used to test for the plausibility of a value for a normal population mean vector, with the same level of significance, considering scores 1 to 2 (jointly), scores 1 to 7 (jointly) and scores 1 to 12 (jointly). Each sample of $\nu = 50$ elements is categorized as follows: (i) number of samples from the healthy wind turbine (healthy sample) which were classified by the hypothesis test as 'healthy' (fail to reject H_0); (ii) faulty sample classified by the test as "faulty" (reject H_0); (iii) samples from the faulty structure (faulty sample) classified as "healthy"; and (iv) faulty sample classified as "faulty". The results for the univariate HT for the first three principal components are described in Table 18. Similarly, the results for the multivariate HT for scores 1 to 2 (jointly), scores 1 to 7 (jointly) and scores 1 to 12 (jointly) are detailed in Table 19. In both tables, the results are organized according to the scheme in Table 4. It can be stressed from these tables that the sum of the columns is constant: 16 samples in the first column (healthy wind turbine) and 8 more samples in the second column (faulty wind turbine).

By examining Tables 18 and 19, it is worth noting that, for a fixed level of significance $\alpha = 10\%$, all decisions are correct only when the first twelve scores are considered jointly. Although the results for the score 1 in Table 18 are quite accept-

Table 18 Categorization of the samples with respect to the presence or absence of a fault and the result of the test considering the first score, the second score and the third score, when the size of the samples to diagnose is $\nu = 50$ and the level of significance is $\alpha = 10\%$

	Score 1		Score 2		Score 3	
	H_0	H_1	H_0	H_1	H_0	H_1
Fail to reject H_0	16	1	13	7	16	8
Reject H_0	0	7	3	1	0	0

Table 19 Categorization of the samples with respect to the presence or absence of a fault and the result of the test considering scores 1–2 (jointly), scores 1–7 (jointly), and scores 1–12 (jointly), when the size of the samples to diagnose is $\nu = 50$ and the level of significance is $\alpha = 10\%$

	Scores 1–2		Scores 1–7		Scores 1–12	
	H_0	H_1	H_0	H_1	H_0	H_1
Fail to reject H_0	12	0	13	0	16	0
Reject H_0	4	8	3	8	0	8

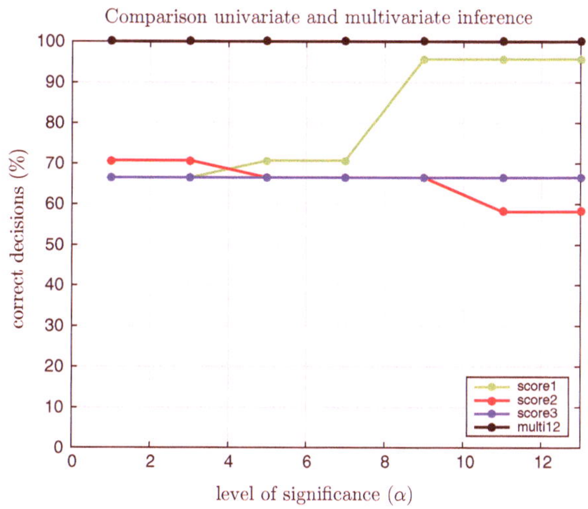

Fig. 29 Percentage of correct decisions using the multivariate hypothesis testing fault detection strategy (scores 1 to 12, jointly) and the univariate hypothesis testing (for the first, second and third score), as a function of the level of significance α

able and it can be improved by increasing the level of significance, it is also important to try to keep α as small as possible since it is related to the probability of committing a type I error. In this sense, it can be observed from Fig. 29 that for very small values of the level of significance α, the percentage of correct decisions in the multivariate HT—considering scores 1 to 12 jointly—is 100%, while in the rest of the univariate HT cases, the correct decisions are about to 65–75%.

5 Concluding Remarks

Two different problems have been addressed in this chapter: early detection of damage in structures, and detection of faults in a wind turbine. In both cases, the proposed methodology, based on PCA plus hypothesis testing, proved to be effective.

In particular, for the experimental set-up, the univariate test showed that the results related to the first score are not as good as those related to the rest of scores. Thus, it is important to note that the projection on the first component is not always the best option to detect damage. Finally, it is shown that multivariate tests improve significantly the results with respect to univariate HT.

On the other hand, for the numerical simulations of the benchmark wind turbine, the univariate HT results using the first score has an excellent performance as all decisions are correct when the used level of significance is $\alpha = 0.36$. However, recall that it is advisable to use small values of significance as this will reduce the number of type I errors. In this case, for $\alpha \in (0, 0.12]$, the multivariate HT obtains the best results with a 100% of correct decisions (while in the univariate HT cases the correct decisions are about 65–75%).

Acknowledgements This work has been partially funded by the Spanish Ministry of Economy and Competitiveness through the research project DPI2014-58427-C2-1-R, and by the Catalonia Government (Generalitat de Catalunya) through the research project 2014 SGR 859.

References

1. Jolliffe, I.T., Cadima, J.: Principal component analysis: a review and recent developments. Phil. Trans. R. Soc. A 374(2065):20150,202 (2016)
2. Mujica, L.E., Rodellar, J., Fernández, A., Güemes, A.: q-statistic and t^2-statistic PCA-based measures for damage assessment in structures. Struct. Health Monit. **10**(5), 539–553 (2011)
3. Mujica, L.E., Ruiz, M., Pozo, F., Rodellar, J., Güemes, A.: A structural damage detection indicator based on principal component analysis and statistical hypothesis testing. Smart Mater. Struct. **23**, 1–12 (2014)
4. Pozo, F., Vidal, Y.: Wind Turbine Fault Detection through Principal Component Analysis and Statistical Hypothesis Testing. Energies **9**(1):3, https://doi.org/10.3390/en9010003, http://www.mdpi.com/1996-1073/9/1/3/htm (2015)
5. Pozo. F., Arruga, I., Mujica, L.E., Ruiz, M., Podivilova, E.: Detection of structural changes through principal component analysis and multivariate statistical inference. Structural Health Monitoring, https://doi.org/10.1177/1475921715624504, http://shm.sagepub.com/cgi/content/abstract/1475921715624504v1 (2016a)
6. Ostachowicz, W., Kudela, P., Krawczuk, M., Zak, A.: Guided Waves in Structures for SHM: The Time-Domain Spectral Element Method. Wiley, (2012)
7. Kaldellis, J.K., Zafirakis, D.: The wind energy (r) evolution: A short review of a long history. Renewable Energy **36**(7), 1887–1901 (2011)
8. Jonkman, J.M., Butterfield, S., Musial, W., Scott, G.: Definition of a 5-MW reference wind turbine for offshore system development. Tech. rep., National Renewable Energy Laboratory, Golden, Colorado, nREL/TP-500-38060, (2009)
9. Kelley, N., Jonkman, B. (Last modified 30-May-2013) NWTC computer-aided engineering tools (Turbsim). http://wind.nrel.gov/designcodes/preprocessors/turbsim/
10. Odgaard, P., Johnson, K.: Wind turbine fault diagnosis and fault tolerant control - an enhanced benchmark challenge. In: Proceeding of the 2013 American Control Conference–ACC,(Washington DC, USA), pp. 1–6, (2013)
11. Odgaard, P.F., Stoustrup, J., Kinnaert, M.: Fault-tolerant control of wind turbines: a benchmark model. Control Sys. Technol. IEEE Trans. **21**(4), 1168–1182 (2013)
12. Liniger, J., Pedersen, H.C., Soltani, M.: Reliable fluid power pitch systems: A review of state of the art for design and reliability evaluation of fluid power systems. In: ASME/BATH 2015 Symposium on Fluid Power and Motion Control, American Society of Mechanical Engineers, pp. V001T01A026–V001T01A026, (2015)
13. Chaaban, R., Ginsberg, D., Fritzen, C.P.: Structural load analysis of floating wind turbines under blade pitch system faults. In: Vidal, Y., Acho, L. Luo N., (eds.) Wind Turbine Control and Monitoring, pp. 301–334. Springer,(2014)
14. Chen, L., Shi. F., Patton, R.: Active FTC for hydraulic pitch system for an off-shore wind turbine. In: Control and Fault-Tolerant Systems (SysTol), 2013 Conference on, IEEE, pp. 510–515 (2013)
15. Jolliffe, I.T.: Discarding Variables in a Principal Component Analysis. II: Real Data. J. R. Stat. Soc. Ser. C (Applied Statistics) **22**(1):21–31, https://doi.org/10.2307/2346300, http://www.jstor.org/stable/2346300, (1973)
16. Pozo, F., Vidal, Y., Serrahima, J.: On Real-Time Fault Detection in Wind Turbines: Sensor Selection Algorithm and Detection Time Reduction Analysis. Energies **9**(7):520, https://doi.org/10.3390/en9070520, http://www.mdpi.com/1996-1073/9/7/520/htm (2016b)

17. Anaya, M., Tibaduiza, D., Pozo, F.: A bioinspired methodology based on an artificial immune system for damage detection in structural health monitoring. Shock Vib. **2015**, 1–15 (2015)
18. Anaya, M., Tibaduiza, D., Pozo, F.: Detection and classification of structural changes using artificial immune systems and fuzzy clustering. Int. J, Bio-Inspired Comput (2016)
19. Odgaard, P.F., Lin, B., Jorgensen, S.B.: Observer and data-driven-model-based fault detection in power plant coal mills. Energy Conversion, IEEE Trans. **23**(2), 659–668 (2008)
20. Tibaduiza, D.A., Mujica, L.E., Rodellar, J.: Damage classification in structural health monitoring using principal component analysis and self-organizing maps. Struct. Control Health Monit. **20**(10), 1303–1316 (2013). https://doi.org/10.1002/stc.1540
21. Gharibnezhad, F., Mujica, L.E., Rodellar, J.: Applying robust variant of Principal Component Analysis as a damage detector in the presence of outliers. Mechanical Systems and Signal Processing **50–51**:467–479, https://doi.org/10.1016/j.ymssp.2014.05.032, http://linkinghub.elsevier.com/retrieve/pii/S0888327014002088 (2015)
22. Sierra-Perez, J., Guemes, A., Mujica, L.E., Ruiz, M.: Damage detection in composite materials structures under variable loads conditions by using fiber Bragg gratings and principal component analysis, involving new unfolding and scaling methods. J. Intell. Material Sys. Struct. **26**(11):1346–1359, (2015) https://doi.org/10.1177/1045389X14541493, http://jim.sagepub.com/cgi/doi/10.1177/1045389X14541493
23. Ugarte, M.D., Militino, A.F., Arnholt, A.: Probability and Statistics with R. CRC Press (Taylor & Francis Group), (2008)
24. Rencher, A.C., Christensen, W.F.: Confirmatory factor analysis. Methods of Multivariate Analysis, 3edn. pp. 479–500, (2012)
25. Korkmaz, S., Goksuluk, D., Zararsiz, G.: MVN: An R package for assessing multivariate normality. R. J. **6**(2), 151–162 (2014)
26. Mardia, K.V.: Applications of some measures of multivariate skewness and kurtosis in testing normality and robustness studies. Sankhyā: Indian J. Statist. Ser. B. 115–128, (1974)
27. Henze, N., Zirkler, B.: A class of invariant consistent tests for multivariate normality. Commun. Statist.-Theory Methods **19**(10), 3595–3617 (1990)
28. Royston, P.: Approximating the Shapiro-Wilk W-test for non-normality. Statist. Computing **2**(3), 117–119 (1992)
29. DeGroot, M.H., Schervish, M.J.: Probability and Statistics. Pearson, (2012)
30. Yinghui, L., Michaels, J.E.: Feature extraction and sensor fusion for ultrasonic structural health monitoring under changing environmental conditions. IEEE Sensors J. **9**(11), 1462–1471 (2009)

Principal Component Analysis for Exponential Family Data

Meng Lu, Kai He, Jianhua Z. Huang and Xiaoning Qian

Principal component analysis (PCA) is a powerful and widely-used dimension reduction tool, which seeks for low-rank approximation to the original data, either achieving the minimum deviation or preserving the maximum variation [14]. The derivation of PCA based on either criterion inherently assumes that the original data are real-valued and follow a multivariate Gaussian distribution. However, such inherent assumptions may not be appropriate when analyzing other data types, for example, with binary, categorical, or count values. Exponential family PCA (ePCA) generalizes traditional PCA for real-valued data to the data belonging to the exponential family. With diverse data types collected in this modern big data era, ePCA is increasingly attracting research attention.

1 A Probabilistic Model for Exponential Family Principal Component Analysis (ePCA)

PCA can be formulated as a maximum likelihood estimation (MLE) problem [35]. From the probabilistic modeling perspective, PCA maximizes the data likelihood, assuming each data point is sampled from a multivariate Gaussian distribution in a low-dimensional subspace. The multivariate Gaussian distribution is suitable for modeling real-valued data, but not appropriate for other data types, especially for discrete data types. For example, binary data are often modeled with Bernoulli distributions; and other types of discrete data can be modeled by the corresponding distributions in the exponential family. Thus, appropriate distributions should be assumed according to the data types to analyze in the MLE framework. The ePCA

M. Lu
Department of Information Management and Management Science,
Institute of Data Science, Tianjin University, Tianjin 300072, China

K. He · J. Z. Huang · X. Qian (✉)
Department of Electrical and Computer Engineering,
Texas A&M University, College Station, Texas 77843, USA
e-mail: xqian@ece.tamu.edu

G. R. Naik (ed.), *Advances in Principal Component Analysis*,
https://doi.org/10.1007/978-981-10-6704-4_8

model generalizes PCA as a MLE problem for the general exponential family of distributions.

1.1 A Probabilistic View of PCA

Given a set of data samples $\mathbf{x}_1, \ldots, \mathbf{x}_n \in \mathscr{R}^d$, PCA projects the data into a principal-component subspace with a lower dimension $L(\leq d)$. A probabilistic interpretation of PCA assumes that the data points can be approximated by linear projections of low-dimensional latent variables plus a Gaussian noise. For each sample $\mathbf{x}_i$ $(1 \leq i \leq n)$, given its corresponding vector of latent variables $\mathbf{z}_i$ lying in the principal-component subspace, we assume

$$\mathbf{x}_i = W\mathbf{z}_i + \mathbf{b} + \boldsymbol{\varepsilon},$$

where W is a principal loading matrix whose columns span the principal-component (PC) subspace; $\mathbf{b}$ is a bias vector and $\boldsymbol{\varepsilon}$ follows a Gaussian distribution $N(0, \sigma^2 I)$. Assuming a vector of canonical parameters $\boldsymbol{\theta}_i = W\mathbf{z}_i + \mathbf{b}$, the conditional probability of $\mathbf{x}_i$ given $\boldsymbol{\theta}_i$ is then represented as:

$$p(\mathbf{x}_i|\boldsymbol{\theta}_i) \sim \mathscr{N}(\mathbf{x}_i|\boldsymbol{\theta}_i, \sigma^2 I)$$

and the conditional probability of $\mathbf{x}_i$ given $\mathbf{z}_i$ is:

$$p(\mathbf{x}_i|\mathbf{z}_i) \sim \mathscr{N}(\mathbf{x}_i|W\mathbf{z}_i + \mathbf{b}, \sigma^2 I).$$

PCA is formulated as an optimization problem of maximizing the log-likelihood of the given data set with respect to the model parameters $\mathbf{z}_i$, W, and $\mathbf{b}$, which is equivalent to maximizing the following objective function:

$$\sum_i -||\mathbf{x}_i - (W\mathbf{z}_i + \mathbf{b})||^2 \quad s.t. \quad W^T W = I. \tag{1}$$

Obviously, this problem is equivalent to minimizing the sum of Euclidean distances from the original data points to their projections in the principal-component subspace, which is exactly the minimum deviation interpretation of PCA [29].

1.2 Exponential Family PCA

From a probabilistic perspective, it is natural to generalize PCA to the exponential family. In the exponential family, a probabilistic latent variable model representing the conditional distribution of a data sample $\mathbf{x}_i$ has a general form as follows [5]:

$$p(\mathbf{x}_i|\boldsymbol{\theta}_i) = \exp(\boldsymbol{\theta}_i^T \mathbf{x}_i + \log q(\mathbf{x}_i) - A(\boldsymbol{\theta}_i)), \tag{2}$$

where $\boldsymbol{\theta}_i$ denotes the vector of canonical parameters corresponding to the data sample $\mathbf{x}_i$. $A(\mathbf{x}_i)$ is the log-normalization factor with the form based on the base measure $q(\mathbf{x}_i)$: $\log \int \exp(\boldsymbol{\theta}_i^T \mathbf{x}_i) q(\mathbf{x}_i) d\mathbf{x}_i$, ensuring that the sum of the conditional probabilities over the domain of $\mathbf{x}_i$ equals 1. The probability distribution functions for different members in the exponential family have different $A(\cdot)$ functions. The $A(\cdot)$ functions for several well-known distributions in the exponential family are provided in Table 1. The resulting data log-likelihood function with respect to the canonical parameters may be of a quadratic form (for Gaussian) or more complicated forms for other exponential family members.

To achieve dimension reduction, the canonical parameters $\boldsymbol{\theta}_i$ are further parameterized with a form of $W\mathbf{z}_i + \mathbf{b}$ using lower-dimensional latent variables $\mathbf{z}_i$, principal loading matrix W and a bias vector $\mathbf{b}$, as similarly done in traditional PCA. In general, ePCA can be achieved by maximizing the data likelihood based on a general form of the probability function shown by (2). After substituting $\boldsymbol{\theta}_i$ by the low-rank representation from $\mathbf{z}_i$, W, and $\mathbf{b}$ into the data likelihood, ePCA is formulated as the following problem:

$$\min_{Z,\mathbf{b}} \min_{W: W^T W = I} \sum_i A(W\mathbf{z}_i + \mathbf{b}) - tr((ZW^T + \mathbf{1}\mathbf{b}^T)X^T), \tag{3}$$

where Z is the $n \times L$ principal component score matrix whose ith row is $\mathbf{z}_i$. In some cases, people might consider imposing the orthonormal constraints on Z instead of W to obtain orthogonal principal components, due to computational considerations.

Taking Gaussian for instance, $A(\boldsymbol{\theta}_i)$ takes a form of $\boldsymbol{\theta}_i^T \boldsymbol{\theta}_i / 2$ to ensure (2) to be a Gaussian distribution function. Then, its data log-likelihood function given $\boldsymbol{\theta}$ is equivalent to

$$\sum_i -||\mathbf{x}_i - \boldsymbol{\theta}_i||^2 \tag{4}$$

up to a constant. After substituting $\boldsymbol{\theta}_i$ into (4), we arrive at (1), which is exactly the objective function of PCA.

Table 1 The $A(\cdot)$ functions and its first derivatives for several well-known exponential family members of scalar variables

Distribution	$A(\theta)$	$\frac{\partial A(\theta)}{\partial \theta}$
Gaussian	$\frac{\theta^2}{2}$	θ
Bernoulli	$\log(1 + \exp(\theta))$	$\frac{\exp(\theta)}{1+\exp(\theta)}$
Poisson	$\exp(\theta)$	$\exp(\theta)$
Exponential	$-\log(-\theta)$	$-\frac{1}{\theta}$

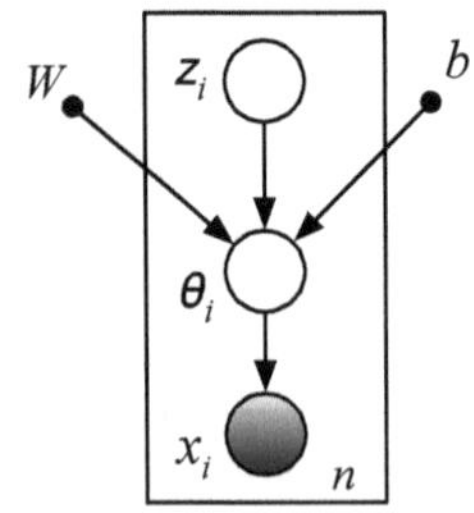

Fig. 1 Probabilistic graphical model for ePCA

A probabilistic graphical model to illustrate ePCA is shown in Fig. 1. Note that the principal component subspace is derived for canonical parameters instead of data samples directly. The low-rank representation of canonical parameters is related to the data through a link function, depending on the data type and assumed distribution function.

2 Two Computational Algorithms for ePCA

Depending on the exponential family distribution functions, solving the ePCA problem (3) can lead to an optimization problem of minimizing a non-jointly convex objective function with non-convex and non-smooth constraints. It is unsatisfactory to utilize the classical gradient descent or block coordinate algorithms to solve this problem, mainly due to the orthogonal constraints. Alternatively, Collins et al. [5] proposed to sequentially update the loading vectors by solving a corresponding set of simple subproblems resembling generalized linear models (GLMs) [26] iteratively. Although the problem complexity is simplified, such a method can not guarantee the joint optimality and joint orthogonality when multiple principal components are required. Different from this method, Guo et al. [10] proposed to solve the ePCA problem by transforming a regularized ePCA to an equivalent problem by exploiting the convex duality of the optimization subproblems. The solutions are obtained by solving the new equivalent problem that admits an efficient optimization procedure. In the following, the above two methods will be reviewed in detail. Other promising methods that solve the special cases of ePCA will be discussed in the Sect. 3.

2.1 Sequential Optimization

Before we study the sequential optimization technique to update the unknown variables for solving (3), we first focus on the simple case of the problem when $L = 1$. The loss function is then reduced to:

$$\ell(\mathbf{z}, \mathbf{w}, \mathbf{b}) = \sum_{i=1}^{n}\sum_{j=1}^{d} A(z_i w_j + b_j) - (z_i w_j + b_j)x_{ij},$$
$$s.t. \quad \|\mathbf{w}\|_2 = 1. \tag{5}$$

This problem can be solved by alternately updating **z**, **w** and **b**, each time minimizing one of them while keeping the others fixed. Let $\tilde{\mathbf{x}}_j$ be a vector containing the value of the jth feature of all the samples. Specifically, in each iteration,
for i = 1, …, n

$$z_i^{(t+1)} = \min_{z_i} \ell(z_i, \mathbf{w}^t, \mathbf{b}^t) = \min_{z_i} A(z_i \mathbf{w}^t + \mathbf{b}^t) - z_i \mathbf{x}_i^T \mathbf{w}^t;$$

for j = 1, …, d

$$w_j^{(t+1)} = \min_{w_j} \ell(w_j, \mathbf{z}^t, \mathbf{b}^t) = \min_{w_j} A(w_j \mathbf{z}^t + b_j^t \mathbf{1_n}) - w_j \tilde{\mathbf{x}}_j^T \mathbf{z}^t$$
$$\mathbf{w}^{(t+1)} = \mathbf{w}^{(t+1)}/\|\mathbf{w}^{(t+1)}\|_2;$$

for j = 1, …, d

$$b_j^{(t+1)} = \min_{b_j} \ell(b_j, \mathbf{z}^t, \mathbf{w}^t) = \min_{b_j} A(w_j \mathbf{z}^t + b_j^t \mathbf{1_n}) - b_j^t \tilde{\mathbf{x}}_j^T \mathbf{1_n}.$$

One can see that there are $n + 2d$ optimization subproblems with only one parameter for each of them to be optimized over in each iteration. Each subproblem is essentially identical to a simple GLM problem, in which z_i or w_j corresponds to the coefficient to be estimated; $\mathbf{x}_i$ or $\tilde{\mathbf{x}}_j$ corresponds to the outcome of the dependent variable; and the other fixed variables correspond to the independent variables.

Now, get back to the original ePCA problem (3) with more than one principal components to be estimated. Considering the canonical parameter matrix $\Theta = \sum_{l=1}^{L} \mathbf{z}_l \mathbf{w}_l^T$, one can estimate each component sequentially by holding the other components and their resulting canonical parameters fixed. For each lth component, $\mathbf{z}_l$ and $\mathbf{w}_l$ can be estimated by solving the similar subproblems as minimizing (5). The procedure can be described as below:

Initialize Z, W and **b**.
Repeat
{
For each component $l = 1, \ldots L$,
{
$s_{ij} = \sum_{c \neq l} z_{ic}^t w_{jc}^t$.

Solve the following problem similarly as minimizing (5):
$z_{il}^{(t+1)}, w_{jl}^{(t+1)} = \arg\min_{z_{il}, w_{jl}} \sum_i \sum_j A(z_{il} w_{jl} + b_j^t + s_{ij}) - z_{il} w_{jl} x_{ij}$.
}

$\mathbf{b}^{(t+1)} = \arg\min_{\mathbf{b}} \sum_i A(W^t \mathbf{z}_i^t + \mathbf{b}) - tr((Z^t W^{tT} + \mathbf{1}\mathbf{b}^T)X^T)$.

}

until convergence.

The convergence is not guaranteed to reach the global minimum since the objective function in each subproblem is not jointly convex in $\mathbf{z}_l$ and $\mathbf{w}_l$ though it is convex in either one of them when the other one is fixed. Moreover, the joint orthogonality constraints for the components might be violated while implementing the sequential updates for the components one by one, which is the price that the sequential technique has to pay.

2.2 *Transformation by Convex Conjugate*

In order to address the difficulties in solving the ePCA problem, an alternative optimization strategy [10] was proposed to add a quadratic regularizer on the loading vectors to the ePCA problem in order to formulate a new problem that is easier to solve. This regularized version of the ePCA problem was proven to be equivalent to a new problem wth a much nicer structure so that the transformed problem can be solved by alternating the updates of Z and W based on closed-form updating rules. There is no worry about the violation of joint orthogonality constraints here because the subproblems under the orthogonality constraints have closed-form solutions according to the Procrustes rotation theorem [25].

Recall the ePCA problem (3). Without loss of generality, a regularized extension can be formulated as

$$\min_{Z:Z^T Z=I} \min_{W} \sum_i A(W\mathbf{z}_i) - tr(ZW^T X^T) + \frac{\lambda}{2} tr(W^T W), \tag{6}$$

by including a quadratic regularizer on W. This regularization term can be interpreted as a zero-mean Gaussian prior on W with a diagonal covariance matrix. The addition of such a regularization term allows a maximum a posteriori (MAP) formulation of the problem and also makes the subsequent optimization procedure simpler. The orthogonality is maintained by the constraints $Z^T Z = I$ for the model simplicity.

It is natural to consider directly optimizing Z and W by gradient decent or block coordinate descent methods; but the constraints on Z make the optimization procedure more complex, which results in low computational efficiency. The main difficulty in solving the ePCA problem is the minimization over Z under the orthogonality constraints. Instead of directly tackling the original problem (6), the problem transformation strategy transforms the original optimization problem based on the convex conjugate of the $A(\cdot)$ function to introduce an optimization function over a new variable U, which enables the minimization over Z in the new problem has closed-form solutions.

There are three main steps to transform the regularized ePCA problem (6) to a new equivalent problem that can be efficiently solved.

Step 1 Replacing the $A(\cdot)$ function by introducing its convex conjugate.

Let U be a $n \times d$ matrix and $A^*(\mathbf{u}_i)$ is the Fenchel conjugate of $A(W\mathbf{z}_i)$. Note that $A(\cdot)$ and $A^*(\cdot)$ are both convex functions. Then, $A(W\mathbf{z}_i)$ can be rewritten as

$$A(W\mathbf{z}_i) = \max_{\mathbf{u}_i} \mathbf{u}_i^T W\mathbf{z}_i - A^*(\mathbf{u}_i).$$

Thus, the inner minimization of (6) becomes

$$\min_W \max_U -\sum_i A^*(\mathbf{u}_i) + tr(ZW^T(U-X)^T) + \frac{\lambda}{2} tr(W^T W). \tag{7}$$

Step 2 Exchange the order of minimization on W and maximization on U in (7).

Let $F(W; U)$ denote the objective function in (7). Guo et al. [10] claimed that one can verify that F satisfies the conditions of the strong minmax theorem [3, 30], which allows the order of the minimization and maximization to be reversed. That is, based on the strong minmax theorem one can conclude that (7) is equivalent to

$$\max_U \min_W -\sum_i A^*(\mathbf{u}_i) + tr(ZW^T(U-X)^T) + \frac{\lambda}{2} tr(W^T W). \tag{8}$$

Then, the inner minimization on W can be easily solved since the objective function of (8) is convex in W for fixed U. By substituting $W = \frac{1}{\lambda} Z^T(X-U)$, the transformed problem (8) turns into

$$\max_U -\sum_i A^*(\mathbf{u}_i) - \frac{1}{2\lambda} tr((X-U)(X-U)^T ZZ^T). \tag{9}$$

After adding back the outer minimization over Z to (9), one arrives at the following problem equivalent to the regularized ePCA problem (6):

$$\min_{Z:Z^TZ=I} \max_U -\sum_i A^*(\mathbf{u}_i) - \frac{1}{2\lambda} tr((X-U)(X-U)^T ZZ^T). \tag{10}$$

Step 3 Exchange the order of minimization on Z and maximization on U in (10).

First, rewrite the outer minimization of Z as a minimization in terms of a square matrix M with constraints: $\{M : I \succeq M \succeq 0; tr(M) = L\}$. The problem (10) is then relaxed as follows:

$$\min_{Z:Z^TZ=I}\max_{U} -\sum_i A^*(\mathbf{u}_i) - \frac{1}{2\lambda}tr((X-U)(X-U)^T ZZ^T)$$
$$\geq \min_{M:I\succeq M\succeq 0;tr(M)=L}\max_{U} -\sum_i A^*(\mathbf{u}_i) - \frac{1}{2\lambda}tr((X-U)(X-U)^T M), \quad (11)$$

where the equivalence holds when $M^2 = M$, which implies $M = ZZ^T$ for some Z such that $Z^T Z = I$.

In the relaxed optimization problem (11), the outer minimization problem with respect to M is convex since the maximum of linear functions is convex and the constraints are convex. After relaxation, the objective function of (11) satisfies the conditions of the strong minmax theorem, which allows another order change of a minimization and maximization in the relaxed optimization problem shown as below:

$$\max_{U}\min_{M:I\succeq M\succeq 0;tr(M)=L} -\sum_i A^*(\mathbf{u}_i) - \frac{1}{2\lambda}tr((X-U)(X-U)^T M). \quad (12)$$

Since solving a semidefinite problem

$$\min_{M:I\succeq M\succeq 0;tr(M)=L} tr(MA)$$

is equivalent to solve

$$\min_{Z:Z^TZ=I} tr(ZZ^T A),$$

the problem (12) can be further transformed to the equivalent problem:

$$\max_{U}\min_{Z:Z^TZ=I} -\sum_i A^*(\mathbf{u}_i) - \frac{1}{2\lambda}tr((X-U)(X-U)^T ZZ^T). \quad (13)$$

Denote (U^*, Z^*) as the solutions to (13). Then (U^*, M^*) are also the solutions to (12) where $M^* = Z^* Z^{*T}$. Because of $M^{*2} = M^*$, the problem (11) is equivalent to (10), suggesting that (U^*, Z^*) are also the solutions to (10).

After a series of equivalent transformations, one ends up at an efficiently solvable problem (13), enabling efficient solutions for the ePCA problem (6). The solutions can be achieved by alternately updating Z and U until convergence. According to the Procrustes rotation theorem [25], Z can be updated by

$$Z^{t+1} = Q^L_{max}((X-U^t)(X-U^t)^T),$$

where $Q^L_{max}(A)$ denotes the first L eigenvectors of A. U can be updated based on the gradient descent methods where the gradient is given by

$$-\frac{\partial}{\partial U}\sum_i A^*(\mathbf{u}_i) - \frac{1}{\lambda}ZZ^T(U-X).$$

Unlike the sequential optimization technique, this transformation strategy allows utilizing closed-form updating rules to search for the optimal Z that satisfies the orthogonal constraints, which thus results in effective and efficient solutions. Apart from these two general approaches to solve the ePCA problem, there are also other methods [18–20] proposed to solve the special cases of ePCA such as logistic PCA, which will be discussed in the following section.

3 A Special Case: Logistic PCA

We have previously introduced the general methods for solving the ePCA problem. In this section, we will discuss some other methods proposed to solve the special cases of ePCA, which are easier to implement than the general methods. A special case of ePCA refers to the applications of ePCA on a particular type of data that follows a certain distribution in the exponential family. For example, when the ePCA model is applied on a binary data set, this is a special case of ePCA called *logistic PCA* since the model exploits the log-odds as the natural parameter of the Bernoulli distribution and the logistic function as the canonical link.

In the applications of ePCA, the choice of appropriate form of $A(\cdot)$ function results in the appropriate distribution to model the given data, which shall intuitively lead to the best model performance. For example, the $A(\theta)$ function should be chosen as a quadratic function $\frac{\theta^2}{2}$ corresponding to the Gaussian distribution assumed for real-valued data. In this case, ePCA reduces to the standard PCA problem. Its corresponding objective function is also a quadratic function that allows the solution can be easily found by solving the corresponding singular value decomposition (SVD). Compared to the quadratic form, other more complex non-quadratic $A(\cdot)$ functions chosen for the distributions other than Gaussian make the general ePCA problem computationally challenging in the applications with non-real-valued data. Next, we will focus on logistic PCA, where $A(\theta) = \log(1 + e^{\theta})$ in the ePCA model for applications with binary data. The way PCA generalized to logistic PCA is analogous to the way linear regression generalized to logistic regression. In the following, we will discuss a strategy [20] proposed by Leeuw to solve logistic PCA which is easier to implement than the general methods.

The general ePCA problem (3) becomes the logistic PCA with $A(\theta) = \log(1 + e^{\theta})$:

$$
\begin{aligned}
\min_{\mathbf{b},Z,W} \ell(\mathbf{b}, Z, W) &= \min_{W:W^TW=I} \min_{\mathbf{b},Z} \sum_i A(W\mathbf{z}_i + \mathbf{b}) - tr((ZW^T + \mathbf{1}\mathbf{b}^T)X^T) \\
&= \min_{W:W^TW=I} \min_{\mathbf{b},Z} \sum_i \sum_j \log(1 + e^{\mathbf{z}_i^T \mathbf{w}_j + b_j}) - (\mathbf{z}_i^T \mathbf{w}_j + b_j)x_{ij} \\
&= \min_{W:W^TW=I} \min_{\mathbf{b},Z} \sum_i \sum_j -\log \frac{e^{\theta_{ij} x_{ij}}}{1 + e^{\theta_{ij}}}
\end{aligned}
$$

$$= \min_{W:W^TW=I} \min_{\mathbf{b},Z} \sum_i \sum_j -\log \pi(q_{ij}\theta_{ij}), \tag{14}$$

in which the canonical parameters are still assumed to take a low-rank representation $\theta_{ij} = \mathbf{z}_i^T \mathbf{w}_j + b_j$; $\pi(a) = \frac{e^a}{1+e^a}$; and

$$q_{ij} = 2x_{ij} - 1 = \begin{cases} -1, & x_{ij} = 0 \\ 1, & x_{ij} = 1. \end{cases}$$

As we observed, the objective function (the log-likelihood function) of (14) is non-quadratic. Instead of directly dealing with the non-quadratic objective function, the majorization-minimization (MM) algorithm [11, 17] is employed by Leeuw [20] to minimize the objective function by iteratively minimizing a suitably defined quadratic surrogate function. The MM algorithm guarantees that the objective function decreases in each iteration and converges to a local minimum of the original objective function. When applying the MM algorithm, the minimization of the surrogate function in each iteration is easier to solve than solving the original optimization problem but the surrogate function should be carefully chosen to satisfy several conditions for the performance guarantee of the MM algorithm.

Let $f(\theta)$ be the objective function to be minimized. At the kth iteration of the algorithm, a constructed convex function $g(\theta|\theta_k)$ will be a qualified surrogate function (the majorized version of the objective function) at θ_k if

$$g(\theta|\theta_k) \geq f(\theta) \quad for \quad all \quad \theta$$
$$g(\theta_k|\theta_k) = f(\theta_k).$$

The MM algorithm minimizes $g(\theta|\theta_k)$ instead of $f(\theta)$. In each iteration, θ is updated until convergence based on the following iteration rule:

$$\theta_{k+1} = \arg\min_{\theta} g(\theta|\theta_k).$$

The minimization of the surrogate function will drive $f(\theta)$ to converge to a local optimum as k goes to infinity, which can be proven as follows:

$$f(\theta_{k+1}) \leq g(\theta_{k+1}|\theta_k) \leq g(\theta_k|\theta_k) = f(\theta_k).$$

The majorization algorithm is also known as the variational methods, or variational bounding [12] in the machine learning literature.

In order to apply the MM algorithm to solve the logistic PCA problem (14), one need to find a surrogate function for $-\log \pi(\alpha)$. The surrogate function $g(\alpha|\alpha_k)$ is constructed by bounding the second-order derivative of $-\log \pi(\alpha)$ according to Leeuw [20]. Assume there exists a function $w \geq 0$ such that

$$-\log \pi(\alpha) \leq g(\alpha|\alpha_k)$$
$$= -\log \pi(\alpha_k) + \frac{-d \log \pi(\alpha)}{d\alpha}(\alpha - \alpha_k) + \frac{1}{2}w(\alpha_k)(\alpha - \alpha_k)^2 \quad (15)$$

for all α and α_k. Then, this bounding function w leads to a quadratic majorization of $-\log \pi(\alpha)$. One can choose

$$w = \frac{1}{4}$$

to achieve the uniform majorization or

$$w(\alpha_k) = \frac{1 - 2\pi(\alpha_k)}{2\alpha_k}$$

to achieve the non-uniform majorization according to [12]. By completing the square, the surrogate function of $-\log \pi(\alpha)$ can be rewritten as

$$g(\alpha|\alpha_k) = -\log \pi(\alpha_k) + \frac{1}{2}w(\alpha_k)\big(\alpha - (\alpha_k - \frac{h(\alpha_k)}{w(\alpha_k)})\big)^2 - \frac{1}{2}\frac{h^2(\alpha_k)}{w(\alpha_k)},$$

where $h(\alpha_k)$ denotes the first-order derivative of $-\log \pi(\alpha)$ at α_k. By substituting α by $q_{ij}\theta_{ij}$ and α_k by $q_{ij}(\theta_{ij})_k$ to $g(\alpha|\alpha_k)$, one will arrive at the surrogate function of $-\log \pi(q_{ij}\theta_{ij})$.

To solve the logistic PCA problem (14), the MM algorithm works by iteratively minimizing the surrogate function $g(q_{ij}\theta_{ij}|q_{ij}(\theta_{ij})_k)$ until convergence. In each iteration,

$$(Z_{k+1}, W_{k+1}, \mathbf{b}_{k+1}) = \arg \min_{Z,W,\mathbf{b}} \sum_i \sum_j g(q_{ij}\theta_{ij}|q_{ij}(\theta_{ij})_k)$$
$$= \arg \min_{Z,W,\mathbf{b}} \sum_i \sum_j \{q_{ij}\theta_{ij} - \big[q_{ij}(\theta_{ij})_k - \frac{h(q_{ij}(\theta_{ij})_k)}{w(q_{ij}(\theta_{ij})_k)}\big]\}^2$$
$$= \arg \min_{Z,W,\mathbf{b}} \sum_i \sum_j \{\theta_{ij} - \big[(\theta_{ij})_k - q_{ij}\frac{h(q_{ij}(\theta_{ij})_k)}{w(q_{ij}(\theta_{ij})_k)}\big]\}^2$$
$$s.t. \quad W^T W = I,$$

where $\theta_{ij} = \mathbf{z}_i^T \mathbf{w}_j + b_j$.

Let

$$M_{k+1} = \{(M_{ij})_{k+1}\} = (\theta_{ij})_k - q_{ij}\frac{h(q_{ij}(\theta_{ij})_k)}{w(q_{ij}(\theta_{ij})_k)}.$$

In the $(k+1)$th iteration, one can first calculate M_{k+1} based on Z_k, W_k and $\mathbf{b}_k$, and then update $(Z_{k+1}, W_{k+1}, \mathbf{b}_{k+1})$ by solving the following problem:

$$(Z_{k+1}, W_{k+1}, \mathbf{b}_{k+1}) = \arg\min_{Z,W,\mathbf{b}} ||ZW^T + \mathbf{1}\mathbf{b}^T - M_{k+1}||_F^2$$
$$s.t. \quad W^T W = I. \tag{16}$$

This is a least-square matrix approximation problem that can be easily solved by solving SVD. Thus, closed-form solutions are available for updating $(Z_{k+1}, W_{k+1}, \mathbf{b}_{k+1})$ in the $(k+1)$th iteration, which makes the MM algorithm easier to implement and more computationally efficient than the aforementioned general methods that involve computing gradients.

In summary, the algorithm works as follows. Start with some initial Z_0, W_0 and $\mathbf{b}_0$. Suppose Z_k, W_k and $\mathbf{b}_k$ are the current best solutions. We keep calculating M_{k+1} and updating $(Z_k, W_k, \mathbf{b}_k)$ to find a better solution $(Z_{k+1}, W_{k+1}, \mathbf{b}_{k+1})$ based on (16) until the log likelihood function converges. This is similar to the iterations between the E-step and the M-step in the EM-algorithm.

4 An Alternative Formulation: Projection of Saturated Model Parameters

The previous logistic PCA is formulated by the low-rank approximation of the canonical parameters for Bernoulli distributions, motivated by the interpretation of the standard PCA from the low-rank approximation perspective. The standard PCA assumes that the data follow Gaussian distributions and maximizes the log-likelihood function, resulting in the following problem:

$$\min \sum_i \sum_j (x_{ij} - E(x_{ij}))^2$$
$$= \min_{W:W^TW=I} \min_Z ||X - (ZW^T + \mathbf{1}\mathbf{b}^T)||_F^2,$$

where the expectation $E(x_{ij})$ equals the canonical parameter θ_{ij} for the Gaussian distribution which is approximated by a low-rank decomposition. The logistic PCA generalizes PCA by replacing the Gaussian distribution by Bernoulli and retaining the low-rank factorization of the canonical parameters. In this way, each sample has its own associated latent factor and the number of parameters increases with the number of samples. When applying logistic PCA to new data for prediction, one needs to carry out another matrix factorization, which is prone to overfit. Thus, the previous formulation of logistic PCA may not get satisfactory performance if applied for prediction on binary data. In contrast, the standard PCA predicts the principal component scores for the new data based only on the linear combinations of the observed values of the variables. It only requires to know the principal component loading matrix for prediction.

To maintain this property for other members in the exponential family, a new formulation was proposed by Landgraf and Lee to generalize PCA in such a way that the generalized principal component scores are linear functions of the data variables [15]. The Pearson's interpretation of PCA aims to find the optimal low-dimensional representation of multivariate data with the minimum squared error. Along this line, PCA is formulated as:

$$\min_{W:W^TW=I} \sum_{i=1}^{n} ||\mathbf{x}_i - \mathbf{b} - WW^T(\mathbf{x}_i - \mathbf{b})||_F^2,$$

which should also be equivalent to minimizing the Gaussian deviance $D(X, \Theta)$:

$$\sum_{i=1}^{n} ||\mathbf{x}_i - \boldsymbol{\theta}_i||_F^2$$

proportional to the negative log-likelihood. This guarantees that the Pearson's interpretation of PCA is essentially equivalent to the interpretation from maximizing the data log-likelihood perspective. According to Landgraf and Lee [15], the equivalence suggests that

$$\begin{aligned}\theta_i &= \mathbf{b} + WW^T(\mathbf{x}_i - \mathbf{b}) \\ &= \mathbf{b} + WW^T(\tilde{\theta}_i - \mathbf{b}),\end{aligned} \tag{17}$$

where $\tilde{\theta}_i$ denotes the canonical parameter of the saturated model that is the best possible fit to the data. Compared to the low-rank approximation of Θ, (17) is a different representation that treats Θ as the function of the canonical parameters of the saturated model. In this way, PCA is interpreted as a projection of the canonical parameter of the saturated model that minimizes the Gaussian deviance (or maximizes the likelihood).

To generalize PCA to binary data, one should minimize the Bernoulli deviance (or maximizes the likelihood) with respect to W that projects the canonical parameters of the corresponding saturated model to a lower dimensional space. The calculation of the canonical parameter of a saturated model is discussed below.

For a random variable x from a one-parameter exponential family distribution, the first-order derivative of $A(\theta)$ equals its expectation $E(x)$ where θ is the canonical parameter for x. The canonical link function $g(\cdot)$ is the inverse of the derivative of $A(\cdot)$ function, which suggests

$$\theta = g(E(x)).$$

Let $\tilde{\theta}$ denote the canonical parameter for the saturated model. As the value of each variable equals its expectation in the saturated model, one will have

$$\tilde{\theta} = g(x).$$

For example, $\tilde{\theta} = x$ for the Gaussian distribution; $\tilde{\theta} = \text{logit}(x)$ for Bernoulli distribution; and $\tilde{\theta} = \log(x)$ for Poisson distribution.

Since $\tilde{\Theta}$ contains constant values for a multivariate random vector, an alternative formulation of generalized PCA is an optimization problem with respect to W and $\mathbf{b}$:

$$\min_{W:W^TW=I} \min_{\mathbf{b}} \sum_i A(\mathbf{b} + WW^T(\tilde{\theta}_i - \mathbf{b})) - tr((\mathbf{1b}^T + (\tilde{\Theta} - \mathbf{1b}^T)WW^T)X^T).$$

When it is applied to binary data, one should assume correspondingly that x_{ij} is sampled from Bernoulli(p_{ij}) where p_{ij} is the success probability that is also the expectation $E(x_{ij})$. The natural parameter for the Bernoulli distribution is $\theta_{ij} = \text{logit}(p_{ij})$. The saturated model occurs when $p_{ij} = x_{ij}$, giving

$$\tilde{\theta}_{ij} = logit(x_{ij}) = \begin{cases} -\infty, & x_{ij} = 0 \\ \infty, & x_{ij} = 1. \end{cases}$$

As the value of $\tilde{\theta}_{ij}$ in this case is not finite, Landgraf and Lee [15] suggest that one can approximate $\tilde{\theta}_{ij}$ by mq_{ij} for a large number m where q_{ij} is defined as $2x_{ij} - 1$ which equals $\{-1, 1\}$ when x_{ij} takes values $\{0, 1\}$. Given $\tilde{\Theta} = \{\tilde{\theta}_{ij}\}$, the logistic PCA can be formulated as:

$$\min_{W:W^TW=I} \min_{\mathbf{b}} \sum_i \mathbf{1}^T \log(1 + \exp(\mathbf{b} + WW^T(\tilde{\theta}_i - \mathbf{b}))) \\ - tr((\mathbf{1b}^T + (\tilde{\Theta} - \mathbf{1b}^T)WW^T)X^T), \qquad (18)$$

where the objective function can be re-expressed as:

$$\sum_i \sum_j -\log \frac{e^{\theta_{ij}x_{ij}}}{1 + e^{\theta_{ij}}} \\ = \sum_i \sum_j -\log \pi(q_{ij}\theta_{ij}),$$

with

$$\theta_{ij} = [\mathbf{1b}^T + (\tilde{\Theta} - \mathbf{1b}^T)WW^T]_{ij}.$$

As we can observe, the objective function of (18) has the same form as that of the previous logistic PCA problem (14) when their objective functions are expressed as functions of canonical parameters θ_{ij}. From the optimization perspective, the only difference between these two formulations of logistic PCA lies in the representation of θ_{ij}, which thus leads to two different optimization problems after θ_{ij} is substituted. Therefore, all the properties we have discussed on $-\log \pi(q_{ij}\theta_{ij})$ still apply here.

Similarly, one can also employ the MM algorithm to solve (18). As we discussed, the surrogate function $g(q_{ij}\theta_{ij}|q_{ij}(\theta_{ij})_k)$ for $-\log\pi(q_{ij}\theta_{ij})$ is

$$\{\theta_{ij} - \big[(\theta_{ij})_k - q_{ij}\frac{h(q_{ij}(\theta_{ij})_k)}{w(q_{ij}(\theta_{ij})_k)}\big]\}^2 + C,$$

where C is a constant term irrelevant with θ_{ij}. Let

$$\begin{aligned}(M_{ij})_{k+1} &= (\theta_{ij})_k - q_{ij}\frac{h(q_{ij}(\theta_{ij})_k)}{w(q_{ij}(\theta_{ij})_k)},\\ \text{and} \quad M_{k+1} &= \{(M_{ij})_{k+1}\}.\end{aligned}$$

Then, the surrogate function of $\sum_i \sum_j -\log\pi(q_{ij}\theta_{ij})$ is

$$\begin{aligned}&\sum_i \sum_j \{\theta_{ij} - (M_{ij})_{k+1}\}^2 + C\\ &= ||\Theta - M_{k+1}||_F^2 + C.\end{aligned}$$

Remember that the new formulation of logistic PCA differs from the previous formulation in the representation of Θ which equals $\mathbf{1b}^T + (\tilde{\Theta} - \mathbf{1b}^T)WW^T$ instead of $ZW^T + \mathbf{1b}^T$. Thus, one will have the surrogate function of the objective function in (18) as

$$||\mathbf{1b}^T + (\tilde{\Theta} - \mathbf{1b}^T)WW^T - M_{k+1}||_F^2 + C,$$

where M_{k+1} is correspondingly calculated based on W_k and $\mathbf{b}_k$ by substituting Θ_k.

The MM algorithm works by iteratively minimizing this surrogate function. In the $(k+1)$th iteration, one can first calculate M_{k+1} based on W_k and $\mathbf{b}_k$, and then update W_{k+1} and $\mathbf{b}_{k+1}$ by solving the following problem:

$$\begin{aligned}(W_{k+1}, \mathbf{b}_{k+1}) &= \arg\min_{W,\mathbf{b}} ||\mathbf{1b}^T + (\tilde{\Theta} - \mathbf{1b}^T)WW^T - M_{k+1}||_F^2\\ &s.t. \quad W^TW = I, \end{aligned} \tag{19}$$

which will simultaneously drive the objective function to decrease for each iteration. The solutions to (19) can be achieved by alternately updating the unknown variables until convergence. Let $\tilde{\mathbf{b}}^{t+1}$ and $\tilde{W}^{t+1}$ denote the $(t+1)$th step estimates when solving problem (19). The updating rules for them are shown as follows.

Update **b**

$$\tilde{\mathbf{b}}^{t+1} = \frac{1}{n}(M_{k+1}^t - \tilde{\Theta}\tilde{W}^t\tilde{W}^t)^T\mathbf{1}.$$

Update W

Denote

$$P = \tilde{\Theta} - \mathbf{1}(\tilde{\mathbf{b}}^t)^T$$

and

$$Q = M_{k+1} - \mathbf{1}(\tilde{\mathbf{b}}^t)^T.$$

Solve

$$\min_W ||PWW^T - Q||_F^2$$
$$s.t. \quad W^T W = I.$$

$\tilde{W}^{t+1}$ is updated as the first L eigenvectors of

$$P^T Q + Q^T P - P^T P$$

by solving the eigen-decomposition [8]. If it takes r iterations to converge when solving problem (19), then W_{k+1} and $\mathbf{b}_{k+1}$ will be estimated as $\tilde{W}^r$ and $\tilde{\mathbf{b}}^r$ respectively.

In summary, the MM algorithm works as follows. Start with some initial W_0 and $\mathbf{b}_0$. Suppose W_k and $\mathbf{b}_k$ are the current best solutions. One need to keep calculating M_{k+1} and updating $(W_k, \mathbf{b}_k)$ to find a better solution $(W_{k+1}, \mathbf{b}_{k+1})$ based on (19) until the log-likelihood function converges. The above MM algorithm and updating rules can also be employed for solving the alternative formulations of ePCA problems generalized to other types of data [16], in which the corresponding surrogate function has to be re-defined and $\tilde{\theta}_{ij}$ should be replaced by the value of the corresponding canonical link function for the given data x_{ij}.

5 Applications: Dimension Reduction and Aggregate Association Study

The generalization of PCA to exponential family enables the extension of PCA to many modern applications with various types of data frequently appearing in biomedicine, finance, and electronic commerce. Compared with PCA, the performance of dimension reduction will be enhanced by applying the corresponding ePCA suitable to the data types. In the recommender systems built by the e-commerce sites, such as Amazon.com, the ratings from customers for each item are in unary or finer continuous scales. By assuming the data distribution as Bernoulli (unary case) distribution or multinomial distribution (integer case) in the ePCA model, the user and/or item characteristics are extracted by the top principal components for further calculation of user and/or item similarity for recommendations. In this case, ePCA is applied as a latent factor analysis tool for unsupervised study.

Apart from unsupervised applications, ePCA can also be embedded in a regression framework for aggregate association analysis. For example, in the applications of identifying the causal factors for diseases, millions of omics variables measured using different high-throughput profiling technologies may contain continuous, binary, or count data. The corresponding data distribution in the ePCA model is thus assumed as Gaussian, Bernoulli, or Poisson respectively when applying ePCA to analyze such diverse omics data mapped to pathways or functional regions. The resulting principal component scores are regarded as the surrogate aggregate signals for the pathways to analyze their associations with the disease. The analysis of association between pathways and disease provides a better way to understand disease etiology from a systematic perspective, which also allows relevant robust results compared to the association analysis of millions of individual genetic variables. Typically, the aggregate association analysis of pathways are accomplished by performing regression analysis between the disease outcome and the aggregate signals for the pathways [21–24].

How to derive good aggregate signals for a pathway is a critical problem that will affect the results of consequent association analyses. With appropriate assumptions of the data distributions tailored to the data types, ePCA is a promising tool that can generate better aggregate signals to capture more accurate genetic variations. Moreover, ePCA should be performed on the subset of genetic variables (GVs) containing the most influential information regarding the disease outcome because the irrelevant GVs will dilute the aggregate signal for the given pathway. The selection of such a subset can be done by two means: (1) automatic selection from the integrated supervised framework involving sparse penalization on the PC loading vectors. (2) heuristic selection guided by the performance of statistical association with disease outcome. The second way is simple and widely applied in the literature [1, 2, 4, 21, 23, 24]. In each pathway, its mapped GVs are first ranked based on their statistical association with disease outcome and grouped as candidate units by gradually increasing the size as typically done in forward feature selection. With the ranked list, ePCA is then implemented to derive multiple potential summary statistics for the formed candidate groups respectively. The final statistical significance of each pathway is the best value derived from the candidate group with the most discriminating power. The heuristic procedure of applying ePCA for aggregate association analysis is summarized as follows:

(1) *Generate candidate GV groups for each pathway*
For each individual GV assigned to a given pathway, its significance (p-value) can be computed by fitting a logistic regression model. Given all GVs belonging to a pathway, we generate several (for example, 20) incremental candidate groups by setting 20 thresholds at each increment of 5 percentiles of p-values for those GVs. Hence, for each pathway, 20 groups of GVs $\{S_1, \ldots, S_{20}\}$ are formed by sequentially grouping GVs with p-values less than each corresponding threshold.
(2) *ePCA on candidate GV groups*
ePCA can be implemented to compute the first PC scores for 20 candidate groups respectively in each pathway.
(3) *Calculate M statistics for candidate groups*
For each candidate group $S_\ell (1 \le \ell \le 20)$, we fit a logistic regression model using the corresponding first PC scores as regressors and estimate the t-statistic t_ℓ. Let $M = \{t_\ell : |t_\ell| = max_{1 \le \ell \le 20} |t_\ell|\}$.

(4) *Estimate the null distribution of M statistics*
For each pathway, we perform the permutation test by generating random disease status for each sample from a Bernoulli distribution with the success probability set to the disease prevalence. Based on randomly generated outcomes, the corresponding M statistic for each pathway can be calculated by repeating steps (1) to (3) as a random sample from the null distribution of M.

(5) *Calculate p-value for each pathway*
Given the M value for each pathway based on true disease status and its corresponding null distribution of M statistic, an empirical p-value for each pathway can be calculated to estimate the pathway significance. This provides a self-contained test which compares pathways to the non-associated genomic background.

By embedding ePCA in a heuristic supervised framework, it has the potential to aggregate weak signals from individual GVs with the explicit modeling of categorical GV data considering outcome. The supervised ePCA is expected to provide more effective association analysis for high-dimensional modern data and is more likely to produce reproducible results.

6 Sparse ePCA Through Penalization

As we have discussed in Sect. 1, ePCA is an attractive method to reduce the dimensionality of exponential family data by making respective appropriate model assumptions based on the specific data types. In addition to deriving low-dimensional projections for model complexity reduction and reproducibility of learning results, people often want to know the physical meanings of the original variables and how they contribute to these projections. For example, when analyzing images, it is of much interest to know which image regions are crucial to represent or capture the essential information contained in the given images. Identifying variables expressing the maximum data variation will also be of interest for next-generation sequencing data analysis in bioinformatics since it would help greatly reduce the profiling cost for biomarker discovery. This is one of the motivations for deriving the sparse model extension of ePCA, named as sparse ePCA, to better interpret the obtained principal components based only on a subset of contributing variables. When the exponential family data are in high-dimensional spaces with $n \ll d$, the model consistency and effectiveness of ePCA can be questionable, for which the sparse models can help greatly improve the performance. The theoretical discussion of the model inconsistency of standard PCA when d is comparable or larger than n can be found in the literature [13, 27, 28].

The sparse ePCA model enforces sparse non-zero entries in the principal component loading matrix by adding a penalty term of the loading vectors into the ePCA model formulation discussed in Sect. 1. This idea is similar to the way of formulating sparse generalized linear models such as sparse logistic regression and sparse loglinear (Poisson) regression. Both of them impose the assumptions of exponential family distributions and sparsity regularization simultaneously in their models; however, sparse ePCA encounters even more difficult challenges in finding its solutions. The difficulties are mainly caused by the non-joint convexity of its objective function and the non-convex and non-smooth constraints in the resulting optimization problem.

A general form of the sparse ePCA problem can be formulated as follows:

$$\min_{Z:Z^TZ=I} \min_{W,\mathbf{b}} \sum_{i=1}^{n} A(W\mathbf{z}_i + \mathbf{b}) - tr((ZW^T + \mathbf{1b}^T)X^T) + P(W, \lambda), \tag{20}$$

where $P(W, \lambda)$ denotes the penalty term to enforce sparse non-zero entries of W. A non-zero entry w_{jl} indicates that the jth variable is selected contributing to the lth PC. There are various choices of the penalty function $P(.,.)$. For example, $\sum_{j=1}^{d}\sum_{l=1}^{L} P(|w_{jl}|, \lambda)$ and $\sum_{j=1}^{d} P(\|\mathbf{w}_j\|_2, \lambda)$ to achieve the element-wise sparsity and row-sparsity respectively. The tuning parameter λ controls the level of sparsity. A larger value of λ will result in fewer non-zero entries in W. This penalized maximum-likelihood estimator attempts to estimate the optimal PC loading vectors that are sparse and meanwhile maintain high accuracy in low-rank approximation. The resulting parsimonious model enables meaningful interpretation of the derived PCs and is expected to alleviate the inconsistency of ePCA in applications with high-dimensional data. In other words, sparse ePCA is an enhanced ePCA model with the aim to improve the performance and model interpretation. It is encouraged to be applied for reducing dimensionality of high-dimensional modern data of various data types, including real-valued data, binary data, categorical data, and count data.

7 Two Computational Algorithms for Sparse ePCA

As we mentioned previously, it is not easy to solve the sparse ePCA problem (20) though the objective function is marginally convex with respect to those unknown variables. One of the big challenges lies in the non-convex and non-smooth constraints on Z. As we discussed for the ePCA problem (3), utilizing variants of gradient descent methods to alternately update $\mathbf{b}$, W and Z is unsatisfactory and computationally slow to solve sparse ePCA problem under such constraints. This problem can be phrased as a Stiefel manifold optimization one, for which packages are already available [32, 37]. These methods have been claimed to be valid yet still computationally expensive for high-dimensional data sets [38].

An alternative way is to employ the MM algorithm to optimize a surrogate function for the original objective function, which is guaranteed to reach a local minimum. The MM algorithm has already been employed by Lee et al. [19] to solve sparse logistic PCA problem that performs dimension reduction for binary data. The sparse logistic PCA problem is a specific case of sparse ePCA tailored to binary data. For the binary data following Bernoulli distributions, the corresponding data log-likelihood can be written as a summation of a set of log inverse logit functions, for which the quadratic upper bounds for the negative log inverse logit functions specified by Jaakkola and Jordan [12]; de Leeuw [20] are chosen as the surrogate functions. Finding an appropriate surrogate function is crucial to ensure that the MM algorithm works. Sometimes it is quite challenging when the objective function is complex for other different types of data, for example, categorical and count data. Recently, a general form of surrogate functions were proposed by Zhang and She [38] to solve the sparse ePCA problem.

Instead of employing the MM algorithm, Lu et al. [22] have proposed another way to solve sparse ePCA problem by transforming it into an equivalent problem that can be solved more effectively and efficiently. Closed-form updating rules are available for alternately updating the unknown variables in the new problem to achieve high computational efficiency.

Next, the above two strategies: the MM algorithm and transformation by convex conjugate will be discussed in detail.

7.1 *Majorization-Minimization*

To address the difficulty in solving (20) caused by non-quadratic objective function with non-convex constraints, the MM algorithm can help by solving a rather easier minimization problem instead with a quadratic surrogate function as the objective function. Compared to directly optimizing the original objective function, minimizing a quadratic function under the orthogonal constraints can be efficiently solved using closed-form updating rules according to Procrustes rotation theorem [25]. Then, the big concern we have now is how to construct the quadratic surrogate function to guarantee the MM algorithm works. A qualified surrogate function has to be convex and meet some inequality conditions, as we discussed earlier.

For the sparse ePCA problem, one can rewrite the objective function in (20) as

$$f(\Theta) = \sum_{i=1}^{n} A(\boldsymbol{\theta}_i) - tr(\Theta X^T) + P(W, \lambda),$$

by assuming that $\Theta = ZW^T + \mathbf{1b}^T$, taking the usual low-rank representation.

Let $l(\Theta) = \sum_{i=1}^{n} A(\boldsymbol{\theta}_i) - tr(\Theta X^T)$. Zhang and She [38] define

$$\begin{aligned} g(\Theta|\Theta_k) &= l(\Theta_k) + \nabla_{\Theta} l(\Theta_k)(\Theta - \Theta_k)^T + \frac{\rho_k}{2}\|\Theta - \Theta_k\|_F^2 + P(W, \lambda) \\ &= l(\Theta_k) + \Big(\sum_{i=1}^{n} \frac{\partial A(\boldsymbol{\theta}_i)}{\partial \Theta}|_{\Theta=\Theta_k} - X\Big)(\Theta - \Theta_k)^T + \frac{\rho_k}{2}\|\Theta - \Theta_k\|_F^2 + P(W, \lambda) \end{aligned}$$

as the surrogate function. It is clear that $g(\mathbf{b}_k, Z_k, W_k|\mathbf{b}_k, Z_k, W_k) = f(\mathbf{b}_k, Z_k, W_k)$. Moreover, they claimed that $g(\mathbf{b}, Z, W|\mathbf{b}_k, Z_k, W_k) \geq f(\mathbf{b}, Z, W)$ can be realized by Taylor expansion when setting a large enough value for ρ_k. Details in choosing the appropriate ρ_k for each iteration are discussed in [38]. With these conditions satisfied, minimizing this quadratic surrogate function will guarantee that $f(\Theta)$ can converge to a local minimum by applying the MM algorithm. Thus, one can approximate the solutions by iteratively updating $\mathbf{b}$, Z, W based on

$$(\mathbf{b}_{k+1}, Z_{k+1}, W_{k+1}) = \arg\min_{\mathbf{b},Z,W} g(\mathbf{b}, Z, W|\mathbf{b}_k, Z_k, W_k). \quad (21)$$

This problem (21) can be rewritten as

$$\begin{aligned}(\mathbf{b}_{k+1}, Z_{k+1}, W_{k+1}) &= \arg\min_{\mathbf{b},Z,W} \Big(\sum_{i=1}^{n} \frac{\partial A(\boldsymbol{\theta}_i)}{\partial \Theta}|_{\Theta=\Theta_k} - X\Big)\big(ZW^T + \mathbf{1}\mathbf{b}^T - \Theta_k\big)^T \\ &\quad + \frac{\rho_{k+1}}{2}\|ZW^T + \mathbf{1}\mathbf{b}^T - \Theta_k\|_F^2 + P(W, \lambda) \\ &= \arg\min_{\mathbf{b},Z,W} \frac{1}{2}\|ZW^T + \mathbf{1}\mathbf{b}^T - M_{k+1}\|_F^2 + \frac{1}{\rho_{k+1}} P(W, \lambda), \quad (22)\end{aligned}$$

where $M_{k+1} = \Theta_k + \frac{1}{\rho_{k+1}}\big(X - \sum_{i=1}^{n} \frac{\partial A(\boldsymbol{\theta}_i)}{\partial \Theta}|_{\Theta=\Theta_k}\big)$. The resulting optimization problem (22) is a quadratic minimization problem with orthogonal constraints, which can be easily solved by alternately updating $\mathbf{b}$, Z, W based on the closed-form updating rules shown below. Let $\tilde{\mathbf{b}}^{t+1}$, $\tilde{Z}^{t+1}$ and $\tilde{W}^{t+1}$ denote the corresponding $(t+1)$th step estimates of $\mathbf{b}$, Z and W respectively.

Update $\mathbf{b}$

$$\tilde{\mathbf{b}}^{t+1} = \frac{1}{n}(M_{k+1}^T - \tilde{W}^t(\tilde{Z}^t)^T)\mathbf{1}.$$

Update Z

Given $\tilde{\mathbf{b}}^t$ and $\tilde{W}^t$, the minimization problem with respect to Z is:

$$\min_{Z:Z^TZ=I} ||M_{k+1}^T - \mathbf{1}(\tilde{\mathbf{b}}^t)^T - Z(\tilde{W}^t)^T||_F^2.$$

This can be identified as a Procrustes rotation problem that has closed-form solutions. Denote Q as $M_{k+1} - \tilde{\mathbf{b}}^t\mathbf{1}^T$. One can first compute the SVD of $Q(\tilde{W}^t)^T = R\Lambda V^T$ and then update $\tilde{Z}^{t+1}$ by $R[, 1 : L]V^T$.

Update W

Given $\tilde{\mathbf{b}}^t$ and $\tilde{Z}^t$, the minimization problem with respect to W is a penalized least square problem:

$$\min_W ||M_{k+1}^T - \mathbf{1}(\tilde{\mathbf{b}}^t)^T - \tilde{Z}^t W^T||_F^2 + \frac{2}{\rho_{k+1}} P(W, \lambda).$$

To estimate $\tilde{W}^{t+1}$, one can refer to many existing work studying this type of problems which adopt various penalty functions including such as ℓ_1, ℓ_0, ℓ_1/ℓ_2, SCAD [7], and $\ell_p(0 < p < 1)$. Proximal gradient method is attractive to solve such convex problems with non-smooth norm penalizations. Iterative reweighted ℓ_1 and ℓ_2 minimization are also two efficient schemes that help produce more focal estimates as optimization progresses. The non-separable iterative reweighing algorithms can even enforce row-sparsity on the presentation of W that makes variable selection occur in a more natural way [6]. In addition, thresholding-based iterative selection procedure (TISP) [31] can be used to solve a P-penalized problem for any penalty term P associated with a thresholding rule [33].

If it takes r iterations to converge when solving problem (22), then $\mathbf{b}_{k+1}$, Z_{k+1} and W_{k+1} will be estimated as $\tilde{\mathbf{b}}^r$, $\tilde{Z}^r$ and $\tilde{W}^r$ respectively.

7.2 Transformation by Convex Conjugate

The reformulation of the sparse ePCA problem (20) is achieved via replacing the term $A(W\mathbf{z}_i + \mathbf{b})$, which is not jointly convex in Z and W, by introducing its convex conjugate. The convex conjugate for a function $h(\boldsymbol{\alpha})$ is defined as:

$$h^*(\mathbf{u}) = \sup_{\boldsymbol{\alpha} \in M} <\mathbf{u}, \boldsymbol{\alpha}> - h(\boldsymbol{\alpha}),$$

where $h^*(\mathbf{u})$ is always convex since the maximum of a linear function is convex. Let $A^*(\cdot)$ denote the convex conjugate of $A(\cdot)$. The explicit form of $A(\cdot)$ and $A^*(\cdot)$ specific to each distribution in the exponential family are listed in the following table 2.

Let Θ be a $n \times d$ matrix whose ith row is $\boldsymbol{\theta}_i$. By introducing linear constraints, the problem (20) is first rewritten as follows:

$$\min_{Z:Z^TZ=I} \min_{W,\mathbf{b}} \min_{\Theta} \sum_i A(\boldsymbol{\theta}_i) + g(Z, W, \mathbf{b})$$
$$s.t. \;\; \boldsymbol{\theta}_i = W\mathbf{z}_i + \mathbf{b}, \quad 1 \le i \le n, \tag{23}$$

where $g(Z, W, \mathbf{b}) = -tr((ZW^T + \mathbf{1}\mathbf{b}^T)X^T) + P(W, \lambda)$. Due to the potentially complex $A(\cdot)$ function forms, it is difficult to directly solve (23). In order to replace

Table 2 Convex conjugate duals corresponding to $A(\cdot)$ for several well-known exponential family members of scalar variables

Distribution	$A(\theta)$	$A^*(u)$	$\frac{\partial A^*(u)}{\partial u}$
Gaussian	$\frac{\theta^2}{2}$	$\frac{u^2}{2}$	u
Bernoulli	$\log(1+\exp(\theta))$	$u \log u + (1 - u)\log(1-u)$	$\log \frac{u}{1-u}$
Poisson	$\exp(\theta)$	$u \log u - u$	$\log u$
Exponential	$-\log(-\theta)$	$-1 - \log u$	$-\frac{1}{u}$

the complex $A(\cdot)$, the minimization of $A(\cdot)$ is first transformed to its equivalent dual problem based on the following Lemma 1 proposed by Lu et al. [22].

Lemma 1 *Let U be the $n \times d$ matrix whose ith row is $\mathbf{u}_i$. The inner minimization of* (23) *with respect to Θ is equivalent to solving the dual problem:*

$$\max_{U} -\sum_{i} A^*(-\mathbf{u}_i) - <\mathbf{u}_i, W\mathbf{z}_i + \mathbf{b}> + g(Z, W, \mathbf{b}).$$

Proof The Lagrangian of (23) is defined as

$$\sum_{i} A(\boldsymbol{\theta}_i) + <\mathbf{u}_i, (\boldsymbol{\theta}_i - W\mathbf{z}_i - \mathbf{b})> + g(Z, W, \mathbf{b}).$$

The inner minimization of (23) on Θ can be reformulated as the saddle point problem

$$\min_{\Theta} \max_{U} \sum_{i} A(\boldsymbol{\theta}_i) + <\mathbf{u}_i, \boldsymbol{\theta}_i> - <\mathbf{u}_i, W\mathbf{z}_i + \mathbf{b}> + g(Z, W, \mathbf{b}). \qquad (24)$$

Since the inner minimization of (23) on Θ is a convex problem with feasible linear constraints, it satisfies Slater's conditions for strong duality and the order of minimization and maximization in (24) can be exchanged:

$$\begin{aligned}
&\max_{U} \min_{\Theta} \sum_{i} A(\boldsymbol{\theta}_i) + <\mathbf{u}_i, \boldsymbol{\theta}_i> - <\mathbf{u}_i, W\mathbf{z}_i + \mathbf{b}> + g(Z, W, \mathbf{b}) \\
&= \max_{U} -(\max_{\Theta} \sum_{i} -A(\boldsymbol{\theta}_i) - <\mathbf{u}_i, \boldsymbol{\theta}_i>) - <\mathbf{u}_i, W\mathbf{z}_i + \mathbf{b}> + g(Z, W, \mathbf{b}) \\
&= \max_{U} -\sum_{i} A^*(-\mathbf{u}_i) - <\mathbf{u}_i, W\mathbf{z}_i + \mathbf{b}> + g(Z, W, \mathbf{b}),
\end{aligned}$$

which completes the proof for the lemma. □

Then, based on Lemma 1, the original optimization problem (20) can be transformed to an equivalent problem by Theorem 1 according to Lu et al. [22].

Theorem 1 *The optimization problem* (20) *is equivalent to*

$$\min_{Z:Z^TZ=I} \min_{W,\mathbf{b}} \max_{U} -\sum_{i} A^*(-\mathbf{u}_i) - tr((ZW^T + \mathbf{1}\mathbf{b}^T)(U + X)^T) + P(W, \mathbf{b}). \qquad (25)$$

Proof It suffices to show that (23) is equivalent to (25). From Lemma 1, it is straightforward to prove that (23) is equivalent to the following problem:

$$\min_{Z:Z^TZ=I}\min_{W,\mathbf{b}}\max_{U} -\sum_i A^*(-\mathbf{u}_i) - <\mathbf{u}_i, W\mathbf{z}_i + \mathbf{b}> + g(Z, W, \mathbf{b})$$
$$= \min_{Z:Z^TZ=I}\min_{W,\mathbf{b}}\max_{U} -\sum_i A^*(-\mathbf{u}_i) - tr((ZW^T + \mathbf{1b}^T)(U + X)^T) + P(W, \mathbf{b}),$$

which leads to (25) and completes the proof for the theorem. □

Closed-form updating rules

Despite of the non-quadratic objective function and non-convex constraints, one can still find the closed-form updating rules to solve (25) with good solution quality. The algorithm based on these updating rules will converge much faster than gradient descent methods. The solutions are achieved by alternately updating the unknown variables based on the closed-form solutions, which are given below.

Let $f(Z, W, \mathbf{b}, U)$ denote the objective function of the min-max problem (25). Obviously, $f(\cdot, \cdot, \cdot, U)$ is concave in U. In each iteration, one can update U by solving the following optimization problem:

$$\max_{U} -\sum_i A^*(-\mathbf{u}_i) - tr((ZW^T + \mathbf{1b}^T)U^T). \tag{26}$$

According to Lu et al. [22], we have:

Theorem 2 *The optimal* $\hat{\mathbf{u}}_i$ *to the problem* (26) *is the negative mean vector of the sample* $\mathbf{x}_i$*:* $-E_{\theta_i}[\mathbf{x}_i]|_{\theta_i = W\mathbf{z}_i + \mathbf{b}}$*, which is also equal to* $-\frac{\partial A(\theta_i)}{\partial \theta_i}|_{\theta_i = W\mathbf{z}_i + \mathbf{b}}$.

Proof To solve (26), one need the following result as proposed by [36]: For all canonical parameters $\boldsymbol{\gamma}$ of the exponential family distribution of random variables $\mathbf{y} \in \mathcal{Y}$, $\sup_{\boldsymbol{\mu} \in \mathcal{M}}\{\langle \boldsymbol{\gamma}, \boldsymbol{\mu}\rangle - A^*(\boldsymbol{\mu})\}$ is attained uniquely at the mean vector $\boldsymbol{\mu}^*$ specified by the moment matching condition:

$$\boldsymbol{\mu}^* = \int_{\mathcal{Y}} \mathbf{y} p(\mathbf{y}|\boldsymbol{\gamma}) d\mathbf{y} = E_{\boldsymbol{\gamma}}[Y].$$

Similarly, consider an optimization problem: $\max_{\mathbf{v} \in \mathcal{M}'} - \langle \boldsymbol{\gamma}, \mathbf{v}\rangle - A^*(-\mathbf{v})$, where $\mathcal{M}' = \{m : -m \in \mathcal{M}\}$. Its maximum is attained at the vector $\mathbf{v}^* = -\boldsymbol{\mu}^* = -E_{\boldsymbol{\gamma}}[Y]$.

According to this result, the optimal $\hat{\mathbf{u}}_i$ in (26) is obtained as the negative mean vector of the sample $\mathbf{x}_i$: $-E_{\theta_i}[\mathbf{x}_i]|_{\theta_i = W\mathbf{z}_i + \mathbf{b}}$. Since the mean vector is further shown to be equal to the first-order derivative of the log-normalization factor $A(\cdot)$ according to Proposition 1 given in [36], we have $\hat{\mathbf{u}}_i = -\frac{\partial A(\theta_i)}{\partial \theta_i}|_{\theta_i = W\mathbf{z}_i + \mathbf{b}}$. One can also verify this solution by setting the first-order derivative of the objective function in (26) with respect to $\mathbf{u}_i$ equal to 0. □

For the outer minimization problem on Z, W and $\mathbf{b}$, Lu et al. [22] consider the penalty term $P(W, \mathbf{b})$ using $\lambda_0||ZW^T + \mathbf{1b}^T||_F^2 + \sum_{l=1}^{L} \lambda_l |W_{:l}|$ where $W_{:l}$ denote the lth column of W. The l_2-norm penalty term is involved here to ensure the stable reconstruction of principal components when $n < d$ and X is not a full rank matrix. It could also be interpreted as a Gaussian prior for canonical parameters to ensure the stability of the model. Then, the objective function $f(Z, W, \mathbf{b}, \cdot)$ is quadratic as shown below.

$$
\begin{aligned}
& f(Z, W, \mathbf{b}, .)|_{Z^T Z=I} \\
&= -tr((ZW^T + \mathbf{1}\mathbf{b}^T)(U + X)^T) + \lambda_0 ||ZW^T + \mathbf{1}\mathbf{b}^T||_F^2 + \sum_{l=1}^{L} \lambda_l |W_{:l}| + C_0 \\
&= \lambda_0 ||\frac{1}{2\lambda_0}(X + U) - ZW^T - \mathbf{1}\mathbf{b}^T||_F^2 + \sum_{l=1}^{L} \lambda_l |W_{:l}| + C_1,
\end{aligned}
$$

where C_0 and C_1 are constant terms unrelated to Z, W, or $\mathbf{b}$. Although the minimization problem with respect to Z, W, and $\mathbf{b}$ involves non-convex constraints, an efficient solution can be achieved owing to the elegant problem structure.

Specifically, in the $(t + 1)$-th iteration, given an optimal U^t, $\mathbf{b}^{t+1}$ can be updated as:

$$
\mathbf{b}^{t+1} = \frac{1}{N}\big(\frac{1}{2\lambda_0}(X + U^t) - Z^t W^{tT}\big)^T \mathbf{1}.
$$

To update Z, the minimization problem with respect to Z is:

$$
\begin{aligned}
& \min_{Z:Z^T Z=I} ||\frac{1}{2\lambda_0}(X + U) - \mathbf{1}\mathbf{b}^T - ZW^T||_F^2 \\
\equiv & \min_{Z:Z^T Z=I} ||\frac{1}{2\lambda_0}(X + U)^T - \mathbf{b}\mathbf{1}^T - WZ^T||_F^2.
\end{aligned}
$$

Denote Q as $\frac{1}{2\lambda_0}(X + U) - \mathbf{1}\mathbf{b}^T$. One need first compute the SVD of $Q^t W^t = R\Lambda V^T$ and then update Z^{t+1} by $R[, 1 : L]V^T$ according to the Procrustes rotation approach [25].

To update W, the minimization problem with respect to W is a LASSO problem:

$$
\begin{aligned}
& \min_W ||\frac{1}{2\lambda_0}(X + U) - ZW^T - \mathbf{1}\mathbf{b}^T||_F^2 + \sum_{l=1}^{L} \frac{\lambda_l}{\lambda_0} |W_{:l}| \\
&\equiv \min_W ||Q - ZW^T||_F^2 + \sum_{l=1}^{L} \frac{\lambda_l}{\lambda_0} |W_{:l}|.
\end{aligned}
$$

The optimal $W_{:l}^{t+1}$ for $l = 1, \ldots, L$ is given by $\big(|Q^{tT} Z_{:l}^t| - \frac{\lambda_l}{2\lambda_0}\big)_+ Sign(Q^{tT} Z_{:l}^t)$, where $Z_{:l}$ denotes the lth column of Z corresponding to the lth PC.

Computational complexity

In the initialization step, it takes $O(nd^2)$ computational operations to compute the SVD. Our algorithm contains two main steps: maximization of U and minimization of Z, W, and $\mathbf{b}$. Computing U has the computational complexity of $O(ndL)$ in each iteration. In each iteration of optimizing Z, computing QW^T and the corresponding SVD has the complexity $O(ndL)$ and $O(nL^2)$, respectively. The estimation of W using the soft-thresholding operation has the complexity of $O(ndL)$ in each iteration. In total, the computational complexity is $O(nd^2) + rO(ndL)$ if it takes r iterations to converge. If $n \ll d$, the cost of SVD in the initialization step can be reduced to n^2d and the total computational complexity is $O(n^2d) + rO(ndL)$. The algorithm usually takes only a few iterations to converge according to our experience.

8 Connection of Sparse ePCA with ePCA and Sparse PCA

The sparse ePCA model is a generalization of both ePCA model and sparse PCA model. It reduces to ePCA or sparse PCA in special cases. The connections can be illustrated in the context of two aforementioned solution methods.

Majorization-Minimization

The sparse ePCA problem reduces to the standard ePCA problem (3) exactly when the penalty term $P(\cdot, \cdot)$ is removed. Different from the solution methods discussed in Sect. 2, the MM algorithm helps approximate the solutions to ePCA by minimizing a quadratic surrogate function. The resulting closed-form updating rules lead to rather higher computational efficiency than the sequential optimization.

When the data are assumed to be sampled from Gaussian distributions, the sparse ePCA has some connections with the following sparse PCA problem:

$$\min_{Z:Z^TZ=I} \min_{W} \|X - ZW^T - \mathbf{1}\mathbf{b}^T\|_F^2 + P(W, \lambda). \tag{27}$$

Assume the MM algorithm takes r iterations to converge. We have $A(\boldsymbol{\theta}_i) = \boldsymbol{\theta}_i^T\boldsymbol{\theta}_i/2$. Thus, at $\mathbf{b}_r$, Z_r, and W_r, the surrogate function reaches the optimum value

$$\begin{aligned} g^* = g(\Theta_r|\Theta_r) &= \frac{1}{2}\|\frac{1}{\rho_r}(X - \sum_{i=1}^{n}\frac{\partial A(\boldsymbol{\theta}_i)}{\partial \Theta}|_{\Theta=\Theta_r})\|_F^2 + \frac{1}{\rho_r}P(W_r, \lambda) \\ &= \frac{1}{2\rho_r^2}\big(\|X - Z_rW_r^T - \mathbf{1}\mathbf{b}_r^T\|_F^2 + P(W_r, 2\rho_r\lambda)\big). \end{aligned}$$

This is equivalent to solving the above sparse PCA problem (27) with a scaled λ.

Transformation by convex conjugate

When λ_l is set to 0 and the bias term $\mathbf{b}$ is dropped, sparse ePCA reduces to an optimization problem with the same objective function and constraints as the ePCA problem presented by Guo [10] but with a different alternating order of optimization with respect to Z, W and U. Fortunately, these two problems are shown to be equivalent irrespective of the optimization order by [10]. Without the l_1-norm penalty term on W, the algorithm updates W and Z by $\frac{1}{2\lambda_0}Z^T(X+U)$ and the first L left vectors of matrix $X+U$ respectively, which conforms to the updates given by [10]. As for the U update step, it directly updates U by a closed-form solution instead of the gradient ascent method used in [10]. Eventually, Guo's gradient ascent approach will find the same solution since the objective function is concave with respect to U; however it will take longer time than the above algorithm discussed.

When the data set X is assumed to be sampled from a Gaussian distribution, $A(\boldsymbol{\theta}_i) = \boldsymbol{\theta}_i^T\boldsymbol{\theta}_i/2$ and $A^*(\mathbf{u}_i) = \mathbf{u}_i^T\mathbf{u}_i/2$. Consequently, U is estimated as $-ZW^T - \mathbf{1}\mathbf{b}^T$. After substituting the estimated U into the objective function in problem (25), one will arrive at the sparse PCA problem (27) with an elastic net penalty term.

9 Application: Clustering of Facial Images

Sparse ePCA is a promising dimension reduction tool in many applications with high-dimensional modern data such as various types of genetic data measuring millions of omics variables in bioinformatics, millions of customer ratings for commercial items in e-commerce, and high-resolution images in computer vision. Take the digital image data for example, it contains millions of integer-valued pixels which requires an appropriate assumption of the data distribution and sparse representations of the loading vectors to alleviate the model inconsistency caused by high-dimensionality.

Here, sparse ePCA will be applied to address a clustering problem that categorizes the given images to several clusters based on their visual appearance. Each image is treated as a sample in the data set while the pixel intensity at each position is treated as a variable. As image data are usually high-dimensional and involve redundant information caused by locally related pixels, sparse ePCA is applied to reduce the dimension and redundancy with variable selection to pursue better performance and interpretation. The clustering performance is investigated based on the obtained PCs lying in a lower-dimensional space instead of the original high-dimensional space using k-means clustering. Standard PCA, ePCA, and sparse ePCA are compared in this application.

We consider a subset of images from the Yale image database [9] in this experiment. There are 11 different images for each of 15 distinct subjects in the Yale database. The images of each subject vary with different facial expressions or configurations: center-light, w/glasses, happy, left-light, w/no glasses, normal, right-light, sad, sleepy, surprised, and wink. A data set containing 44 images from 4 subjects: 1, 4, 6 and 8, corresponding to 4 ground-truth clusters are randomly selected for our experiment. Consider a center region of 128 $\times$ 128 (=16,384) pixels from the original images by removing the redundant white background pixels. These images are shown in Fig. 2. This data set can be represented by a 44$\times$16,384 matrix with each row corresponding to one image. The goal is to cluster these images into 4 clusters corresponding to the four selected subjects. Due to the high dimensionality of of these face images, one can perform the clustering in a lower L-dimensional subspace constructed by the generalized principal components obtained by perform-

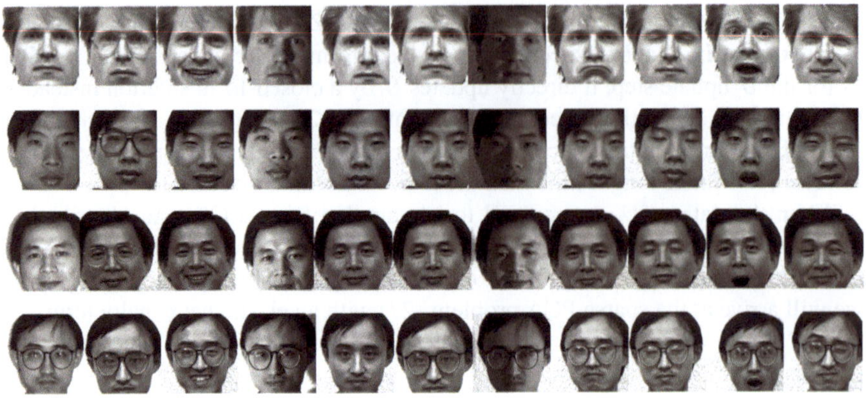

Fig. 2 Visualization of the 44 Yale images from subjects 1, 4, 6 and 8 shown by four rows correspondingly

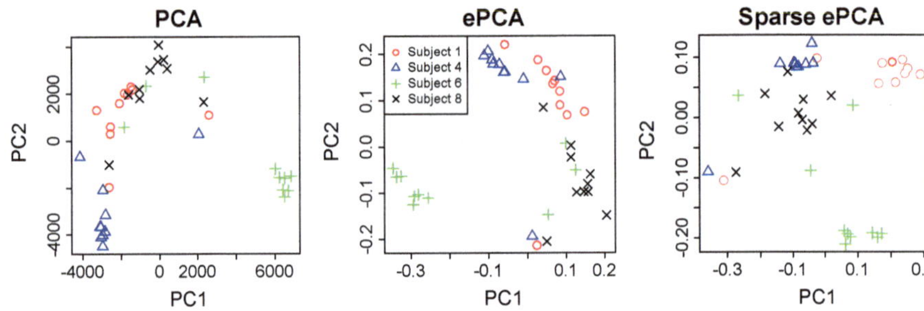

Fig. 3 Visualization of the distribution of the 44 Yale images from 4 subjects in the 2-PC subspaces constructed by PCA, ePCA, and sparse ePCA respectively

ing sparse ePCA on the images, where the pixel intensities are considered as count data following Poisson distributions. The irrelevant or redundant pixels are taken care by sparsity regularization of the PC loading vectors. In this experiment, we consider L as 2. The problem transformation strategy is utilized to solve sparse ePCA, where λ_0 is set to $1E+4$ and (λ_1, λ_2) is set to $(60, 0)$. Standard PCA and ePCA are also applied on this data set for comparison. Figure 3 shows the two-dimensional projections obtained from PCA, ePCA, and sparse ePCA respectively, from which one can see that the PCs of sparse ePCA shows more obvious clustering boundary and smaller within-cluster distance. This demonstrates that sparse ePCA can obtain more accurate results by alleviating the model inconsistency of ePCA when applied on high-dimensional data.

10 Conclusions and Future Directions

This chapter introduces two dimension reduction tools: ePCA and sparse ePCA, suitable for large-scale modern data with various data types other than the real-valued Gaussian data. They generalize PCA and sparse PCA respectively and extend their applications to various types of data that can be modeled by the distributions in the exponential family. The ePCA and sparse ePCA extensions are promising dimension reduction tools that can be applied in many areas such as computer vision, bioinformatics, e-commerce, finance, and social science.

Both ePCA and sparse ePCA models lead to non-jointly convex optimization with non-convex constraints, which makes finding efficient optimization solutions a challenging problem. Solution strategies including the MM algorithm and problem transformation strategy based on convex conjugate are the current two popular ways more efficient than those gradient methods. Both solutions are favorable in computational efficiency due to the closed-form updating rules for alternating updates of the unknown variables. There is more space to study efficient optimization strategies for solving ePCA and sparse ePCA problems since the scalability of the algorithms is still a concern when applying ePCA and sparse ePCA on large-scale modern data.

The sparse model of exponential family PCA—sparse ePCA—enables variable selection in low-dimensional analysis of exponential family data for better systematic interpretation in real-world applications. Sparse ePCA also improves the reconstruction accuracy of ePCA when the data dimension is larger than the sample number or the data has a latent sparse structure.

Finally, the choice of the form of exponential family distributions can be made by examining the data distribution before running any PCA methods. When the data follow the Gaussian distribution, the specific form of ePCA or sparse ePCA naturally reduces to standard PCA or the sparse PCA [34]. It is convenient to use the Bernoulli distribution for binary data and Poisson distribution for count data when applying ePCA or sparse ePCA method. Sparse ePCA is generally preferred over ePCA when there is a large number of variables under consideration because of its model consistency and good interpretation of results.

Sparse ePCA is flexible and highly extensible. With integration of additional label information into the current framework, it can be extended to a supervised-learning model to solve classification or regression problems involving high dimensional exponential family data. Compared to the heuristic supervised ePCA, the appropriately integrated supervised sparse ePCA allows the optimal subset of variables containing the most discriminant information automatically selected by the model itself. The problem transformation strategy by convex conjugate is still applicable for this problem due to the similar form of models shared by ePCA and GLMs. Similar closed-form updating rules can be derived for efficient supervised learning. Based on such supervised sparse ePCA models, hierarchical analysis based on several dominant variables can also be derived to simultaneously estimate the principal component effects as well as the individual effects of the dominant variables in association analyses. Moreover, one can introduce different regularization terms on principal component loadings such as involving prior network structure to achieve smoothness or perform graph-regularized learning. The network structure regularization can be widely considered to involve the relations between customers or users for user behavior analysis in social science or to take care of the gene-gene interactions for association analysis in bioinformatics.

References

1. Bair, E., Hastie, T., Paul, D., Tibshirani, R.: Prediction by supervised principal components. J. Am. Stat. Assoc. **101**(473), 119–137 (2006)
2. Bair, E., Tibshirani, R.: Semi-supervised methods to predict patient survival from gene expression data. PLoS Biol. **2**(4), e108 (2004)
3. Borwein, J., Lewis, A.: Convex Analysis and Nonlinear Optimization. Springer (2000)
4. Chen, X., Wang, L., Hu, B., Guo, M., Barnard, J., Zhu, X.: Pathway-based analysis for genome-wide association studies using supervised principal components. Genet. Epidemiol. **34**, 716–724 (2010)
5. Collins, M., Dasgupta, S., Schapire, R.E.: A generalization of principal component analysis to the exponential family. Adv. Neural Inf. Process. Syst. **14**, 617–642 (2002)
6. David, W., Srikantan, N.: Iterative reweighted l1 and l2 methods for finding sparse solutions. IEEE J. Sel. Top. Sig. Process. **4**(2), 317–329 (2010)
7. Fan, J., Li, R.: Variable selection via nonconcave penalized likelihood and its oracle properties. J. Am. Stat. Assoc. **96**(456), 1348–1360 (2001)
8. Fan, K.: On a theorem of weyl concerning eigenvalues of linear transformations: II. Proc. Natl. Acad. Sci. U. S. A. **35**(11), 652–655 (1949)
9. Georghiades, A.S., Belhumeur, P.N.: From few to many: illumination cone models for face recognition under variable lighting and pose. IEEE Trans. Pattern Anal. Mach. Intell. **23**(6), 643–660 (2001)
10. Guo, Y., Schuurmans, D.: Efficient global optimization for exponential family PCA and low-rank matrix factorization. In: Proceedings of the 46th Annual Allerton Conference on Communication, Control, and Computing, pp. 1100–1107 (2008)
11. Hunter, D.R., Lange, K.: A tutorial on MM algorithms. Am. Stat. **58**(1), 30–37 (2004)
12. Jaakkola, T., Jordan, M.I.: Bayesian parameter estimation via variational methods. Stat. Comput. **10**, 25–37 (2000)
13. Johnstone, I.M., Lu, A.Y.: On consistency and sparsity for principal components analysis in high dimensions. J. Am. Stat. Assoc. **104**(486), 700 (2009)
14. Jolliffe, I.T.: Principal Component Analysis. Springer, New York (2002)
15. Landgraf, A.J., Lee, Y.: Dimensionality reduction for binary data through the projection of natural parameters. Technical Report No. 890, Department of Statistics, The Ohio State University (2015)
16. Landgraf, A.J., Lee, Y.: Generalized principal component analysis: projection of saturated model parameters. Technical Report No. 892, Department of Statistics, The Ohio State University (2015)
17. Lange, K., Hunter, D.R., Yang, I.: Optimization transfer using surrogate objective functions (with discussion). J. Comput. Graphical Stat. **9**, 1–20 (2000)
18. Lee, S., Huang, J.Z.: A coordinate descent MM algorithm for fast computation of sparse logistic PCA. J. Comput. Stat. Data Anal. **62**, 26–38 (2013)
19. Lee, S., Huang, J.Z., Hu, J.: Sparse logistic principal components analysis for binary data. Ann. Appl. Stat. **4**(3), 1579–1601 (2010)
20. Leeuw, J.D.: Principal component analysis of binary data by iterated singular value decomposition. J. Comput. Stat. Data Anal. **50**(1), 21–39 (2006)

21. Lu, M., Huang, J.Z., Qian, X.: Supervised logistic principal component analysis for pathway based genome-wide association studies. In: ACM Conference on Bioinformatics, Computational Biology and Biomedicine (ACM BCB), pp. 52–59 (2012)
22. Lu, M., Huang, J.Z., Qian, X.: Sparse exponential family principal component analysis. Pattern Recogn. **60**, 681–691 (2016)
23. Lu, M., Lee, H.S., Hadley, D., Huang, J.Z., Qian, X.: Logistic principal component analysis for rare variants in gene-environment interaction analysis. IEEE/ACM Trans. Comput. Biol. Bioinf. **11**(6), 1020–1028 (2014)
24. Lu, M., Lee, H.S., Hadley, D., Huang, J.Z., Qian, X.: Supervised categorical principal component analysis for genome-wide association analyses. BMC Genomics **15**, (S10) (2014)
25. Mardia, K., Kent, J., Bibby, J.: Multivariate Analysis. Academic Press, New York (1979)
26. McCullagh, P., Nelder, J.A.: Generalized Linear Models, 2nd ed. CRC (1990)
27. Nadler, B.: Finite sample approximation results for principal component analysis: a matrix perturbation approach. Ann. Stat. **36**(6), 2791–2817 (2008)
28. Paul, D.: Asymptotics of sample eigenstructure for a large dimensional spiked covariance model. Stat. Sinica **17**(4), 1617 (2007)
29. Pearson, K.: On lines and planes of closest fit to systems of points in space. Lond. Edinb. Dublin Phylos. Mag. J. Sci. Sixth Ser. **2**, 559–572 (1901)
30. Rockafellar, R.: Convex Analysis. Princeton University Press (1970)
31. She, Y.: Thresholding-based iterative selection procedures for model selection and shrinkage. Electron. J. Stat. **3**, 384–415 (2009)
32. She, Y., Li, S., Wu, D.: Robust orthogonal complement principal component analysis. J. Am. Stat. Assoc. **111**(514), 763–771 (2016)
33. She, Y., Owen, A.B.: Outlier detection using nonconvex penalized regression. J. Am. Stat. Assoc. **106**(494), 626–639 (2011)
34. Shen, H., Huang, J.Z.: Sparse principal component analysis via regularized low rank matrix approximation. J. Multivar. Anal. **101**, 1015–1034 (2008)
35. Tipping, M.E., Bishop, C.M.: Probabilistic principal component analysis. J. R. Stat. Soc. B **6**(3), 611–622 (1999)
36. Wainwright, M.J., Jordan, M.I.: Graphical models, exponential families, and variational inference. Found. Trends Mach. Learn. **1**, 1–305 (2008)
37. Wen, Z., Yin, W.: A feasible method for optimization with orthogonality constraints. Math. Program. **142**(1–2), 397–434 (2013)
38. Zhang, Q., She, Y.: Sparse generalized principal component analysis for large-scale applications beyond gaussianity. arXiv:1512.03883 (2016)

Application and Extension of PCA Concepts to Blind Unmixing of Hyperspectral Data with Intra-class Variability

Yannick Deville, Charlotte Revel, Véronique Achard and Xavier Briottet

Abstract The most standard blind source separation (BSS) methods address the situation when a set of signals are available, e.g. from measurements, and all of them are linear memoryless combinations, with unknown coefficient values, of the same limited set of unknown source signals. BSS methods aim at estimating these unknown source signals and/or coefficients. This generic problem is e.g. faced in the field of Earth observation (where it is also called "unsupervised unmixing"), when considering the commonly used *(over)simplified model* of hyperspectral images. Each pixel of such an image has an associated reflectance spectrum derived from measurements, which is defined by the fraction of sunlight power reflected by the corresponding Earth surface at each wavelength. Each source signal is then the *single* reflectance spectrum associated with one of the classes of pure materials which are present in the region of Earth corresponding to the overall considered hyperspectral image. Besides, the associated coefficients define the surfaces on Earth covered with each of these pure materials in each sub-region corresponding to one pixel of the considered image. However, *real* hyperspectral data e.g. obtained in urban areas have a much more complex structure than the above basic model: each class of pure materials (e.g. roof tiles, trees or asphalt) has so-called spectral or intra-class variability, i.e. it yields a somewhat *different* spectral component in each pixel of the image. In this complex framework, this chapter shows that Principal Component Analysis (PCA) and its proposed extension are of high interest at three stages of our investigation. First, PCA allows us to analyze the above-mentioned spectral variability

Y. Deville (✉) · C. Revel
IRAP (Institut de Recherche en Astrophysique et Planétologie),
Université de Toulouse, UPS-CNRS-OMP, 31400 Toulouse, France
e-mail: Yannick.Deville@irap.omp.eu

C. Revel
e-mail: Charlotte.Revel@irap.omp.eu

V. Achard · X. Briottet
Department of Theoretical and Applied Optics,
ONERA The French Aerospace Lab, 31400 Toulouse, France
e-mail: Veronique.Achard@onera.fr

X. Briottet
e-mail: Xavier.Briottet@onera.fr

G. R. Naik (ed.), *Advances in Principal Component Analysis*,
https://doi.org/10.1007/978-981-10-6704-4_9

of real high-dimensional hyperspectral data and to propose an extended data model which is suited to these complex data. We then develop a versatile extension of BSS methods based on Nonnegative Matrix Factorization, which adds the capability to handle arbitrary forms of intra-class variability by transposing PCA concepts to this original version of the BSS framework. Finally, PCA again proves to be very well suited to analyzing the high-dimensional data obtained as the result of the proposed BSS method.

1 Introduction

Standard Blind Source separation (BSS) methods [4–8, 11, 20], also called unsupervised (or blind) unmixing methods by the Earth observation community, may be briefly defined as follows. They aim at estimating a set of "source signals" (which have unknown values but some known properties), using a set of available signals which are "mixtures" of the source signals to be restored, without knowing (or with very limited knowledge about) the "mixing transform", i.e. the transform of source signals which yields their mixtures. The term "signal" is here to be understood in a broad sense: the considered problem not only concerns monodimensional functions (especially time-dependent functions), but also images and various types of data.

BSS is thus a generic signal processing problem, which may be represented as shown in Fig. 1, where $s_1(t)$ to $s_M(t)$ are the M unknown source signals to be restored, $x_1(t)$ to $x_P(t)$ are the P available (or "observed") mixed signals used for restoring the source signals and $y_1(t)$ to $y_M(t)$ are the M estimated source signals. Various practical configurations corresponding to this generic model may then be derived, depending on the considered application domain. In particular, the BSS problem has been studied in the field of audio and acoustics. The associated typical configuration is shown in Fig. 2. In that case, each source signal is an emitted acoustic signal (speech signal or "noise" e.g. generated by a car) and each observed signal is provided by a

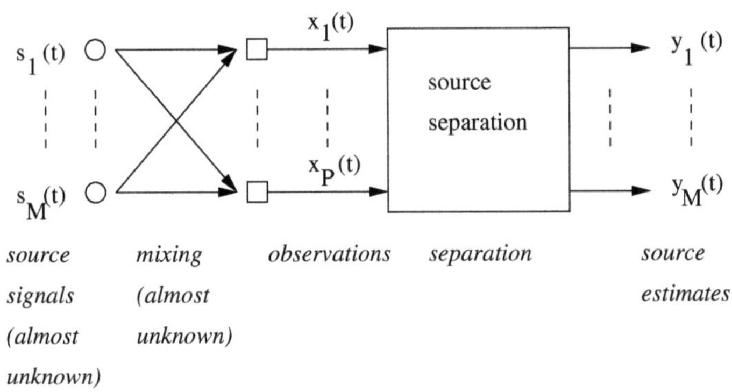

Fig. 1 General configuration for the blind source separation problem

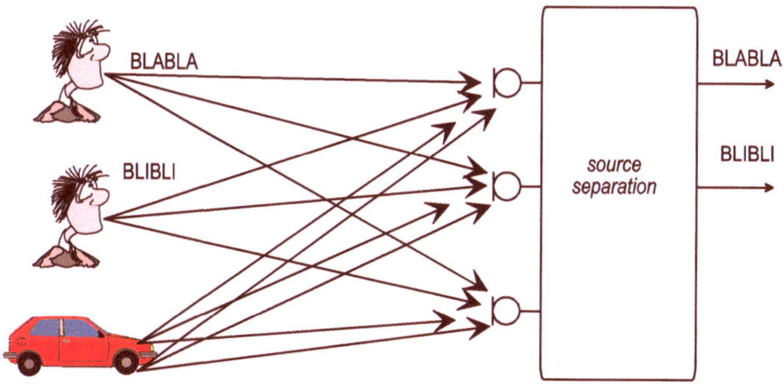

Fig. 2 Application of blind source separation methods to acoustic signals

corresponding sensor, which is a microphone. Besides, the mixing phenomenon then results from the simultaneous propagation of all source signals to each microphone. Another application field, which is of major interest in the framework of this chapter, is Earth observation, also referred to as "remote sensing". The definition of the corresponding configuration requires a more detailed analysis than above and is therefore postponed to Sect. 2.1. That analysis shows that the *basic version* of that Earth observation problem falls in the scope of the above type of BSS configurations.

A major feature of the above generic mixing model, corresponding to *standard* BSS configurations, is that all observed signals are mixtures of the *same* set of source signals. This assumption is relevant in many applications. For instance, in the audio application considered above (see Fig. 2), all microphones receive combinations of the same speech signals, which may be considered as emitted by a limited set of point-like sources in various configurations.[1] More generally speaking, using the terminology provided in Fig. 1 for this standard mixing model, this model is such that any observed signal $x_i(t)$ with index i is influenced by any source with index j by depending on the same source signal $s_j(t)$ whatever the value of i. However, that standard model may be too simplified for other application fields. This especially includes applications where the source signals have so-called "variability" from one observed signal to another. Still using the terminology of Fig. 1, this means that any observed signal $x_i(t)$ with index i is here influenced by any source with index j by depending on a source signal which is not exactly the same for observed signals with different indices i, although these observation-dependent versions of "the jth source signal" share some similarities in all observed signals. As detailed in Sect. 2.2, this phenomenon especially appears in Earth observation data and is the main motivation of the investigation reported in this chapter.

The above source variability then leads us to use and extend Principal Component Analysis (PCA) concepts at the following three levels:

1. This source variability should first be detected, and its magnitude should then be characterized. This cannot easily be performed by using a plain visualization

[1] "noise source signals", when non-negligible, may correspond to more complex phenomena.

of the considered data, since they consist of a large number of spectra extracted from hyperspectral images (defined in Sect. 2.1) and therefore lie in very high-dimensional spaces. A dimensionality reduction method is therefore needed to analyze these data, and the tool used to this end is PCA, as shown in Sect. 3.
2. Due to the existence of source variability in the considered data, standard BSS techniques do not apply to them. Therefore, we then develop original BSS methods aiming at handling sources which have arbitrary variability patterns. As shown in Sect. 4, this is achieved by extending PCA concepts to the considered original version of the BSS framework.
3. Finally, the performance of the proposed BSS methods should be assessed. This essentially consists of determining with which accuracy the estimated source spectra approximate the actual ones. Again, the visual analysis of these high-dimensional data requires one to first reduce their dimensionality, by using PCA, as detailed in Sect. 5.

Conclusions are eventually drawn from this investigation in Sect. 6.

2 Mixing Models for Earth Observation

2.1 Standard Model

The BSS problem investigated in this chapter concerns the analysis of hyperspectral images of parts of the Earth surface. These images are provided by sensors onboard a satellite or an aircraft, as shown in Fig. 3. Whereas each "observation" (i.e. each observed signal) was a function of time in the configurations shown in Sect. 1, it is a function of wavelength in the approach considered here [14]. More precisely, the data typically derived for each pixel of an image sensor consist of a reflectance spectrum corresponding to the part of Earth surface seen by this pixel. This reflectance spectrum is a monodimensional series of values, which forms a vector, where each value corresponds to a given wavelength (more precisely, to a narrow spectral band), situated or not in the visible part of the spectrum of sunlight. At each considered wavelength, the corresponding value of the above reflectance spectrum is equal to the fraction of light power which is reflected by the Earth and sent to the considered sensor pixel, with respect to the light power received from the sun by the considered surface on Earth. In the specific case when a pixel receives light from a flat and homogeneously illuminated part of Earth surface only covered with a single pure material, the observed reflectance spectrum derived for that pixel is directly equal to the reflectance spectrum specific to this pure material. But in practice, a pixel often corresponds to a part of Earth surface which is composed of several pure materials (see Fig. 3). For flat and homogeneously illuminated surfaces on Earth (and without intimate mixing), the observed reflectance spectrum obtained for a pixel may then be shown to be well approximated by a linear combination of the reflectance spectra

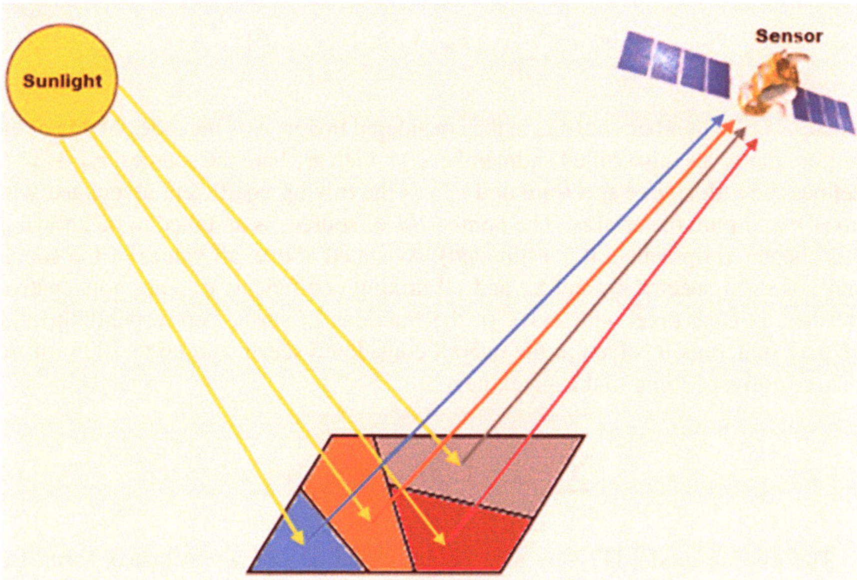

Fig. 3 Linear mixing in Earth observation in the spectral reflective domain, [0.4 μm, 2.5 μm]: the light emitted by the sun is reflected by the surface of the Earth and then reaches the sensing device (courtesy of S. Karoui)

of the pure materials present in the associated surface on Earth [3, 15], and the coefficients of that combination are referred to as the "mixing coefficients". Moreover, the following simplifying assumption is most often made: any given type of pure materials (e.g. roof tile, tree or asphalt) is considered to yield exactly the *same* pure reflectance spectrum in any pixel of the considered scene which is partly or fully covered with this pure material. This yields the standard linear (memoryless [8]) mixing model, where each source signal to be estimated is the reflectance spectrum of a pure material, each observation is the reflectance spectrum obtained for one pixel and, for each observation, the mixing coefficients associated with all sources are respectively equal to the fractions of surface on Earth associated with all pure materials, within the overall surface corresponding to the considered pixel. In this Earth observation framework, the mixing coefficients are therefore often called "abundance fractions" or "abundances". It should be noted that all source values and mixing coefficients are nonnegative in this configuration.

More precisely, the mixing model resulting from the above analysis may be formally expressed as follows. Let us denote as L the number of wavelengths at which the considered hyperspectral image is recorded. Each pixel, with index p, of this image then yields a recorded reflectance spectrum defined by a column vector $\mathbf{x_p} \in \mathbb{R}^{L \times 1}$. Due to the above mixing model, this vector reads

$$\mathbf{x_p} = \sum_{m=1}^{M} c_{pm}\mathbf{r_m} \quad \forall p \in \{1, \ldots, P\} \tag{1}$$

where P is the number of pixels in the considered image, m is the index of one of the M pure materials (also called endmembers[2]) which are here the sources, $\mathbf{r_m} \in \mathbb{R}^{L\times 1}$ defines the mth source spectrum and c_{pm} is the mixing coefficient associated with pixel p and pure material m. The number M of sources is assumed to be known in this chapter (in practice, it is estimated). As stated above, all values of the source and observed spectra $\mathbf{r_m}$ and $\mathbf{x_p}$ and all mixing coefficients c_{pm} are nonnegative. Besides, in each pixel with index p, the fractions of surface corresponding to all M pure materials involved in the overall considered scene sum up to 100% of the surface corresponding to this pixel, i.e.

$$\sum_{m=1}^{M} c_{pm} = 1 \quad \forall p \in \{1, \ldots, P\}. \tag{2}$$

The mixing model (1) may also be expressed as follows in matrix form. Let $\mathbf{X} = [\mathbf{x_1}, \ldots, \mathbf{x_P}]^T$, with ^{T} standing for transpose, denote the matrix of recorded reflectance spectra, and let $\mathbf{R} = [\mathbf{r_1}, \ldots, \mathbf{r_M}]^T$ denote the matrix of pure material reflectance spectra. Besides, let us denote as $\mathbf{c_p} = [c_{p1}, \ldots, c_{pM}]^T$ the M-element column vector containing the set of mixing coefficients associated with the pth observed spectrum. Finally, $\mathbf{C} = [\mathbf{c_1}, \ldots, \mathbf{c_P}]^T$ is the mixing coefficient matrix. The mixing model (1) then yields

$$\mathbf{X} = \mathbf{CR}. \tag{3}$$

Only knowing matrix $\mathbf{X}$, the blind unmixing problem then consists of estimating the pure material spectra (i.e. source signals) which compose matrix $\mathbf{R}$ and/or the abundance fractions (i.e. mixing coefficients) which compose matrix $\mathbf{C}$, under the above-defined nonnegativity and sum-to-one constraints. For reviews of (blind or non-blind) unmixing methods dedicated to this mixing model, the reader may e.g. refer to [3, 15].

2.2 Extended Model: Intra-class Variability

The above mixing model (1) or (3) represents a complete hyperspectral image with a very limited number M of pure material spectra. For example, it may aim at describing a complete urban scene by using a single spectrum for all tiles which cover roofs, together with a single spectrum for all areas covered with trees, a single spectrum for all areas covered with asphalt and so on. This model is widely used because of its simplicity, but it is only a coarse approximation of most actual data.

[2]The term "endmembers" is also used for the reflectance spectra of these pure materials.

For instance, the reflectance spectra of the roof tiles of a scene are not the same in all pixels of the considered scene, e.g. because (i) the overall magnitude of these spectra depends on the illumination of the considered tiles, (ii) these tiles may be more or less weathered due to aging or (iii) they may have slightly different mineral compositions. These spectra are stated to have some variability. Yet, they may be hoped to remain rather similar to one another, so that the "distances" between them may be hoped to often remain lower than the distances between such tile spectra and, say, spectra corresponding to trees or asphalt. This means that, although each "type of materials" (e.g. tiles) cannot then be reduced to a single spectrum, it may hopefully still be relevant to consider that it yields a class of spectra, which has limited intra-class variability, and which can thus be distinguished from another class of spectra corresponding to another type of materials (e.g. trees), since these classes have no or limited overlap in data representation domains.

A survey of spectral variability problems and of associated unmixing methods is available in [29]. These methods most often require prior knowledge about the considered data. However, the shape of the subspace or, more generally, manifold that may be spanned by the spectra belonging to the same class in a real scene is not always well characterized, and it may be quite general as will be shown by the example provided in Sect. 3. To handle such varied situations, we hereafter consider a versatile extended version of the standard mixing model (1)–(3), that we started to introduce in [23], and we investigate associated general blind unmixing methods. In this extended model, a separate set of M pure material spectra $\mathbf{r_m}(p)$, with m ranging from 1 to M, is associated with each pixel p. Each recorded spectrum $\mathbf{x_p}$ is then expressed with respect to this *pixel-dependent* set of pure material spectra, still considering a linear model for combining them, that is

$$\mathbf{x_p} = \sum_{m=1}^{M} c_{pm} \mathbf{r_m}(p) \quad \forall p \in \{1, \ldots, P\}. \tag{4}$$

Extracting this large number of source spectra ($M \times P$ spectra for the complete image) from such observations without further constraining them would yield an ill-posed problem. Therefore, further in this chapter, we will show that PCA concepts are particularly suited to introducing constraints which are relevant for the considered application. Moreover, we here keep the type of constraints defined above for the standard mixing model, that is (i) the nonnegativity of all spectra $\mathbf{r_m}(p)$ and all mixing coefficients c_{pm}, and (ii) the sum-to-one constraint (2).

The extended mixing model (4) may then be expressed in matrix form as follows. We introduce $\mathbf{R}(p) = [\mathbf{r_1}(p), \ldots, \mathbf{r_M}(p)]^T \in \mathbb{R}^{M \times L}$, which contains the set of M source (i.e. pure) spectra associated with the observed spectrum $\mathbf{x}_p$, then $\tilde{\mathbf{R}} = \begin{bmatrix} \mathbf{R}(1) \\ \cdots \\ \mathbf{R}(P) \end{bmatrix} \in \mathbb{R}^{PM \times L}$, the matrix containing all the source spectra of the complete scene and $\tilde{\mathbf{C}} \in \mathbb{R}^{P \times PM}$ the block-diagonal extended mixing coefficient matrix:

$$\tilde{\mathbf{C}} = \begin{bmatrix} \mathbf{c_1}^T & 0\dots0\dots0\dots0 \\ 0\dots0 & \mathbf{c_2}^T \;\; \dots 0\dots0 \\ & \ddots \\ 0\dots0\,0\dots0\dots & \mathbf{c_P}^T \end{bmatrix} \tag{5}$$

with $\mathbf{c_p}$ defined as in Sect. 2.1. Equation (4) then yields the matrix expression

$$\mathbf{X} = \tilde{\mathbf{C}}\tilde{\mathbf{R}}. \tag{6}$$

It should be noted that, for any given class of pure materials with index m, where $m \in \{1, \dots, M\}$, the pure spectra corresponding to that class and to pixels $p = 1, \dots, P$ are equal to the rows with indices $(m + (p-1)M)$ of matrix $\tilde{\mathbf{R}}$.

Only knowing matrix $\mathbf{X}$, we aim at estimating matrices $\tilde{\mathbf{C}}$ and $\tilde{\mathbf{R}}$ under the non-negativity and sum-to-one constraints, and the PCA-related constraint introduced further (and possibly up to the usual scale and permutation indeterminacies of BSS).

A different and much more restricted mixing model, which includes some aspects of intra-class variability, has also been used in the literature (see e.g. [27]). It may be seen as a subset of model (4), obtained as follows: separately for each class of pure materials, i.e. separately for each value of m, the pure spectra involved in all pixels p for that class are constrained to be proportional, i.e.

$$\forall m \in \{1, \dots, M\}, \;\; \forall p \in \{1, \dots, P\}, \qquad \exists\, \gamma_{pm} \in \mathbb{R}^{*+}, \qquad \mathbf{r_m}(p) = \gamma_{pm}\mathbf{r_m}, \tag{7}$$

where $\mathbf{r_m}$ is the single prototype reflectance spectrum associated with the considered class m of pure materials and each parameter γ_{pm} defines with which scale factor the above prototype appears in each pixel p. This model mainly describes the intra-class variability which is due to illumination variations, that e.g. result from landscape slope variations in non-flat areas, such as slope differences from one part of a roof to another. However, this model does not take into account other types of variability, e.g. due to the above-mentioned aging or composition variations. The resulting limitations are shown in the next section.

3 Experimental Characterization of Spectral Variability with PCA

The first step of the standard procedure for developing a BSS method [7, 8] is the definition of the considered mixing model (and associated sources). Therefore, we here analyze which of the above-defined models is relevant for the considered Earth observation application, especially focusing on urban scenes. To this end, we use a part of a hyperspectral image which was recorded over the center of Toulouse city, France (see Fig. 4). This image contains 405 bands covering the wavelength domain ranging from 414 to 2498 nm. It was converted to a 1.8 by 1.8 m pixel spatial

Fig. 4 Part of image used for characterizing spectral variability [2]

resolution. In this part of image, we manually selected a set of pixels which belong to homogeneous areas and which are assumed to be pure. The corresponding reflectance spectra are thus supposed to belong to three classes, namely tile, vegetation (trees) and asphalt. It should be clear that this pre-characterization of some features of the considered type of data is not performed in the same conditions as the final operation of the BSS methods addressed further in this chapter: whereas we here use pure (i.e. unmixed) observed spectra, each belonging to a known class, we will eventually aim at using *blind* source separation methods operating with observed mixtures of unknown types of pure spectra. The non-blind non-mixed framework considered in the first investigation reported in the current section only aims at choosing the mixing model to be later used.

The most natural approach for analyzing the features of the above recorded spectra consists of representing each of them in a two-dimensional figure, by plotting the variations of the considered reflectance with respect to wavelength. Gathering all these plots on the same graph yields Fig. 5, where we removed the reflectance values corrresponding to the bands where they have poor quality, due to atmospheric absorption entailed by water vapor, CO_2... The remaining data thus consists of 214 spectral bands. This figure reveals part of the important features of the considered data. First, if these spectra could be described with the standard model of Sect. 2.1, all spectra recorded for the same class of materials would be identical and therefore superimposed in this figure. Clearly, this is not the case, which means that these classes of spectra exhibit intra-class variability. We then need to analyze the structure of this variability, which cannot easily be performed with Fig. 5, especially because most

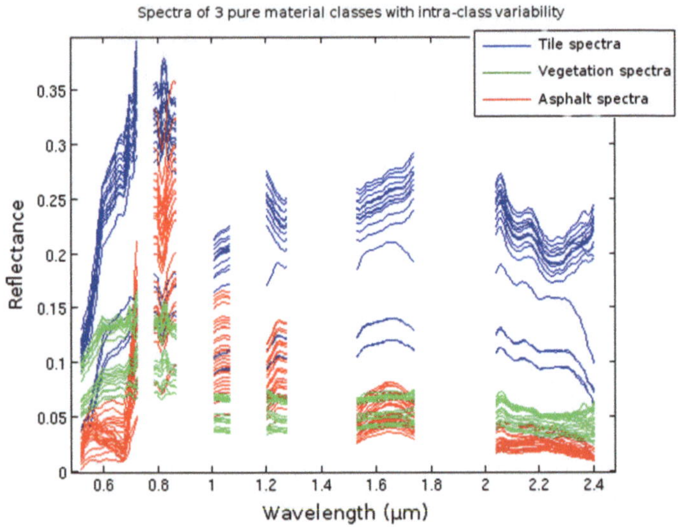

Fig. 5 Examples of reflectance spectra of three classes of materials: tile, vegetation (trees) and asphalt

spectra from the same class, and some from different classes, are close to one another or even interleaved. Besides, as discussed in Sect. 2.2, we should here determine if it is relevant to associate a separate class with each type of materials. To this end, we should determine if the above classes do not highly overlap, i.e. if the intra-class variabilities are lower than class-to-class variabilities. Although Fig. 5 may suggest that this condition is actually met for the considered data, this type of representation does not allow one to analyze this phenomenon in detail.

To avoid the limitations of the above plain graphical representation, we then need a much more powerful data processing and visualization tool. This tool is requested to be able to process the overall set of recorded spectra, without taking their above user-defined classes into account during this processing, so as to check if classes with different features naturally emerge from the processed data. Moreover, this data processing tool should allow one to display a transformed version of each recorded spectrum as a point in a two-dimensional subspace, to allow one to plot the scatter plot of these points. We will then create this scatter plot by using a different pictogram for each of the above user-defined classes, to check if the spectra as labeled above do belong to classes with different features from the point of view of the considered representation : we will check if the points belonging to the same user-defined class tend to be close to one another but further from the points belonging to other user-defined classes, thus leading to no or limited overlap between the subsets of points corresponding to different pictograms in the considered representation. This means that the transform used to map the original high-dimensional-space spectra to the very-low-dimensional-space (2 dimensions) scatter plot to be visually analyzed should be selected so as to make this scatter plot readable. To this end, the points in the scatter plot of the transformed data should be kept far enough from one another, and ideally "as far as possible". Although the above requests might be considered to be quite demanding, a data analysis tool which meets all these constraints does exist: these constraints are precisely those which lead to Principal Component Analysis (PCA). PCA has been used in many application fields and the reader may e.g. refer to [1, 9, 11, 13, 25, 26] for its description.[3] Therefore, we here do not detail the well-known practical procedure which results from the data processing concepts that we presented above, and we directly focus on how the resulting data interpretation principles defined above may be applied to the Earth observation problem tackled in this chapter. PCA uses a set of "objects" or "individuals". Each of them is here represented by a recorded reflectance spectrum. PCA considers a set of variables as its inputs. Each of these variables here corresponds to the reflectance value (for each object) measured at a given wavelength. PCA creates new variables, which are linear combinations of the original variables, i.e. here linear combinations of reflectances measured at different wavelengths. Its defines new directions in the data space, called principal axes, and it measures the components of the considered data along these axes. The latter components are called principal components. Keeping only the first two of these components for each object allows one to reduce these objects to a two-dimensional representation (by projecting the original data on the subspace spanned by the first two principal axes), while preserving "as much as possible" the shape of the scatter plot of the original high-dimensional data. This scatter plot of the projected objects (i.e. spectra) in the above two-dimensional subspace is the main result of PCA considered hereafter. It is shown in Fig. 6, where the data corresponding to the "tile" class are moreover split into two sub-classes, namely "sunny tiles" and "shaded tiles", to analyze these data in more detail.

This figure should be analyzed keeping in mind the two mixing models with intra-class variability that we introduced in Sect. 2.2:

- If spectral variability in a given class, with index m, is reduced to scale factors, as defined by (7), then all reflectance vectors $\mathbf{r_m}(p)$ associated with that class are collinear. Therefore, all corresponding points in the PCA projection are situated on the same line, which goes through the origin and which has a direction defined by the value of the prototype vector $\mathbf{r_m}$ of that class. Each such point then moves further from the origin when the scale factor γ_{pm} is increased.
- On the contrary, if spectral variability can lead to "any" spectral vector $\mathbf{r_m}(p)$ in any pixel p, then the corresponding points in the two-dimensional PCA representation may be situated anywhere in the projection plane.

Applying these considerations to the data of Fig. 6 shows that the scale-based mixing model (7) is of interest but not sufficient for describing the complexity of

[3]In all this chapter, the metric used for PCA is defined by the identity matrix.

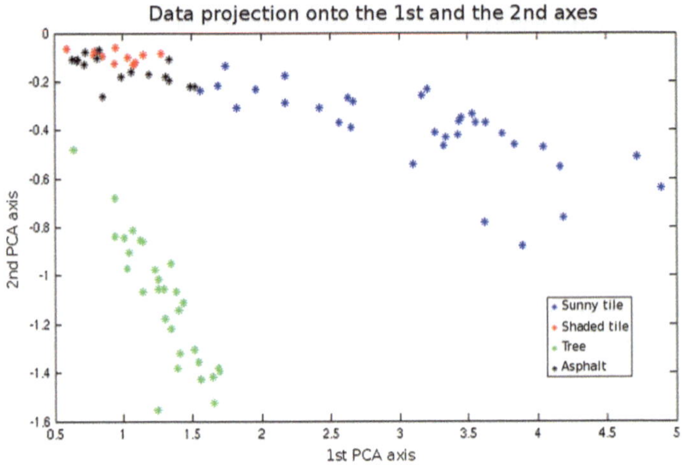

Fig. 6 Projection of pure reflectance spectra on first and second principal axes

these data. For each (sub)class, two effects appear in the corresponding part of the scatter plot:

1. All points more or less tend to be close to the same line, hereafter called the "main class axis", which goes through the origin (and which may approximately be considered as the line which contains the center of gravity of that part of the scatter plot), as expected from the scale-based mixing model (7). The data have a large variance along that main class axis. Moreover, for the two tile sub-classes, the scale of each reflectance spectrum is highly correlated to the illumination of the considered tile: the shaded tiles yield points in Fig. 6 which are closer to the origin than the points obtained for sunny tiles. Similarly, The two (sub)classes which correspond to the darkest materials in the considered data are shaded tile and asphalt. They are therefore expected to yield projected points which are closer to the origin than the points associated with the other two (sub)classes. Figure 6 confirms that this is actually the case.
2. However, each part of the scatter plot associated with one (sub)class also has significant variance along the "orthogonal class axis", i.e. the axis corresponding to the direction of Fig. 6 which is orthogonal to the above main class axis.[4] To model the considered data accurately enough, this variance along the orthogonal class axis should not be disregarded. Therefore, the scale-based mixing model (7) is not suitable in this situation and we will hereafter instead use the more general model (4), or its matrix form (6), when developing blind source separation methods which aim at handling spectral variability.

[4]Among all lines which have this "orthogonal direction", one may e.g. consider the line which includes the center of gravity of the considered part of the scatter plot. What matters here is not the position of the orthogonal class axis, but its direction.

4 A New PCA-related Blind Source Separation Method for Handling Spectral Variability

4.1 Background: Nonnegative Matrix Factorization (NMF)

One of the main classes of BSS methods is Nonnegative Matrix Factorization (NMF). Under this name, it was introduced by Lee and Seung in [17, 18]. However, related methods were previously reported, especially by Paatero et al. [21], under the name Positive Matrix Factorization (PMF). Since the beginning of the 2000s, many NMF methods have been developed, e.g. including Lin's algorithms [19]. For a detailed overview of NMF, the reader may e.g. refer to [5]. NMF is also described e.g. in [6, 8]. More specifically, NMF has been adapted to remote sensing applications, as discussed e.g. in [3].

The basic aspects of the most standard NMF method which are of importance for the current chapter are as follows. The considered data follow the standard mixing model (3), moreover with nonnegative matrices $\mathbf{R}$ and $\mathbf{C}$, and hence $\mathbf{X}$. The standard NMF method involves two nonnegative matrices $\hat{\mathbf{R}}$ and $\hat{\mathbf{C}}$, which aim at respectively estimating $\mathbf{R}$ and $\mathbf{C}$ (up to scale and permutation indeterminacies, as usual in BSS), so as to achieve $\mathbf{X} \simeq \hat{\mathbf{C}}\hat{\mathbf{R}}$. To this end, this method minimizes the cost function defined as

$$J_{nmf} = \frac{1}{2} \left\| \mathbf{X} - \hat{\mathbf{C}}\hat{\mathbf{R}} \right\|_F^2 \tag{8}$$

where $\|.\|_F$ stands for Frobenius norm. This cost function therefore defines the reconstruction error achieved when deriving an approximation of the observed data matrix $\mathbf{X}$ as the product of the adaptive matrices $\hat{\mathbf{C}}$ and $\hat{\mathbf{R}}$. The "hat" sign, i.e. $\hat{}$, used above in the notations $\hat{\mathbf{C}}$ and $\hat{\mathbf{R}}$ for the *adaptive* variables is often omitted for the sake of simplicity. The above cost function (8) is then expressed as

$$J_{nmf} = \frac{1}{2} \left\| \mathbf{X} - \mathbf{C}\mathbf{R} \right\|_F^2 . \tag{9}$$

We stress that the latter notations should be used with care: in (9), $\mathbf{C}$ and $\mathbf{R}$ *are not the fixed matrices* involved in the mixing model (3), otherwise the cost function (9) would always be equal to zero!

4.2 Proposed Unconstrained Pixel-by-pixel NMF Method

4.2.1 Principle and Cost Function

When considering our extended mixing model (6), instead of the standard model (3), we are led to introduce a natural extension of the standard NMF method based on the cost function (9). Instead of the actual matrices $\mathbf{C}$ and $\mathbf{R}$ and their estimates $\hat{\mathbf{C}}$ and $\hat{\mathbf{R}}$ considered in the above standard NMF method, we here use the extended matrices $\tilde{\mathbf{C}}$ and $\tilde{\mathbf{R}}$ and their estimates, still requesting all of them to be nonnegative. Then, instead of the above cost function (9), we here introduce its extended form

$$J_{upnmf} = \frac{1}{2} \left\| \mathbf{X} - \tilde{\mathbf{C}}\tilde{\mathbf{R}} \right\|_F^2 \tag{10}$$

where we again stress that $\tilde{\mathbf{C}}$ and $\tilde{\mathbf{R}}$ do not represent the *actual* data which lead to the observed reflectance spectra according to (6), but the associated *adaptive* variables used to estimate the actual matrices: the "hat" sign ^ is here again omitted for the sake of readability. The adaptive matrix $\tilde{\mathbf{R}}$ of this new approach thus yields a separate set of estimates of pure spectra for each pixel (together with the associated mixing coefficients in the adaptive matrix $\tilde{\mathbf{C}}$). These spectra are arranged in the adaptive matrix $\tilde{\mathbf{R}}$ as in the corresponding actual matrix involved in the mixing model (6). Therefore, due to the structure of that actual matrix detailed above (just after Eq. (6)), the structure of the adaptive matrix $\tilde{\mathbf{R}}$ considered here is as follows. For any given class of pure materials with index m, where $m \in \{1, \dots, M\}$, the estimated pure spectra corresponding to that class and to pixels $p = 1, \dots, P$ are equal to the rows $(m + (p - 1)M)$ of the adaptive matrix $\tilde{\mathbf{R}}$.

This adaptive matrix $\tilde{\mathbf{R}}$ is here only subject to the nonnegativity constraint of NMF (and sum-to-one constraint), as opposed to the additional constraint introduced in a modified version of this type of approaches, further in this paper. This first proposed method is therefore referred to as Unconstrained Pixel-by-pixel NMF, or UP-NMF, hereafter ("pixel-by-pixel" means that a separate set of spectra is estimated for each pixel; all these sets are *simultaneously* estimated).

4.2.2 Gradient-Based Algorithm

The algorithm used hereafter to minimize the above cost function is based on a gradient descent, combined with

- a projection of all elements of the adaptive matrices $\tilde{\mathbf{C}}$ and $\tilde{\mathbf{R}}$ on $\mathbb{R}^{+*}$, to ensure the nonnegativity of these matrices,
- the normalization of all mixing coefficients, separately for each pixel, to ensure that their sum is equal to one, as in (2).

To perform the required gradient calculations, the cost function (10) is first rewritten as

$$J_{upnmf} = \frac{1}{2} Tr((\mathbf{X} - \tilde{\mathbf{C}}\tilde{\mathbf{R}})(\mathbf{X} - \tilde{\mathbf{C}}\tilde{\mathbf{R}})^T). \tag{11}$$

Standard matrix derivation properties, e.g. available in [22], then yield

$$\frac{\partial J_{upnmf}}{\partial \tilde{\mathbf{R}}} = -\tilde{\mathbf{C}}^T(\mathbf{X} - \tilde{\mathbf{C}}\tilde{\mathbf{R}}) \tag{12}$$

$$\frac{\partial J_{upnmf}}{\partial \tilde{\mathbf{C}}} = -(\mathbf{X} - \tilde{\mathbf{C}}\tilde{\mathbf{R}})\tilde{\mathbf{R}}^T. \tag{13}$$

Denoting $\alpha_{\tilde{\mathbf{R}}}$ and $\alpha_{\tilde{\mathbf{C}}}$ the positive adaptation gains, the associated gradient-descent rules for updating the adaptive matrices $\tilde{\mathbf{C}}$ and $\tilde{\mathbf{R}}$ read

$$\tilde{\mathbf{R}}^{(i+1)} \longleftarrow \tilde{\mathbf{R}}^{(i)} - \alpha_{\tilde{\mathbf{R}}} \frac{\partial J_{upnmf}^{(i)}}{\partial \tilde{\mathbf{R}}} \tag{14}$$

$$\tilde{\mathbf{C}}^{(i+1)} \longleftarrow \tilde{\mathbf{C}}^{(i)} - \alpha_{\tilde{\mathbf{C}}} \frac{\partial J_{upnmf}^{(i)}}{\partial \tilde{\mathbf{C}}} \tag{15}$$

except that only part of the adaptive matrix $\tilde{\mathbf{C}}$ should in fact be thus updated, so that this matrix keeps the same structure as the *actual* matrix of coefficients of the mixing model, defined by (5): only the parts of the *adaptive* matrix $\tilde{\mathbf{C}}$ corresponding to all $\mathbf{c_p}^T$ in (5) should be updated with (15), whereas the other elements of this adaptive matrix $\tilde{\mathbf{C}}$ are kept to zero. The updates of the resulting complete UP-NMF algorithm, including projection and sum-to-one normalization, read as follows (ε is a small positive constant).

1. Update matrix $\tilde{\mathbf{R}}$:
$\tilde{\mathbf{R}}^{(i+1)} \longleftarrow \tilde{\mathbf{R}}^{(i)} + \alpha_{\tilde{R}} \tilde{\mathbf{C}}^{(i)T}(\mathbf{X} - \tilde{\mathbf{C}}^{(i)}\tilde{\mathbf{R}}^{(i)})$
$\tilde{\mathbf{R}}^{(i+1)} \longleftarrow \max(\tilde{\mathbf{R}}^{(i+1)}, \varepsilon)$
2. Update matrix $\tilde{\mathbf{C}}$:
for $p = 1$ *to* P **do**
| $\mathbf{c_p}^{(i+1)T} \longleftarrow \mathbf{c_p}^{(i)T} + \alpha_{\tilde{\mathbf{C}}}(\mathbf{x_p}^T - \mathbf{c_p}^{(i)T}\mathbf{R}(p)^{(i)})\mathbf{R}(p)^{(i)T}$
| $\mathbf{c_p}^{(i+1)} \longleftarrow \max(\mathbf{c_p}^{(i+1)}, \epsilon)$
end
3. Normalize coefficients :
for $p = 1$ *to* P **do**
| $\mathbf{c_p}^{(i+1)} \longleftarrow \mathbf{c_p}^{(i+1)} / \sum_{m=1}^{M} c_{pm}^{(i+1)}$
end

In the above algorithm, the superscripts $^{(i)}$ and $^{(i+1)}$ define the considered values of the adaptive variables, throughout their evolution (these superscripts are not exponents).

The procedures for initializing the adaptive matrices $\tilde{\mathbf{R}}$ and $\tilde{\mathbf{C}}$ are discussed further in this paper, when introducing an extended version of this algorithm (see Sect. 4.3.3).

4.2.3 Limitations

As explained above, after running the UP-NMF algorithm, the estimates of the pure spectra corresponding to a given class with index m and to all pixels are obtained in rows of the adaptive matrix $\tilde{\mathbf{R}}$ which have fixed indices (that is, rows m, $[m + M], \ldots, [m + (P-1) \times M]$). Even if these rows were initialized coherently before running UP-NMF (e.g. with similar spectra, typical of that class, for all pixels), the above UP-NMF update rules then let them evolve so that they may end up with quite different values. In other words, that simple algorithm introduces a very large number of adaptive spectra per class, but does not guarantee that they eventually keep similar enough features to still represent the same class of materials. To avoid this issue, an improved version of that approach is introduced hereafter.

4.3 Proposed Inertia-Constrained Pixel-by-pixel NMF Method

4.3.1 Principle and Cost Function

As explained in Sect. 4.2.3, the above UP-NMF algorithm should be further extended, so as to control to which extent all estimates (one per pixel) of pure spectra corresponding to the same class of materials are allowed to spread away from one another. This is achieved by adding a penalty term, which measures the spread of these spectra, to the cost function of UP-NMF. The question is then how to define such a penalty term. PCA concepts then yield a natural answer to this question, defined as follows. As explained above, throughout the adaptation of matrix $\tilde{\mathbf{R}}$, the estimated pure spectra for any class m consist of all L-element row vectors of $\tilde{\mathbf{R}}$ with indices equal to $m + (p-1)M$. These P vectors may equivalently be seen as a set of P points in an L-dimensional space. From PCA, it is well known (see e.g. [25]) that their spread, as measured by their inertia, is equal to the trace of the covariance matrix of these points associated with class m. The latter quantity is hereafter denoted as $Tr(Cov(\tilde{\mathbf{R}}_{\mathscr{C}_m}))$, where $\tilde{\mathbf{R}}_{\mathscr{C}_m} \in \mathbb{R}^{P \times L}$ is the matrix containing all above-defined estimates of pure material spectra for the mth class and for pixels 1 to P, respectively in its rows 1 to P. We therefore here choose to use the sum of these inertias[5] associated with all M classes as the overall penalty term of our extended method. The resulting cost function reads

[5]We use the quantity $\sum_{m=1}^{M} Tr(Cov(\tilde{\mathbf{R}}_{\mathscr{C}_m}))$, not the inertia which is defined as $Tr(Cov(\tilde{\mathbf{R}}))$ and which gathers the estimated spectra corresponding to *all* classes. This is motivated by the fact that the latter expression tends to aggregate the spectra of all classes, whereas we here aim at aggregating spectra *separately* for each class.

$$J_{ipnmf} = \frac{1}{2} \left\| \mathbf{X} - \tilde{\mathbf{C}}\tilde{\mathbf{R}} \right\|_F^2 + \mu \sum_{m=1}^{M} Tr(Cov(\tilde{\mathbf{R}}_{\mathscr{C}_m})) \tag{16}$$

where its IP-NMF acronym refers to the fact that this algorithm thus performs an Inertia-constrained Pixel-by-pixel NMF, and where the positive parameter μ is selected so as to define the emphasis put on the penalty term.

4.3.2 Gradient-Based Algorithm

As in Sect. 4.2.2, the algorithm used hereafter to minimize the cost function (16) is based on a gradient descent, combined with projection on $\mathbb{R}^{+*}$ and sum-to-one normalization. To perform the required gradient calculations, the cost function (16) is first rewritten as

$$J_{ipnmf} = J_{RE} + \mu J_I \tag{17}$$

with

$$J_{RE} = \frac{1}{2} \left\| \mathbf{X} - \tilde{\mathbf{C}}\tilde{\mathbf{R}} \right\|_F^2 \tag{18}$$

$$J_I = \sum_{m=1}^{M} Tr(Cov(\tilde{\mathbf{R}}_{\mathscr{C}_\mathbf{m}})) \tag{19}$$

$$= \sum_{m=1}^{M} \left(\frac{1}{P} Tr(\tilde{\mathbf{R}}_{\mathscr{C}_\mathbf{m}}^T \tilde{\mathbf{R}}_{\mathscr{C}_\mathbf{m}}) - \frac{1}{P^2} Tr(Q_{\mathscr{C}_m}) \right) \tag{20}$$

$$Q_{\mathscr{C}_m} = \tilde{\mathbf{R}}_{\mathscr{C}_\mathbf{m}}^T \mathbf{1}_{P,P} \tilde{\mathbf{R}}_{\mathscr{C}_\mathbf{m}} \tag{21}$$

where $\mathbf{1}_{P,P}$ is the $P \times P$-dimensional matrix with all elements equal to one. As shown by (18), J_{RE} defines the Reconstruction Error achieved when deriving an approximation of the observed data matrix $\mathbf{X}$ as the product of the adaptive matrices $\tilde{\mathbf{C}}$ and $\tilde{\mathbf{R}}$. Besides, J_I defines the Inertia constraint.

As shown by (18) and (10), J_{RE} is equal to J_{upnmf}. The gradient terms associated with J_{RE} are therefore already available from (12) and (13). Besides, J_I does not depend on $\tilde{\mathbf{C}}$, so the corresponding derivative is equal to zero. Finally, to obtain the derivative of J_I with respect to $\tilde{\mathbf{R}}$, we here partly use a scalar approach. We start by transforming J_I so as to express part of it with respect to $\tilde{\mathbf{R}}$. Equation (20) thus yields

$$J_I = \sum_{m=1}^{M}\left(\frac{1}{P}\sum_{l=1}^{L}\sum_{k=1}^{P}[\tilde{R}_{\mathscr{C}_m}]_{k,l}^2 - \frac{1}{P^2}Tr(Q_{\mathscr{C}_m})\right) \tag{22}$$

$$= \frac{1}{P}\sum_{m=1}^{M}\sum_{l=1}^{L}\sum_{k=1}^{P}[\tilde{R}]_{(k-1)M+m,l}^2 - \frac{1}{P^2}\sum_{m=1}^{M}Tr(Q_{\mathscr{C}_m}) \tag{23}$$

$$= \frac{1}{P}\sum_{\kappa=1}^{PM}\sum_{l=1}^{L}[\tilde{R}]_{\kappa,l}^2 - \frac{1}{P^2}\sum_{m=1}^{M}Tr(Q_{\mathscr{C}_m}) \tag{24}$$

$$= \frac{1}{P}Tr(\tilde{\mathbf{R}}^T\tilde{\mathbf{R}}) - \frac{1}{P^2}\sum_{m=1}^{M}Tr(Q_{\mathscr{C}_m}). \tag{25}$$

The derivative of the first term of (25) can be determined by using the matrix formulas in [22], which yield

$$\frac{\partial}{\partial\tilde{\mathbf{R}}}\left(\frac{1}{P}Tr(\tilde{\mathbf{R}}^T\tilde{\mathbf{R}})\right) = \frac{2}{P}\tilde{\mathbf{R}}. \tag{26}$$

Now focusing on the second term of (25) and taking (21) into account, we introduce

$$\mathbf{A} = \tilde{\mathbf{R}}_{\mathscr{C}_m}^T \tag{27}$$

$$\mathbf{B} = \mathbf{1}_{P,P}\tilde{\mathbf{R}}_{\mathscr{C}_m}. \tag{28}$$

Their elements with indices (i, j) respectively read

$$a_{ij} = [\tilde{R}_{\mathscr{C}_m}]_{ji} \tag{29}$$

$$b_{ij} = \sum_{\beta=1}^{P}[\tilde{R}_{\mathscr{C}_m}]_{\beta j}. \tag{30}$$

As shown by (21),

$$Q_{\mathscr{C}_m} = \mathbf{AB}. \tag{31}$$

Therefore, its element with indices (i, j) reads

$$[Q_{\mathscr{C}_m}]_{ij} = \sum_{\alpha=1}^{P}a_{i\alpha}b_{\alpha j} \tag{32}$$

$$= \sum_{\alpha=1}^{P}\sum_{\beta=1}^{P}[\tilde{R}_{\mathscr{C}_m}]_{\alpha i}[\tilde{R}_{\mathscr{C}_m}]_{\beta j}. \tag{33}$$

Hence

$$Tr(Q_{\mathscr{C}_m}) = \sum_{l=1}^{L} [Q_{\mathscr{C}_m}]_{l,l} \tag{34}$$

$$= \sum_{l=1}^{L} \sum_{\alpha=1}^{P} \sum_{\beta=1}^{P} [\tilde{R}_{\mathscr{C}_m}]_{\alpha\, l} [\tilde{R}_{\mathscr{C}_m}]_{\beta\, l}. \tag{35}$$

From (34) we can calculate the derivative of $\sum_{m=1}^{M} Tr(Q_{\mathscr{C}_m})$ with respect to one element of $\tilde{\mathbf{R}}$, denoted as $[\tilde{R}]_{\gamma\lambda}$. Due to the above-defined structure of $\tilde{\mathbf{R}}_{\mathscr{C}_m}$, a given element $[\tilde{R}]_{\gamma\lambda}$ is present in only one of the matrices $Q_{\mathscr{C}_m}$, i.e. the one with $m = 1 + (\gamma - 1)(\text{mod } M)$, denoted as η hereafter. Therefore

$$\frac{\partial}{\partial[\tilde{R}]_{\gamma\lambda}} \left(\sum_{m=1}^{M} Tr(Q_{\mathscr{C}_m}) \right) = \sum_{m=1}^{M} \frac{\partial}{\partial[\tilde{R}]_{\gamma\lambda}} (Tr(Q_{\mathscr{C}_m})) \tag{36}$$

$$= \frac{\partial}{\partial[\tilde{R}]_{\gamma\lambda}} (Tr(Q_{\mathscr{C}_\eta})) \tag{37}$$

$$= \frac{\partial}{\partial[\tilde{R}]_{\gamma\lambda}} \left(\sum_{l=1}^{L} \sum_{\alpha=1}^{P} \sum_{\beta=1}^{P} [\tilde{R}_{\mathscr{C}_\eta}]_{\alpha\, l} [\tilde{R}_{\mathscr{C}_\eta}]_{\beta\, l} \right) \tag{38}$$

due to (35). In (38), four cases should be distinguished:

$$\frac{\partial \left(\sum_{l=1}^{L} [\tilde{R}_{\mathscr{C}_\eta}]_{\alpha\, l} [\tilde{R}_{\mathscr{C}_\eta}]_{\beta\, l} \right)}{\partial[\tilde{R}]_{\gamma\lambda}} = \begin{cases} 0 & \text{if } \alpha \neq \gamma \text{ and } \beta \neq \gamma, \\ [\tilde{R}_{\mathscr{C}_\eta}]_{\beta\, \lambda} & \text{if } \alpha = \gamma \text{ and } \beta \neq \gamma, \\ [\tilde{R}_{\mathscr{C}_\eta}]_{\alpha\, \lambda} & \text{if } \alpha \neq \gamma \text{ and } \beta = \gamma, \\ 2[\tilde{R}_{\mathscr{C}_\eta}]_{\alpha\, \lambda} & \text{if } \alpha = \beta = \gamma. \end{cases} \tag{39}$$

Equation (38), then yields

$$\frac{\partial}{\partial[\tilde{R}]_{\gamma\lambda}} \left(\sum_{m=1}^{M} Tr(Q_{\mathscr{C}_m}) \right) = 2 \sum_{\alpha=1}^{P} [\tilde{R}_{\mathscr{C}_\eta}]_{\alpha\, \lambda} \tag{40}$$

$$= 2 \sum_{\alpha=1}^{P} [\tilde{R}]_{(\alpha-1)M+\eta, \lambda}. \tag{41}$$

The expression (41) for one element $[\tilde{R}]_{\gamma\lambda}$ can then be extended to all elements of $\tilde{\mathbf{R}}$, which yields

$$\frac{\partial}{\partial \tilde{\mathbf{R}}} \left(\sum_{m=1}^{M} Tr(Q_{\mathscr{C}_m}) \right) = 2\mathbf{U}\tilde{\mathbf{R}} \tag{42}$$

with $\mathbf{U} \in \mathbb{R}^{PM \times PM}$ defined as

$$\mathbf{U} = \left[\begin{array}{c} \overbrace{\begin{array}{ccccc} 1\,0 & \ldots & 0\,1 & \ldots \\ 0\,1 & \ldots & 0\,0 & \ldots \\ \vdots & \ddots & \vdots & \\ 0\,0 & \ldots & 1\,0 & \ldots \end{array}}^{M} \\ \begin{array}{ccccc} 1\,0 & \ldots & 0\,1 & \ldots \\ \vdots & & & \ddots \end{array} \end{array} \right] = \begin{bmatrix} \mathbf{Id}_M & \ldots & \mathbf{Id}_M \\ \vdots & \ddots & \vdots \\ \mathbf{Id}_M & \ldots & \mathbf{Id}_M \end{bmatrix} \tag{43}$$

where the notation $\mathbf{Id}_D$ stands for the D-dimensional identity matrix. Equations (25), (26) and (42) then yield

$$\frac{\partial J_I}{\partial \tilde{\mathbf{R}}} = \frac{2}{P}(\mathbf{Id}_{PM} - \frac{1}{P}\mathbf{U})\tilde{\mathbf{R}}. \tag{44}$$

By combining (12), (13) and (44) we obtain the two partial derivatives of the general cost function (17) with respect to $\tilde{\mathbf{R}}$ and $\tilde{\mathbf{C}}$:

$$\frac{\partial J_{ipnmf}}{\partial \tilde{\mathbf{R}}} = -\tilde{\mathbf{C}}^T(\mathbf{X} - \tilde{\mathbf{C}}\tilde{\mathbf{R}}) + \frac{2\mu}{P}(\mathbf{Id}_{PM} - \frac{1}{P}\mathbf{U})\tilde{\mathbf{R}} \tag{45}$$

$$\frac{\partial J_{ipnmf}}{\partial \tilde{\mathbf{C}}} = -(\mathbf{X} - \tilde{\mathbf{C}}\tilde{\mathbf{R}})\tilde{\mathbf{R}}^T. \tag{46}$$

The resulting gradient-based algorithm is obtained by inserting the above expressions of derivatives in the same update rules as (14) and (15) except that, again, only the parts of the adaptive matrix $\tilde{\mathbf{C}}$ corresponding to all $\mathbf{c_p}^T$ in (5) should be updated with (15), whereas the other terms of this adaptive matrix $\tilde{\mathbf{C}}$ are kept to zero. Again, projection and normalization are then applied. The complete IP-NMF update algorithm thus obtained reads as follows.

1. Update matrix $\tilde{\mathbf{R}}$:

$\tilde{\mathbf{R}}^{(i+1)} \longleftarrow \tilde{\mathbf{R}}^{(i)} + \alpha_{\tilde{R}}(\tilde{\mathbf{C}}^{(i)T}(\mathbf{X} - \tilde{\mathbf{C}}^{(i)}\tilde{\mathbf{R}}^{(i)}) - \frac{2\mu}{P}(\mathbf{Id}_{PM} - \frac{1}{P}\mathbf{U})\tilde{\mathbf{R}}^{(i)})$

$\tilde{\mathbf{R}}^{(i+1)} \longleftarrow \max(\tilde{\mathbf{R}}^{(i+1)}, \varepsilon)$

2. Update matrix $\tilde{\mathbf{C}}$:

for $p = 1$ *to* P **do**

$\quad \mathbf{c_p}^{(i+1)T} \longleftarrow \mathbf{c_p}^{(i)T} + \alpha_{\tilde{C}}(\mathbf{x_p}^T - \mathbf{c_p}^{(i)T}\mathbf{R}(p)^{(i)})\mathbf{R}(p)^{(i)T}$

$\quad \mathbf{c_p}^{(i+1)} \longleftarrow \max(\mathbf{c_p}^{(i+1)}, \varepsilon)$

end

3. Normalize coefficients:

for $p = 1$ *to* P **do**

$\quad \mathbf{c_p}^{(i+1)} \longleftarrow \mathbf{c_p}^{(i+1)} / \sum_{m=1}^{M} c_{pm}^{(i+1)}$

end

4.3.3 Algorithm Initialization

As usual e.g. for NMF algorithms, the proposed UP-NMF and IP-NMF algorithms require one to select the initial values of the adaptive matrices $\tilde{\mathbf{R}}$ and $\tilde{\mathbf{C}}$. Various methods previously reported for initializing standard NMF algorithms (see e.g. [5], [12, 16]) may be extended to our approaches. In particular, we considered the following approaches.

The initial value $\tilde{\mathbf{R}}^{(0)}$ of the adaptive matrix $\tilde{\mathbf{R}}$ may be set as follows. Separately for each class with index m, the same class-dependent value may be assigned to all P adaptive spectra of $\tilde{\mathbf{R}}$ associated with that class. The M pure spectra thus needed to initialize $\tilde{\mathbf{R}}$ may e.g. be obtained by using one of the following alternative methods:

1. Randomly select M mixed spectra from the observations.
2. Process the observed data matrix $\mathbf{X}$ with a classical remote sensing blind source separation method, such as N-FINDR [28], which only extracts a *single* set of M relatively pure spectra from the whole image. This is expected to yield better performance than the random selection of M mixed spectra used in approach no. 1 above.
3. When performing characterization tests by mixing *known* pure spectra, a possible third initialization method consists of using, for each class with index m, the average of all the pure spectra which are available for that class (and which were extracted from the observed image in the present investigation). This is likely to yield a good initialization, which then allows one to investigate how the considered unmixing methods behave in such good conditions. However, it should be clear that this approach cannot then be used in real (i.e. blind) conditions, since the pure (i.e. source) spectra are then unknown and to be estimated.

The matrix $\tilde{\mathbf{C}}$ of mixing coefficients may be initialized by using the following alternative methods:

1. For each pixel, set all M coefficients to the same value, which is therefore equal to $\frac{1}{M}$, due to the sum-to-one constraint.
2. Use the standard Fully Constrained Least Square (FCLS) regression method [10] to derive these coefficients from the observed data matrix $\mathbf{X}$ and from the value $\tilde{\mathbf{R}}^{(0)}$ of the adaptive matrix $\tilde{\mathbf{R}}$ assigned as explained above.

5 Analyzing Source Separation Results with PCA

The tests reported here were performed with the following data:

- Three classes of pure materials were considered, namely tiles, vegetation and asphalt. For each of these classes, various supposedly pure spectra were extracted from the hyperspectral image shown in Fig. 4. These pure spectra are plotted in Fig. 5.

- A semi-synthetic hyperspectral image was created by computing each spectrum associated with a pixel as a linear combination of a tile spectrum, a vegetation spectrum and an asphalt spectrum randomly selected from the above-defined set of pure spectra. The nonnegative coefficients of this linear combination were randomly drawn and then rescaled so as to sum to one in each pixel.

This approach combines several attractive features: it is guaranteed to follow the mixing model (6), it uses realistic pure spectra (including realistic variability) and, especially, it uses *known* pure spectra, which allows one to compare them with the estimated spectra that are extracted by the considered source separation methods, which operate in a *blind* way, so as to assess the performance of these methods in this investigation dedicated to their characterization.

Still, the above approach leaves one question open, which is how to compare the actual pure spectra (hereafter also called constituent spectra), used above to create numerically mixed spectra, with their estimates derived from BSS methods. Plotting all these spectra is not a suitable solution, especially because many such spectra are to be analyzed (three estimated spectra per pixel) and they contain a large number of points. This problem is therefore the same as the one that we faced in Sect. 3, when analyzing the properties of a large set of spectra. Therefore, this problem is here solved by using the same approach as in Sect. 3: here again, PCA is a very attractive tool, because it allows one to project all spectra to be analyzed on a two-dimensional subspace, essentially in order to check if the pure spectra estimated by the considered BSS methods for any given class yield projected points which are close to one another and close to the points associated with the actual pure spectra for that class, whereas they are further from the points associated with estimated and actual spectra for other classes. Moreover, standard (i.e. without the sum-to-one constraint) NMF methods can only estimate the pure spectra up to positive scale factors, and the proposed UP-NMF and IP-NMF methods may be expected to yield related scale indeterminacies because they use limited constraints (thanks to the additional degrees of freedom provided by the use of different estimated pure spectra from one pixel to another). For each point of the two-dimensional PCA representation, changing such a scale factor results in moving that point along the line containing the position of that point before it was moved and the origin. This should be taken into account when analyzing each scatter plot of estimated points associated with a given class of materials in the PCA representation, and when comparing the estimated class scatter plots with the class scatter plots of the actual pure spectra used to create the considered semi-synthetic mixed spectra: even if the estimated points are shifted along the above-mentioned lines, due to scale factors, the main class axes of the actual and estimated scatter plots should remain coherent, and the spreads of these scatter plots along their orthogonal class axes should be interpreted accordingly. Before moving to that discussion of the obtained results, we hereafter define in which conditions the considered BSS methods were operated.

Before applying the update rules of the UP-NMF and IP-NMF algorithms, the adaptive matrices $\tilde{\mathbf{R}}$ and $\tilde{\mathbf{C}}$ were initialized by using the following approach, among those described in Sect. 4.3.3: $\tilde{\mathbf{R}}^{(0)}$ was derived from the N-FINDR method and

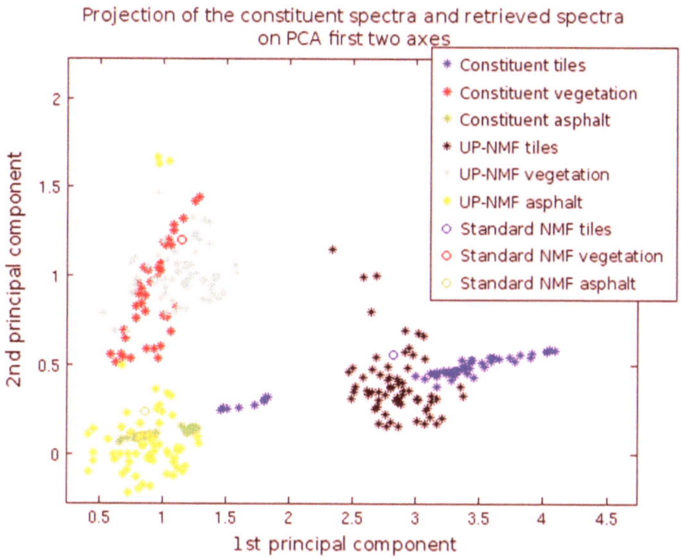

Fig. 7 Projection, on the first two PCA axes, of constituent spectra (blue, red, green stars), UP-NMF spectra (black, cyan, yellow stars) and standard NMF spectra (blue, red, green circles)

all mixing coefficients were set to $\frac{1}{M}$. For more details about the influence of the initialization procedure on the performance of the UP-NMF and IP-NMF methods, the reader is referred to [24], which is dedicated to this topic. Besides, for the IP-NMF method, the constraint parameter μ involved in the cost function (16) was varied from 0 to 100 to assess the impact of its value on algorithm performance. As shown by (16) and (10), for $\mu = 0$ the IP-NMF method becomes identical to UP-NMF. Tests were also performed with the well-known NMF method of [18] (extended to the sum-to-one constraint), which is based on the standard mixing model (3) and therefore provides a single estimated spectrum per class of materials, that is, three spectra for the whole considered image: a tile spectrum, a vegetation spectrum and an asphalt spectrum.

Figure 7 shows the projections, on the first two PCA axes,[6] of (i) the sets of pure spectra used to create the semi-synthetic observations, (ii) the spectra estimated by IP-NMF operated with $\mu = 0$ (that is, by UP-NMF) and (iii) the spectra estimated with standard NMF. Figures 8 and 9 are organized in the same way, except that they respectively correspond to $\mu = 30$ and $\mu = 100$ for IP-NMF. This shows the impact of the inertia constraint of the cost function of IP-NMF on the scatter plots of the estimated spectra.

As expected, the PCA representations in the above-mentioned figures make it possible to easily analyze the behavior of the considered methods. The main outcomes

[6]In Figs. 7, 8 and 9, the PCA axes are determined by applying PCA to the complete set of constituent spectra. These axes are then used to project these constituent spectra and the spectra provided by the considered BSS methods.

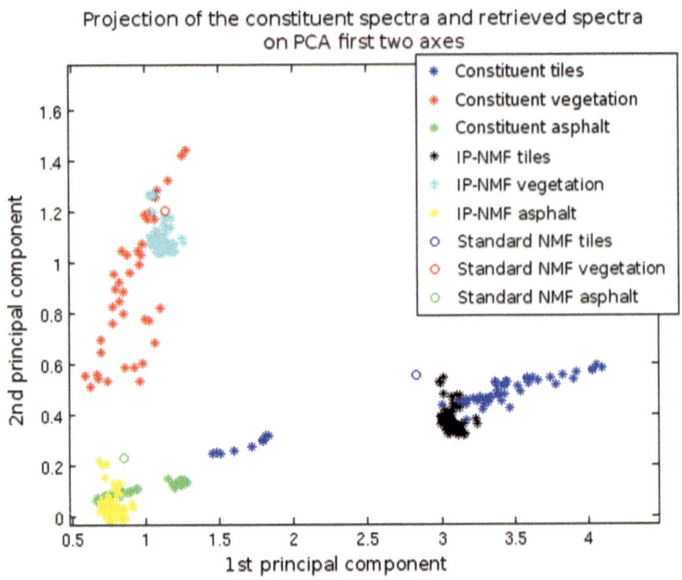

Fig. 8 Projection, on the first two PCA axes, of constituent spectra (blue, red, green stars), IP-NMF spectra with $\mu = 30$ (black, cyan, yellow stars) and standard NMF spectra (blue, red, green circles)

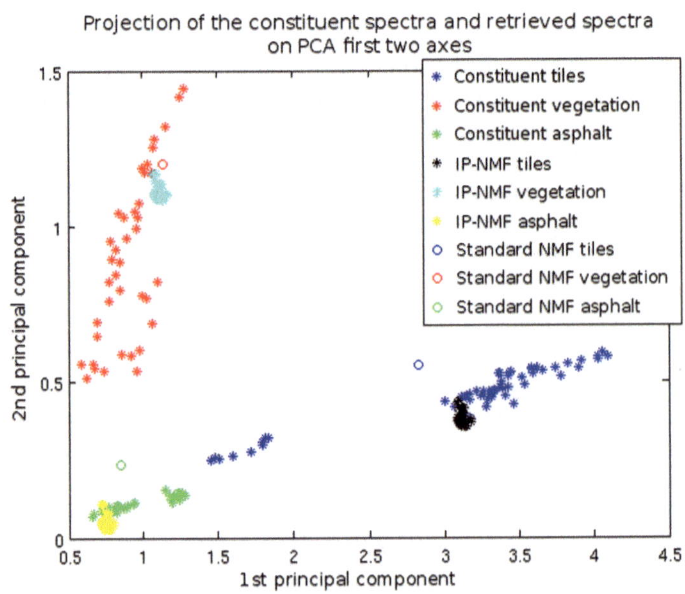

Fig. 9 Projection, on the first two PCA axes, of constituent spectra (blue, red, green stars), IP-NMF spectra with $\mu = 100$ (black, cyan, yellow stars) and standard NMF spectra (blue, red, green circles)

of this analysis are as follows. Of course, the standard NMF method provides a very restrictive view of the actual pure spectra involved in the observations, since it represents each scatter plot associated with a class of such constituent spectra by a *single* estimated spectrum and hence a single "point" (actually, a circle in these figures) in the considered PCA representation. Moreover, even this point is not satisfactory: it is situated significantly outside the corresponding class scatter plot, whereas one would like this NMF method to provide a kind of average of all the projected spectra of the considered class. Both above limitations are strong motivations for moving to the proposed UP-NMF and IP-NMF methods. Among the latter two approaches, we anticipated in Sect. 4.2.3 that UP-NMF is likely to have the drawback of allowing the spectra estimated for one class to spread too far away from one another and from the constituent spectra. For the considered data, this is confirmed by Fig. 7, where the scatter plots (one per class of materials) of estimated spectra tend to have wider projected "extents" (e.g. measured by their variance or by the width of their interval of variation) along the orthogonal class axes than the scatter plots of constituent spectra (as explained above, one should not focus on their extents along the main class axes, in case they would be influenced by scale indeterminacies). This problem of UP-NMF is solved by resorting to the IP-NMF method: as shown by Fig. 8, for intermediate values of μ (e.g. $\mu = 30$), the constituent and estimated spectra yield relatively similar extents along the orthogonal class axes. Other tests, not detailed here, resulted in similar performance for a wide range of values of μ. The performance obtained for a quite large value of μ (that is, $\mu = 100$) is provided in Fig. 9. This PCA representation shows that, by highly increasing this parameter μ and hence the weight of the penalty term in the cost function (16), one can force the extents of the class scatter plots of the estimated spectra to remain quite close to or even somewhat lower than those of the constituent spectra.

The performance improvement achieved by the proposed methods, as compared with the standard one, is obtained at the expense of significantly higher computational times: around 6 s per run for UP-NMF or IP-NMF, instead of around 0.02 s for standard NMF, when running Matlab implementations of these methods on current PCs. However, this is not an issue, since these computational times remain quite low anyway.

6 Conclusion

Principal component analysis (PCA) is a well-known data analysis tool, especially used to project high-dimensional data on a two-dimensional subspace, which then allows one to explore the structure of the resulting scatter plots in detail. The first aspect of this chapter consists of using these capabilities of PCA in two ways:

- PCA is first used to analyze the intra-class variability of observed data faced in the application field of Earth observation: each initial data point processed by PCA here corresponds to a high-dimensional reflectance spectrum. This variability analysis was a required first step for deriving an original model suited to the considered data.

- New blind source separation (BSS) methods (also called unsupervised unmixing methods) based on the above model were developed to tackle intra-class variability. PCA was then needed again to analyze the structure of the data obtained as the *outputs* of these BSS methods, in addition to its above-mentioned application to the observed data which are the *inputs* of these BSS methods.

Whereas the above aspects may be considered as "black-box use" of this PCA tool (in an original application), we also proceeded further in this chapter, by exploiting the internal data processing concepts underlying PCA. More precisely, as explained above, PCA mainly requires one to measure the "spread", related to covariance and inertia, of projections of the considered data, in order to select optimal projection directions. We here reported on detailed mathematical derivations, performed to transpose this concept to a processing function which is different from plain projection, that is, blind source separation: we developed a modified cost function, which aims at controlling the "spread" of the extracted sources in the advanced configuration faced when arbitrary source variability is taken into account.

This investigation shows that, although standard uses of PCA concepts are now well-defined, these concepts are still a source of inspiration for new data processing functions, that we plan to further investigate.

Acknowledgements This work was funded by the French ANR project on "HYperspectral imagery for Environmental urban Planning" (HYEP, no. ANR 14-CE22-0016-01).

References

1. Abdi, H., Williams, L.J.: Principal component analysis. WIREs Comput. Stat. **2**, 433–459 (2010). www.wiley.com/wires/compstats, https://doi.org/10.1002/wics.101
2. Adeline, K.R.M., Le Bris, A., Coubard, F., Briottet, X., Paparoditis, N., Viallefont, F., Rivière, N., Papelard, J.-P., David, N. , Déliot, P., Duffaut, J., Poutier, L., Foucher, P.-Y., Achard, V., Souchon, J.-P., Thom, C., Airault, S., Maillet, G.: Description de la campagne aéroportée UMBRA : étude de l'impact anthropique sur les écosystèmes urbains et naturels avec des images THR multispectrales et hyperspectrales - Urban material characterization in the sun and shade of built-up structures and trees and their retrieval from airborne image acquisitions over two French cities (UMBRA). RFPT, no. 200 (2012)
3. Bioucas-Dias, J.M., Plaza, A., Dobigeon, N., Parente, M., Du, Q., Gader, P., Chanussot, J.: Hyperspectral unmixing overview: geometrical, statistical, and sparse regression-based approaches. IEEE J. Select. Topics Appl. Earth Observations Remote Sens. **5**(2), 354–379 (2012)
4. Cichocki, A., Amari, S.-I.: Adaptive Blind Signal and Image Processing. Learning Algorithms and Applications. Wiley, Chichester, England (2002)
5. Cichocki, A., Zdunek, R., Phan, A.H., Amari, S.-I.: Nonnegative Matrix and Tensor Factorizations. Applications to Exploratory Multi-way Data Analysis and Blind Source Separation. Wiley, Chichester, UK (2009)
6. Comon, P., Jutten, C. (eds.): Handbook of Blind Source Separation. Independent Component Analysis and Applications. Academic Press, Oxford, UK (2010)

7. Deville, Y.: Traitement du signal: signaux temporels et spatiotemporels–analyse des signaux, théorie de l'information, traitement d'antenne, séparation aveugle de sources. Ellipses Editions Marketing, Paris, France (2011)
8. Deville, Y.: Blind source separation and blind mixture identification methods. J. Webster (ed.) Wiley Encyclopedia of Electrical and Electronics Engineering, pp. 1–33. Wiley, Hoboken (2016). https://doi.org/10.1002/047134608X.W8300
9. Gonzalez, R.C., Woods, R.E.: Digital Image Processing. Prentice Hall, Upper Saddle River, New Jersey (2002)
10. Heinz, D., Chang, C.-I.: Fully constrained least squares linear spectral mixture analysis method for material quantification in hyperspectral imagery. IEEE Trans. Geosci. Remote Sens. **39**(3), 529–545 (2001)
11. Hyvärinen, A., Karhunen, J., Oja, E.: Independent Component Analysis. Wiley, New York (2001)
12. Janecek, A., Tan, Y.: Using population based algorithms for initializing nonnegative matrix factorization. In: Tan, Y., Shi, Y., Chai, Y., Wang, G. (eds.) Advances in Swarm Intelligence (ICSI 2011). Lecture Notes in Computer Science, vol. 6729. Springer, Berlin, Heidelberg (2011)
13. Jolliffe, I.T.: Principal Component Analysis. Springer, New York (2002)
14. Karoui, M.S., Deville, Y., Hosseini, S., Ouamri, A.: Blind spatial unmixing of multispectral images: new methods combining sparse component analysis, clustering and non-negativity constraints. Pattern Recogn. **45**, 4263–4278 (2012)
15. Keshava, N., Mustard, J.F.: Spectral unmixing. IEEE Signal Process. Mag. **19**(1), 44–57 (2002)
16. Langville, A.N., Meyer, C.D., Albright, R.: Initializations for the nonnegative matrix factorization. In: Proceedings of the 12th ACM SIGKDD International Conference on Knowledge Discovery and Data Mining (KDD 2006), (2006)
17. Lee, D.D., Seung, H.S.: Learning the parts of objects by non-negative matrix factorization. Nature **401**, 788–791 (1999)
18. Lee, D.D., Seung, H.S.: Algorithms for non-negative matrix factorization. Adv. Neural Info. Proc. Syst. **13**, 556–562 (2001)
19. Lin, C.-J.: Projected gradient methods for nonnegative matrix factorization. Neural Comput. **19**, 2756–2779 (2007)
20. Makino, S., Lee, T.-W., Sawada, H. (eds.): Blind Speech Separation. Springer, Dordrecht, The Netherlands (2007)
21. Paatero, P., Tapper, U., Aalto, P., Kulmala, M.: Matrix factorization methods for analysing difussion battery data. J. Aerosol Sci. **22**(suppl. 1), S273–S276 (1991)
22. Petersen, K.B., Pedersen, M.S.: The Matrix Cookbook. Technical University of Denmark, Nov. 2012, version 20121115 (2012). Available online: http://www2.imm.dtu.dk/pubdb/views/publication_details.php?id=3274
23. Revel, C., Deville, Y., Achard, V., Briottet, X.: A method based on nonnegative matrix factorization dealing with intra-class variability for hyperspectral unmixing. In: Proceedings of the 7th Workshop on Hyperspectral Image and Signal Processing: Evolution in Remote Sensing (WHISPERS 2015), Tokyo, Japan, 2–5 June 2015
24. Revel, C., Deville, Y., Achard, V., Briottet, X.: Impact of the initialisation of a blind unmixing method dealing with intra-class variability. In: Proceedings of the European Symposium on Artificial Neural Networks, Computational Intelligence and Machine Learning (ESANN 2017), pp. 227–232. Bruges, Belgium, 26–28 April 2017
25. Saporta, G.: Probabilités. Analyse des données et statistique. Technip, Paris, France (1990)
26. Theodoridis, S., Koutroumbas, K.: Pattern Recognition, 4th edn. Academic Press, San Diego, California, USA (2009)
27. Veganzones, M.A., Drumetz, L., Tochon, G., Dalla Mura, M., Plaza, A., Bioucas-Dias, J., Chanussot, J.: A new extended linear mixing model to address spectral variability. In: Proceedings of the 6th Workshop on Hyperspectral Image and Signal Processing: Evolution in Remote Sensing (WHISPERS 2014), Lausanne, Switzerland, 24–27 June 2014

28. Winter, M.E.: N-FINDR: an algorithm for fast autonomous spectral end-member determination in hyperspectral data. In: Proceedings of the SPIE Conference on Imaging Spectrometry V, SPIE vol. 3753, pp. 266–275. Denver, Colorado, July 1999
29. Zare, A., Ho, K.C.: Endmember variability in hyperspectral analysis. IEEE Signal Process. Mag. **31**(1), 95–104 (2014)

MIX
Papier aus verantwortungsvollen Quellen
Paper from responsible sources
FSC® C105338

If you have any concerns about our products,
you can contact us on
ProductSafety@springernature.com

In case Publisher is established outside the EU,
the EU authorized representative is:
Springer Nature Customer Service Center GmbH
Europaplatz 3, 69115 Heidelberg, Germany

Printed by Libri Plureos GmbH
in Hamburg, Germany